Sustainable Concrete Materials and Resilient Structures

Sustainable Concrete Materials and Resilient Structures

Guest Editor

Jiping Bai

MDPI

Basel • Beijing • Wuhan • Barcelona • Belgrade • Novi Sad • Cluj • Manchester

Guest Editor
Jiping Bai
Faculty of Computing,
Engineering and Science
University of South Wales
Pontypridd
UK

Editorial Office
MDPI AG
Grosspeteranlage 5
4052 Basel, Switzerland

This is a reprint of the Special Issue, published open access by the journal *Applied Sciences* (ISSN 2076-3417), freely accessible at: https://www.mdpi.com/journal/applsci/special_issues/W5S0DJR0T4.

For citation purposes, cite each article independently as indicated on the article page online and as indicated below:

Lastname, A.A.; Lastname, B.B. Article Title. *Journal Name* **Year**, *Volume Number*, Page Range.

ISBN 978-3-7258-5407-3 (Hbk)
ISBN 978-3-7258-5408-0 (PDF)
https://doi.org/10.3390/books978-3-7258-5408-0

Contents

Jiping Bai
Advancing Sustainable Concrete Materials and Resilient Structures
Reprinted from: *Appl. Sci.* **2025**, *15*, 9683, https://doi.org/10.3390/app15179683 1

Sama Tajasosi, Jalil Barandoust, Ashkan Saradar, Mohammad Mohtasham Moein, Sam E. Rigby and Moses Karakouzian
Investigation of Mechanical and Fresh Properties of Ultra-High-Performance Concrete Incorporating Second-Generation Superplasticizers
Reprinted from: *Appl. Sci.* **2025**, *15*, 5133, https://doi.org/10.3390/app15095133 7

Mthabisi Adriano Nyathi, Jiping Bai and Ian David Wilson
Concrete Crack Width Measurement Using a Laser Beam and Image Processing Algorithms
Reprinted from: *Appl. Sci.* **2023**, *13*, 4981, https://doi.org/10.3390/app13084981 37

Hasan Mostafaei, Niyousha Fallah Chamasemani, Mohammadreza Mashayekhi, Naser Safaeian Hamzehkolaei and Paulo Santos
Sustainability Enhancement and Evaluation of a Concrete Dam Using Recycling
Reprinted from: *Appl. Sci.* **2025**, *15*, 2479, https://doi.org/10.3390/app15052479 50

Hanna Csótár, Szabolcs Szalai, Dmytro Kurhan, Mykola Sysyn and Szabolcs Fischer
Evaluating 3D-Printed Polylactic Acid (PLA)-Reinforced Materials: Mechanical Performance and Chemical Stability in Concrete Mediums
Reprinted from: *Appl. Sci.* **2025**, *15*, 2165, https://doi.org/10.3390/app15042165 68

Darya Anop, Marzhan Sadenova, Nail Beisekenov, Olga Rudenko, Zulfiya Aubakirova and Assel Jexembayeva
Additive Manufacturing as an Alternative to Core Sampling in Concrete Strength Assessment
Reprinted from: *Appl. Sci.* **2025**, *15*, 7737, https://doi.org/10.3390/app15147737 98

Jurga Šeputytė-Jucikė, Sigitas Vėjelis, Viktor Kizinievič, Agnė Kairytė and Saulius Vaitkus
The Effect of Expanded Glass and Crushed Expanded Polystyrene on the Performance Characteristics of Lightweight Concrete
Reprinted from: *Appl. Sci.* **2023**, *13*, 4188, https://doi.org/10.3390/app13074188 126

Helen Negash Shiferaw and Toshiyuki Kanakubo
Structural Performance of Fiber-Reinforced Cementitious Composite Members Reinforced with Fiber-Reinforced Polymer Bars: A Systematic Review
Reprinted from: *Appl. Sci.* **2025**, *15*, 7681, https://doi.org/10.3390/app15147681 146

Tomas Gómez Prieto, María Isabel Prieto, Alberto Leal and Daniel Ferrández
Mechanical Behavior of Carbon Fiber-Reinforced Concrete Structures After Structural Failure
Reprinted from: *Appl. Sci.* **2025**, *15*, 1783, https://doi.org/10.3390/app15041783 168

Farzad Hejazi, Ramin Azarm and Ali Akbar Firoozi
Efficiency of Flange-Bonded CFRP Sheets in Relocation of Plastic Hinge in RC Beam–Column Joints
Reprinted from: *Appl. Sci.* **2023**, *13*, 11870, https://doi.org/10.3390/app132111870 185

Gabriela Rutkowska, Mariusz Żółtowski, Filip Chyliński, Yuliia Trach and Elżbieta Gortych
The Effect of Glass Flour on the Microstructure and Properties of Fiber-Reinforced Concrete: Experimental Studies
Reprinted from: *Appl. Sci.* **2023**, *13*, 11937, https://doi.org/10.3390/app132111937 206

applied sciences

Editorial

Advancing Sustainable Concrete Materials and Resilient Structures

Jiping Bai

Faculty of Computing, Engineering and Science, University of South Wales, Treforest Campus, Pontypridd CF37 1DL, UK; jiping.bai@southwales.ac.uk

1. Introduction

This editorial posits that the defining challenge for 21st-century civil engineering is the synthesis of two historically distinct goals: environmental sustainability and structural resilience. Arguably, these are not competing priorities but rather two faces of the same coin, integral to the creation of infrastructure that is reliable, resilient, robust, and in harmony with our planet. Global cement manufacturing, a cornerstone of modern society, stands at a critical juncture, facing immense pressure to reduce its significant carbon footprint, while our aging infrastructure must be fortified against escalating climate-driven hazards. This editorial begins by outlining the urgent need for change in global infrastructure, followed by a summary of the main findings from the papers featured in this Special Issue. Then, it presents a forward-looking vision for a new generation of materials and structures that are not only low carbon but also intelligent, adaptive, and self-healing. Subsequently, it serves as a formal Call for Papers for the second volume of this Special Issue, inviting the global research community to contribute novel and innovative work that will turn this vision into reality.

It is with immense pleasure that I introduce this volume, a comprehensive collection of innovative research, originally presented as a Special Issue in *Applied Sciences*. This work brings together a diverse array of studies that collectively advance our understanding and capabilities in the critical areas of sustainable concrete materials and the design and assessment of resilient structures. In the face of contemporary economic, social, and environmental challenges, the construction industry is continually driven to innovate, seeking solutions that enhance resilience, reduce costs, and simplify maintenance while mitigating its significant environmental impact [1]. This volume serves as a vital resource for academics, researchers, graduate students, and industry professionals navigating these evolving demands.

2. The Need for Transformation: Global Infrastructure at a Crossroads

As a global community of engineers, scientists, and builders, we are the architects of the modern world. The very fabric of our civilization—our cities, transport networks, and shelters—is built upon material of our own making: cement and concrete. Yet, we must now confront the profound consequences of our success. Let us begin with an uncomfortable fact about our current reality: the worldwide yearly production of concrete is projected to increase from 14 billion cubic meters today to 20 billion cubic meters by the middle of the century, driven by urbanization and rising infrastructure needs [2]. This annual production has built our world, but it has come at a staggering environmental cost. Figure 1 demonstrates that cement production, once a minor source of emissions, has become a significant contributor to global CO_2 levels, underscoring the link between economic development, construction activity, and climate impact. Without intervention, this upward trajectory suggests cement will continue to be a major challenge for decarbonization efforts.

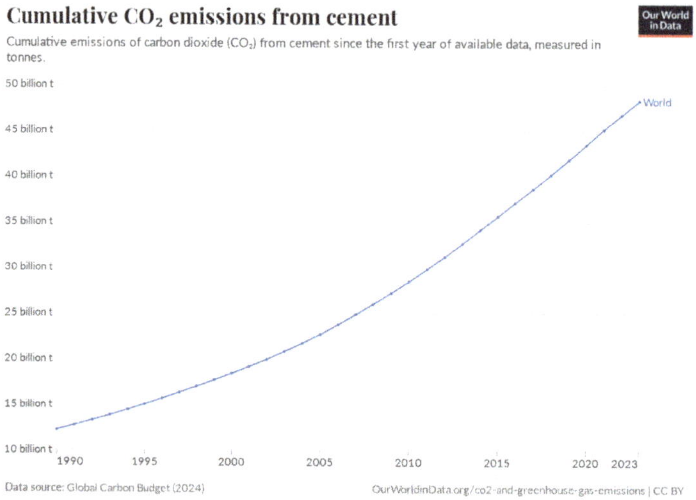

Figure 1. Cumulative CO_2 emissions from cement.

The reliance on Portland cement, an excellent binder, yet one whose creation is both energy- and carbon-intensive, is fundamentally unsustainable at its current scale and trajectory.

This urgency for transformation is further driven by the intensifying climate crisis, with global warming projected to exceed the critical 1.5 °C threshold without immediate reductions [3–5]. The crisis of resources is now colliding with a crisis of risk, creating a perfect storm for our built environment. The increasing frequency and intensity of hurricanes, floods, wildfires, and seismic events expose the deep vulnerabilities of legacy infrastructure [6]. These structures were often excellently engineered in light of the static historical climate data of a world that no longer exists. The traditional focus on singular metrics such as concrete compressive strength or prescriptive code-based detailing, while foundational to safety, is proving insufficient to ensure functional continuity and rapid recovery in the face of these new, dynamic, and often concurrent threats.

It is from this critical point that it is important to realize the prevailing design philosophies, where materials are inert, structures are passive, and lifecycles are linear, are inadequate. The economic and environmental extravagance of building disposable infrastructure, designed only to be demolished and replaced, can no longer be afforded. This moment of challenge, however, is also a moment of opportunity. It compels us to move beyond incremental adjustments and, instead, fundamentally rethink how we design, what we build with, and the very nature of the legacy we leave for generations to come. The imperative for change is not a distant concern; it is here, and it demands our immediate and collective ingenuity.

3. Development of Sustainable and Resilient Infrastructure

The papers in this collection exemplify the multidisciplinary efforts underway to address the need for transformation for concrete construction. They address four key themes crucial for the development of sustainable and resilient infrastructure:

1. Theme 1: innovative manufacturing and material optimization

A significant thrust in modern construction involves the adoption of advanced manufacturing techniques and novel material compositions. Additive manufacturing (3D concrete printing) is a prime example, offering flexibility in material selection and geometric forms and high customization [7]. For 3D-printed concrete to be widely adopted, robust quality control is essential. A non-destructive method has been developed to verify the on-site compressive strength of printed elements, which avoids the need for destructive core sampling [8].

2. Theme 2: enhancing structural resilience and durability through advanced composites

Carbon fiber-reinforced polymer (CFRP) is a highly effective material for enhancing the resilience of concrete structures, particularly against seismic loads or environmental degradation [9]. Its high strength, durability, and corrosion resistance make it ideal for retrofitting. For instance, applying CFRP to beam–column joints can strategically relocate potential failure zones, known as plastic hinges, from the vulnerable joint to the stronger beam section. This significantly boosts the connection's capacity and the structure's overall seismic performance. Additionally, wrapping concrete elements with CFRP provides powerful confinement. This technique can dramatically increase the compressive strength and ductility.

3. Theme 3: sustainable material sourcing and waste utilization

A key aspect of sustainability in concrete is the reduction in its environmental footprint, particularly that associated with cement production [10]. The performance of concrete can be enhanced through the strategic addition of waste materials [11]. Research in this collection indicates that using glass flour as a 10% substitute for traditional cement not only reduces the clinker content but also boosts the final product's compressive strength [12]. Adding polypropylene fibers to this formulation further improves its resilience, creating a composite capable of withstanding negative temperatures. This method provides a pathway to high-performance concrete while simultaneously achieving environmental goals such as waste reduction and lower emissions.

4. Theme 4: life cycle assessment and advanced predictive modeling

Understanding the full environmental and economic impacts of concrete structures is crucial for truly sustainable design. A life cycle analysis of large concrete structures reveals that cement production is the primary source of carbon emissions, but recycling demolished concrete can significantly cut these emissions [13]. This benefit often outweighs the slightly higher demolition costs and must be balanced against design choices, where stronger concrete reduces the material volume but increases emissions due to its higher cement content.

The collective body of work presented herein reflects the dynamic and innovative spirit within the field of concrete engineering. It shows diverse approaches to tackling the complex challenges of modern construction, from developing materials with reduced environmental footprints to enhancing the longevity and resilience of vital infrastructure. The reflections within this volume will undoubtedly inspire further research, foster interdisciplinary collaboration, and accelerate the adoption of sustainable and resilient practices in the global construction industry.

4. A Forward-Looking Vision: The Next Generation of Reliable, Resilient, and Robust Infrastructure

Having achieved significant progress and established the urgent need for a new path, it is time now to turn from the problems of the present to the possibilities of the future.

Our vision is not one of incremental improvement but of a shift in principle as to how we conceive of our fundamental building materials and the structures they form. It is essential to move beyond the passive and inert and begin to engineer materials and systems with integrated life-like functionalities.

First, it is envisioned that the next frontier of concrete is not a static material but one with a metabolism. Imagine concrete that actively contributes to environmental remediation by sequestering atmospheric CO_2 through advanced mineral carbonation, effectively turning our buildings and bridges into carbon sinks [14,15]. Picture bio-inspired concretes with internal vascular networks, analogous to a biological circulatory system, that can transport healing agents to autonomously repair microcracks as they form [16,17]. This would dramatically extend the service life, enhance the durability, and eliminate the cycle of costly carbon-intensive manual repairs, creating structures that heal instead of degrading.

Second, the future of resilience lies in creating a sentient structure. The infrastructure could be integrated with dense networks of self-powered sensors, from embedded fiber optics to piezoelectric nano-sensors, that function as a coherent nervous system [18–20]. This system would continuously stream data to a high-fidelity "digital twin", a dynamic virtual model that evolves with its physical counterpart [21,22]. This would allow for predictive health monitoring and real-time performance assessment during an extreme event, but more importantly, it would enable adaptive responses. A bridge could detect the unique vibrational signature of an approaching seismic shockwave and trigger actuators to momentarily stiffen key structural members, actively mitigating damage before it occurs.

Third, as we look ahead, critical areas for continued investigation include the broader validation of new materials and methods across various environmental conditions and scales, more in-depth analyses of chemical interactions and long-term durability, and the further integration of advanced computational tools such as machine learning for real-time monitoring and optimization [23–25].

This vision is achievable only through the synergy of the digital and the physical. The discovery of novel low-carbon binder chemistries will be radically accelerated by artificial intelligence (AI) and machine learning (ML), allowing us to navigate vast compositional spaces and move beyond laborious trial-and-error experimentation. Concurrently, advanced manufacturing, such as robotic 3D printing, will liberate us from the constraints of conventional formwork. This will enable the creation of structurally optimized and materially efficient geometries that place strength and function precisely where needed, integrating multi-functionality directly into the structural form and heralding a new era of architectural and engineering creativity.

5. The Call to Action: Announcing the Second Special Issue

The concepts outlined above are not distant dreams; their foundational elements are emerging from laboratories around the globe at this very moment. To accelerate this critical transition from laboratory potential to field-deployed reality and to foster the cross-disciplinary collaboration this work demands, we are proud to announce the Call for Papers for Volume II of our Special Issue on "Sustainable Concrete Materials and Resilient Structures".

This is a direct challenge to the community. We seek contributions that transcend incremental improvements and instead redefine the boundaries of what is possible. We challenge our colleagues to submit their most ambitious boundary-pushing research that directly addresses the integration of sustainability and resilience. We are particularly interested in work that demonstrates a clear and viable pathway from fundamental scientific discovery to transformative engineering application.

While we welcome all innovative work in this area, we are especially encouraging submissions that focus on the following high-interest topics:

- Next-Generation Binders and Systems: research on non-Portland systems such as geopolymers, alkali-activated materials, magnesium-based cements, and calcined clay–limestone (LC3) systems with demonstrated low-carbon footprints and robust durability performance.
- The Circular Economy in Practice: advanced research on the valorization of complex industrial and post-consumer waste streams as high-value aggregates and supplementary cementitious materials, including comprehensive life cycle assessments that validate their net environmental benefit.
- Multi-Scale Mechanics of Resilience: novel experimental and computational studies that link material micro-mechanisms of damage and healing to the macro-scale performance and fragility of entire structural systems under multi-hazard conditions.
- Sensing, Monitoring, and Adaptation: papers detailing innovative sensing technologies for embedded structural health monitoring (SHM), data fusion techniques for creating high-fidelity digital twins, and proof-of-concept studies of adaptive structural components.
- Innovative Case Studies: real-world or large-scale laboratory case studies of structures designed and built with an explicit quantifiable integration of sustainability and resilience metrics, providing invaluable data and lessons for the broader community.

6. Concluding Remarks: Building Tomorrow's Legacy, Together

The path forward requires a radical reimagining of the boundaries of our disciplines. The challenges of decarbonization and resilience are far too complex for any single field to solve in isolation. Unprecedented collaboration between materials scientists, structural engineers, architects, data scientists, and policymakers is not just beneficial; it is essential. It is critical to foster an ecosystem of innovation, where a breakthrough in binder chemistry can seamlessly inform a new parameter in a structural design model, and where policy enables, rather than hinders, the adoption of these next-generation solutions.

This Special Issue and the continuing work it represents is more than an academic exercise. It is a collective commitment to our professional and societal responsibility to provide the safe, enduring, and environmentally harmonious infrastructure that future generations deserve. The papers we publish, the research we conduct, and the students we mentor are all cornerstones of this new edifice.

The blueprint for the future is unwritten, and the materials are waiting to be invented. Let us, together, answer this call and begin the vital work of engineering the legacy of tomorrow.

Conflicts of Interest: The author declares no conflict of interest.

References

1. United Nations Environment Programme. *Global Alliance for Buildings and Construction Global Status Report for Buildings and Construction 2024–2025: Key Messages*; United Nations Environment Programme: Nairobi, Kenya, 2025. Available online: https://wedocs.unep.org/20.500.11822/47261 (accessed on 25 August 2025).
2. World Economic Forum. *High-Emitting Sectors: Challenges and Opportunities for Low-Carbon Suppliers*; World Economic Forum: Cologny, Switzerland, 2024. Available online: https://www3.weforum.org/docs/WEF_High_Emitting_Sectors_Challenges_and_Opportunities_for_Low_Carbon_Suppliers_2024.pdf (accessed on 24 August 2024).

3. Al Khaffaf, I.; Hawileh, R.A.; Sahoo, S.; Abdalla, J.A.; Kim, J.H. Toward carbon- neutral construction: A review of zero-carbon concrete. *J. Build. Eng.* **2025**, *99*, 111578. [CrossRef]
4. Camargo-Bertel, A.A.; Hincapie, D.; Pugliese, V.; Gonzalez-Quiroga, A.; Pupo-Roncallo, O. Decarbonizing the cement industry in Latin America and the Caribbean: A comprehensive review of strategies, barriers, and policies. *Energy Convers. Manag. X* **2025**, *26*, 100956. [CrossRef]
5. Nehdi, M.L.; Marani, A.; Zhang, L. Is net-zero feasible: Systematic review of cement and concrete decarbonization technologies. *Renew. Sustain. Energy Rev.* **2024**, *191*, 114169. [CrossRef]
6. Harrison, C.G.; Williams, P.R. A systems approach to natural disaster resilience. *Simul. Model. Pract. Theory* **2016**, *65*, 11–31. [CrossRef]
7. Csótár, H.; Szalai, S.; Kurhan, D.; Sysyn, M.; Fischer, S. Evaluating 3D-Printed Polylactic Acid (PLA)-Reinforced Materials: Mechanical Performance and Chemical Stability in Concrete Mediums. *Appl. Sci.* **2025**, *15*, 2165. [CrossRef]
8. Anop, D.; Sadenova, M.; Beisekenov, N.; Rudenko, O.; Aubakirova, Z.; Jexembayeva, A. Additive manufacturing as an alternative to core sampling in concrete strength assessment. *Appl. Sci.* **2025**, *15*, 7737. [CrossRef]
9. Hejazi, F.; Azarm, R.; Firoozi, A.A. Efficiency of flange-bonded CFRP sheets in relocation of plastic hinge in RC beam–column joints. *Appl. Sci.* **2023**, *13*, 11870. [CrossRef]
10. Barbhuiya, S.; Kanavaris, F.; Das, B.B.; Idrees, M. Decarbonising cement and concrete production: Strategies, challenges and pathways for sustainable development. *J. Build. Eng.* **2024**, *86*, 108861. [CrossRef]
11. Al-Waked, Q.; Bai, J.; Kinuthia, J.; Davies, P. Durability and microstructural analyses of concrete produced with treated demolition waste aggregates. *Constr. Build. Mater.* **2022**, *347*, 128597. [CrossRef]
12. Rutkowska, G.; Żółtowski, M.; Chyliński, F.; Trach, Y.; Gortych, E. The effect of glass flour on the microstructure and properties of fiber-reinforced concrete: Experimental studies. *Appl. Sci.* **2023**, *13*, 11937. [CrossRef]
13. Mostafaei, H.; Chamasemani, N.F.; Mashayekhi, M.; Hamzehkolaei, N.S.; Santos, P. Sustainability enhancement and evaluation of a concrete dam using recycling. *Appl. Sci.* **2025**, *15*, 2479. [CrossRef]
14. Tian, Y.; Xu, J.-C.; Zhang, G.-Y.; Tian, Z.-S.; Li, B.; Zhao, R.-Y.; Zeng, Q.; Huang, B.-T.; Yan, D.-M. Modeling of carbon sink for concrete structures: From hydration kinetics to climate-driven carbonation. *Sustain. Chem. Pharm.* **2025**, *47*, 102148. [CrossRef]
15. Wu, A.; Wang, Z. Exploring urban building carbon sinks: A SHAP-driven machine learning approach. *Sustain. Cities Soc.* **2025**, *126*, 106428. [CrossRef]
16. Wong, P.Y.; Mal, J.; Sandak, A.; Luo, L.; Jian, J.; Pradhan, N. Advances in microbial self-healing concrete: A critical review of mechanisms, developments, and future directions. *Sci. Total Environ.* **2024**, *947*, 174553. [CrossRef]
17. Deeba, S.; Ammasi, A.K. State-of-the-art review on self-healing in mortar, concrete, and composites. *Case Stud. Constr. Mater.* **2024**, *20*, e03298. [CrossRef]
18. Salehi, H.; Burgueño, R.; Chakrabartty, S.; Lajnef, N.; Alavi, A.H. A comprehensive review of self-powered sensors in civil infrastructure: State-of-the-art and future research trends. *Eng. Struct.* **2021**, *234*, 111963. [CrossRef]
19. Zhao, C.; Dong, W.; Liu, J.; Peng, S.; Li, W. Toward intelligent buildings and civil infrastructure: A review on multifunctional concrete through nanotechnology. *Cem. Concr. Compos.* **2025**, *163*, 106165. [CrossRef]
20. Siahkouhi, M.; Rashidi, M.; Mashiri, F.; Aslani, F.; Ayubirad, M.S. Application of self-sensing concrete sensors for bridge monitoring—A review of recent developments, challenges, and future prospects. *Measurement* **2025**, *245*, 116543. [CrossRef]
21. Jeon, C.-H.; Shim, C.-S.; Lee, Y.-H.; Schooling, J. Prescriptive maintenance of prestressed concrete bridges considering digital twin and key performance indicator. *Eng. Struct.* **2024**, *302*, 117383. [CrossRef]
22. Barreto, L.M.; Alves Salvador Filho, J.A. Applications of Digital Twins in Reinforced and Prestressed Concrete Bridge Infrastructure. *Procedia Struct. Integr.* **2024**, *64*, 1168–1175. [CrossRef]
23. Nyathi, M.A.; Bai, J.; Wilson, I.D. Deep Learning for Concrete Crack Detection and Measurement. *Metrology* **2024**, *4*, 66–81. [CrossRef]
24. Barbhuiya, S.; Sharif, M.S. Integrating machine learning with concrete science: Bridging traditional testing and advanced predictive modelling. In Proceedings of the 2024 International Conference on Innovation and Intelligence for Informatics, Computing, and Technologies (3ICT), Sakhir, Bahrain, 17–19 November 2024; pp. 88–93. [CrossRef]
25. Parhi, S.K.; Dwibedy, S.; Patro, S.K. Managing waste for production of low-carbon concrete mix using uncertainty-aware machine learning model. *Environ. Res.* **2025**, *279*, 121918. [CrossRef] [PubMed]

applied sciences

MDPI

Article

Investigation of Mechanical and Fresh Properties of Ultra-High-Performance Concrete Incorporating Second-Generation Superplasticizers

Sama Tajasosi [1], Jalil Barandoust [2], Ashkan Saradar [2,*], Mohammad Mohtasham Moein [3], Sam E. Rigby [4] and Moses Karakouzian [5,*]

1 Department of Water and Environmental Engineering, Faculty of Civil Engineering, Shahrood University of Technology, Shahrood P.O. Box 3619995161, Iran; sama.tajasosi@yahoo.com
2 Department of Civil Engineering, University of Guilan, Rasht P.O. Box 419961377, Iran; j.barandoust@gmail.com
3 Department of Civil Engineering, Allameh Mohaddes Nouri University, Nour P.O. Box 4641859558, Iran; m.mohtasham.moein@gmail.com
4 Arup Resilience, Security & Risk, 3 Piccadilly Pl, Manchester M1 3BN, UK; sam.rigby@sheffield.ac.uk
5 Department of Civil and Environmental Engineering and Construction, University of Nevada, Las Vegas, NV 89154, USA
* Correspondence: ashkan.saradar@gmail.com (A.S.); mkar@unlv.nevada.edu (M.K.); Tel.: +1-702-895-0959 (M.K.)

Abstract: Ultra-high-performance concrete (UHPC) has been following economic and environmental trends for the past two decades. Limited research has been conducted on the significance of superplasticizers in UHPC products, despite the high costs they entail for projects. The current study assesses UHPC based on rheological properties and mechanical characteristics considering different factors. In this study, the effects of different levels of superplasticizer derived from sulfonated naphthalene formaldehyde (SNF: 0.7%, 0.8%, and 0.9%), silica fume (SF: 15%, 20%, and 25%), and the water-to-binder ratio (w/b: 0.18, 0.20, and 0.22) were examined. Fresh tests such as slump flow, Vicat needle, and squeezing, as well as hardened tests like compressive strength, flexural strength, and electrical resistivity, were conducted. In the analysis, an artificial neural network (ANN) model and a fuzzy logic (FL) model were employed to forecast compressive strength results at 7 and 28 days. The results indicated that a higher SF dosage reduced slump flow and set time, whereas the opposite was observed for SNF and the w/b ratio. Three distinct behaviors were identified in the squeezing flow test findings: (1) specific elastic behavior and low plasticity, (2) extensive plastic behavior and significant dilatancy, and (3) heightened responsiveness to compressive flow rate and material ratio. SNF demonstrated promise in enhancing compressive, flexural, and electrical strength. The prediction models suggested that the FL (error range 3.18–4.36%) and ANN (0.74–1.03%) models performed well in predicting compressive strength at 7 and 28 days. The encouraging findings from this study set the stage for further sustainable and cost-effective construction methods.

Keywords: ultra-high-performance concrete; superplasticizers; squeeze flow; mechanical properties; artificial neural networks; fuzzy logic

1. Introduction

Ultra-high-performance concrete (UHPC) stands out among construction materials as an extraordinary marvel that surpasses conventional constraints [1]. UHPC has key characteristics like a high particle packing density (0.825–0.855), a low water-to-binder ratio (0.15–0.25), steel fiber content exceeding 2%, and suitable chemical admixtures [2]. In this respect, UHPC has transformed the construction sector by offering attributes like outstanding strength, excellent durability, acceptable ductility, and crack resistance, enabling the construction of lighter, more flexible, and visually appealing structures [1,3]. The outstanding performance of UHPC has resulted in its use in various construction projects, ranging from bridges and high-rise buildings to architectural marvels [4–6]. Over the past three decades, UHPC has undergone four phases of evolution [2]: (1) laboratory-only production with vacuum mixing and heat curing (before the 1980s), (2) the use of MDF and DSP cement to reach a compressive strength of 200 to 345 MPa (in the early 1980s), (3) the addition of steel fibers to reduce the fragility of this kind of concrete (in the mid-1980s), and (4) the introduction of superplasticizers and the creation of RPC, which marked a significant advancement for UHPC (in the 1990s). The production of UHPC has shown a new trend since 2000, which includes reducing CO_2 emissions and reducing initial costs.

The rheological properties of UHPC, such as viscosity and yield stress, are critical to its performance and are often evaluated using methods like the squeeze flow test. This test is particularly effective for assessing mortars under compression, with studies demonstrating strong correlations between squeeze flow and rotational rheometry results [7,8]. Because of the low water-to-binder (w/c) ratio and the high quantity of fine particles like silica fume, quartz sand, and fibers, UHPC is more viscous than normal concrete [9]. This makes the particles stickier and more resistant to flow, resulting in a stiffer and less fluid mixture [10]. The high viscosity can have negative effects on the workability, fiber distribution, and curing of UHPC, so it is essential to use appropriate additives, such as superplasticizers, to make the mixture more fluid and dispersed [11]. Superplasticizers are generally classified into four groups based on their chemical structure [12]: (1) lignosulfonates (SP), (2) sulfonated naphthalene formaldehyde condensate (SNF), (3) sulfonated melamine formaldehyde condensate (SMF), and (4) polycarboxylates (PCE). Figure 1 depicts details on the mechanism of action, chemical structure, and generation of each Superplasticizer. Beyond superplasticizers, advancements in material science—such as nano-absorbers for electromagnetic wave shielding [13–15] and sustainable low-carbon composites [16,17]—highlight the broader potential of tailored material design in enhancing UHPC performance.

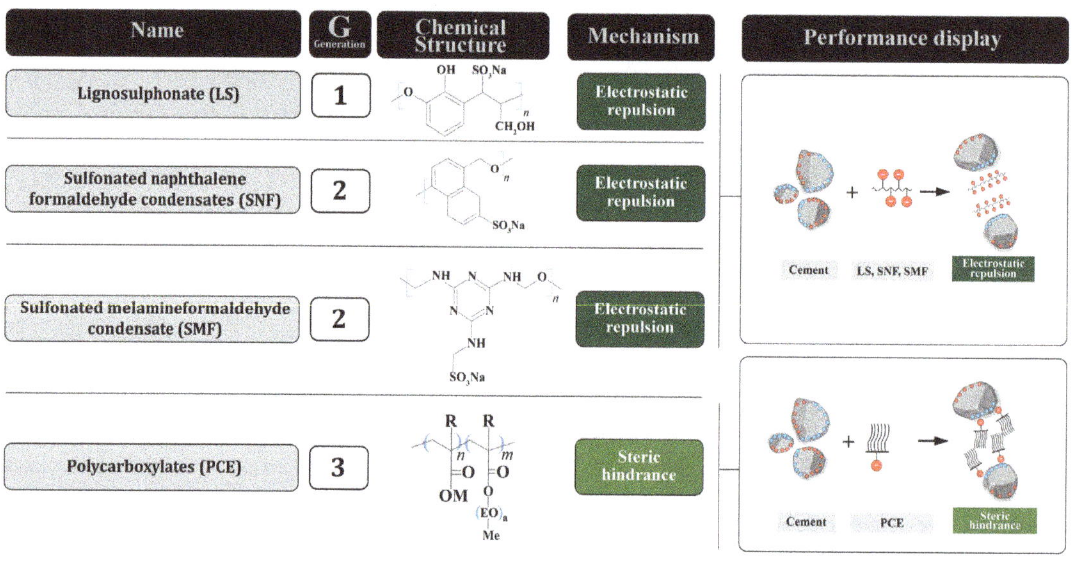

Figure 1. Types of superplasticizers based on chemical structure.

The tests used to assess concrete's mechanical properties can be problematic due to their time-consuming nature, high costs, and reliance on laborious, intensive work [18–20]. Moreover, the results might be uncertain because of varying laboratory conditions or human error [18,21]. In this regard, to more precisely assess the mechanical properties of concrete, the focus has been placed on techniques like artificial neural networks (ANNs) and fuzzy logic (FL). Various researchers have demonstrated the effectiveness of ANNs and FL in studying the mechanical properties of UHPC [22–24].

ANN, inspired by the biological nervous system, aims to process information similarly to the human brain [25–29]. ANN allows the modeling of nonlinear relationships between input and output [30]. The three layers of ANN are the input layer, the hidden layer, and the output layer [30]. In the ANN modeling process, determining the number of hidden layers and neurons is a crucial stage. A high number of hidden layers increases training duration, whereas a low count (insufficient) results in the model lacking resources to tackle intricate, nonlinear issues [30]. Figure 2a illustrates an ANN structure.

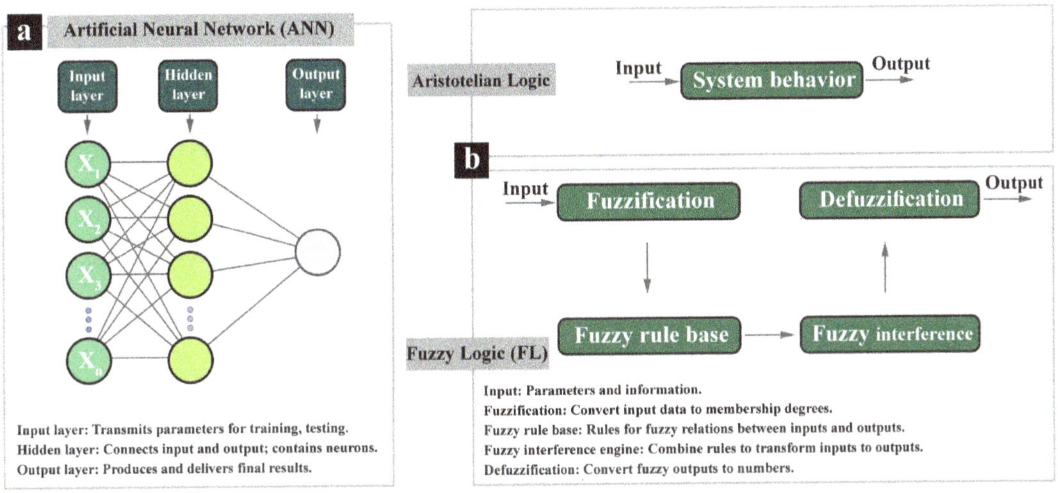

Figure 2. Structure of ANN and FL. (**a**) Artificial Neural Network. (**b**) Fuzzy Logic.

The concept of fuzzy sets was initially introduced by Professor Lotfi A. Zadeh [31]. In contrast to Aristotelian logic, which operates with binary values of 0 or 1, FL allows for the selection of values ranging between 0 and 1 for a given variable [32–35]. Data ambiguity, statistical variance, inadequate data, statistical inference, and limited understanding of variable relationships in a field are some of the factors that highlight the importance of fuzzy logic. Since the inception of FL, this concept has evolved and remains widely utilized across various engineering disciplines. FL effectively represents human knowledge and experience through mathematical equations, enabling the analysis of intricate real-world issues. It operates on 'if-then' rules to facilitate communication between input and output variables [18,36]. FL is mainly used in situations that involve uncertainty. To create fuzzy inference systems, fuzzy implication operators and a combination of fuzzy relations are used. The application of this method in predicting problems related to UHPC has been repeatedly investigated, and positive results have been obtained [22,23,37]. Figure 2b illustrates a schematic comparison between Aristotelian logic and fuzzy logic.

2. Literature Review

2.1. Literature on Superplasticizers

Researchers are currently analyzing how superplasticizers interact with various components of UHPC to address potential challenges arising from inadequate interaction. This includes studying the behavior of superplasticizers concerning cement particles, fine aggregates, and other additives. Most research on superplasticizer use in UHPC focuses on third-generation products. Second-generation products have been overshadowed by third-generation ones, leading to less attention on their potential. Lowering costs and meeting regulatory requirements with second-generation products are positive signs for incorporating UHPC as a target.

Zhang et al. [38] investigated the preferential adsorption of two superplasticizers (one containing only carboxyl groups and one containing both carboxyl and phosphate groups) on cement and silica fume. The research indicated that the carboxyl-containing PCE demonstrated higher adsorption potential on the cement surface, influencing the flow spread of the cement paste. Additionally, the carboxyl- and phosphate-containing PCE showed a higher affinity for adsorption on silica fume at lower concentrations, while at higher concentrations, it exhibited a strong adsorption capacity on both cement and silica fume. Li et al. [39] reported that the dispersing efficacy of PCE-type superplasticizers was intrinsically dependent on their chemical composition and justified the significant effect of chemical structure on the rheological properties of the pastes by an exponential correlation. They attributed the retardation effect (which affects concrete's setting time) to two factors: PCE molecules adsorbed onto cement particles and those within the aqueous phase. It was reported that factors like the type and dosage of SP caused chemical shrinkage of paste only in the initial 24 h. Wang et al. [40] found that enhancing the water-to-binder ratio and superplasticizer dosage increased flowability in fresh cement mortar, resulting in reduced yield stress. They noted that adjusting the superplasticizer dosage had no impact on plastic viscosity. According to Teng et al. [41], combining PCE-SP with nano-clay resulted in enhanced flexural toughness (by 45%), increased flexural strength (by 30%), a slight boost in compressive strength, and decreased porosity in UHPC samples. Other studies, like Yu et al. [42] and Tuan et al. [43], also found that rheological indices increased and free water decreased when nano or micro silica was combined with a superplasticizer [44,45]. Murugesan et al. [46] evaluated the efficacy of three distinct superplasticizer variants (namely SNF, SMF, and PCE) incorporated into UHPC at dosages of 0%, 0.5%, 1%, 1.5%, and 2%. The findings indicated that the mixtures containing PCE-SP outperformed those with sulfonated superplasticizers in terms of water reduction capacity, retention of workability, and mechanical strength.

2.2. Literature on Numerical Methods

ANNs and FL have gained significant popularity as innovative and effective tools in scientific and engineering research, particularly in the analysis and prediction of concrete behavior. Ramkumar et al. [47] investigated the application of ANNs in optimizing the mix design of self-consolidating concrete (SCC). Their study focused on the properties of SCC, particularly reinforcement with fibers and the utilization of agricultural and industrial waste as partial cement replacements. The authors reported that some studies indicate a confidence level of approximately 0.9995 for ANN predictions, with significantly lower errors than linear regression. Balf et al. [48] explored the application of ANNs in conjunction with data envelopment analysis (DEA) for designing SCC mixes containing fly ash. The ANN served as a comparative model to validate the DEA approach, which estimated efficiency based on input parameters such as superplasticizers and aggregates. The study indicated that the ANN model correlated well with the experimental results, enhancing the understanding and optimization of SCC mix design. Ly et al. [49] developed a deep neural network (DNN) model to predict the compressive strength of rubber concrete. A comprehensive database was utilized, with input parameters related to binders and aggregates and compressive strength as the output. They reported that DNN outperformed other neural network architectures across various performance metrics, demonstrating superior accuracy compared to existing machine learning methods. Liu and Zhang [50] compared the performance of two ANN models in predicting explosive spalling in polypropylene fiber-reinforced concrete subjected to elevated temperatures. One model was based on concrete mix proportions, while the other was based on concrete strength. The results, obtained from training datasets of 306 and 300 tests, showed that both models achieved high prediction accuracies, with ANN1 attaining 100% accuracy and ANN2 reaching 90%, indicating the effectiveness of ANNs in assessing the risk of explosive spalling. Biswas et al. [51] developed a hybrid ANN model to predict the compressive strength of SCC, considering factors such as cement replacement and aggregate types. The ANN, optimized using the Runge–Kutta optimization algorithm, achieved high prediction accuracy (R^2 = 0.933 for training and R^2 = 0.9203 for testing), surpassing the performance of traditional models. Feature importance and Taylor diagram analyses further supported the effectiveness of the ANN as a predictive tool for the mechanical properties of SCC. Khan et al. [52] proposed an ANN model optimized using the Levenberg–Marquardt backpropagation algorithm to predict compressive strength in both normal and high-strength concrete. By evaluating 1637 samples with eight input variables, the study optimized the model's architecture. K-fold cross-validation confirmed its reliability, and the statistical results were promising. The ANN demonstrated significant accuracy and efficiency, revealing that cement content and superplasticizers had a notable impact on compressive strength. Shafaie et al. [53] introduced an FL system to predict the shear bond strength of fiber-reinforced self-consolidating concrete (FRSCC). They reported high prediction accuracy, with R^2 values reaching up to 0.96 depending on the applied t-norms. The study highlighted the potential of FL in guiding the precise selection of pozzolan and optimizing mechanical performance. Abbas et al. [54] presented an FL numerical model designed to predict the complete stress–strain behavior of hybrid fiber-reinforced concrete (HFRC). Using data from 27 mixes (including 18 self-made and 9 collected from the literature), the FL model exhibited strong predictive capabilities compared to existing equations. The model was particularly suitable for HFRC, with strengths ranging from 60 to 90 MPa, encompassing both metallic and non-metallic fibers. Demir [55] compared a fuzzy modeling approach with traditional methods for determining the elastic modulus of normal and high-strength concrete. The study demonstrated that the fuzzy modeling approach provides a simpler and more effective way to evaluate the elastic modulus by combining multiple parameters. Al-Swaidani et al. [18] employed both ANNs and FL to predict the efficiency factor (EF) and durability indicator (DI) of nano natural pozzolana (NNP) as a cement replacement. Key input variables included the curing time, NNP content, particle size, water/binder (w/b) ratio, and superplasticizer dosage. They concluded that the ANN model exhibited higher accuracy than FL. For EF prediction via ANN, they achieved R^2 and mean absolute percentage error (MAPE) values of 0.992 and 18.5, respectively, outperforming both FL and multiple linear regression (MLR). The application of ANN and FL methods has shown great promise in improving our understanding of concrete and driving significant performance gains. These techniques offer the potential for substantial time and cost savings, as well as a reduced environmental impact.

3. Significance of the Investigation

In recent years, being economically and environmentally friendly has become one of the key requirements for UHPC. Hence, this study aimed to explore UHPC based on the following aspects:

Economic aspect

- Superplasticizer is a crucial material in the production of UHPC. However, its high cost in large projects has led to a search for more economically efficient approaches. Research has primarily focused on evaluating third-generation superplasticizers, while the potential of second-generation superplasticizers for UHPC has been largely overlooked, with limited studies conducted.
- The results of ANN and FL models were compared with the results of experimental tests (which are costly and time-consuming)

Environmental aspect

- Various percentages (15%, 20%, and 25%) of silica fume were studied as a partial substitute for cement.

Industrial application

- The key practical aspect of this study was the construction workshop for prefabricated parts production. To efficiently create parts in the workshop, quick molding and removal from the mold are essential. Speed (enhancing construction pace) and cost reduction are crucial in such workshops. The second-generation superplasticizer (naphthalene-based) holds promise, with a shorter processing time and lower cost than the third-generation alternative (polycarboxylate-based), offering potential time and cost savings.

Figure 3 delineates the methodological framework employed in the present study.

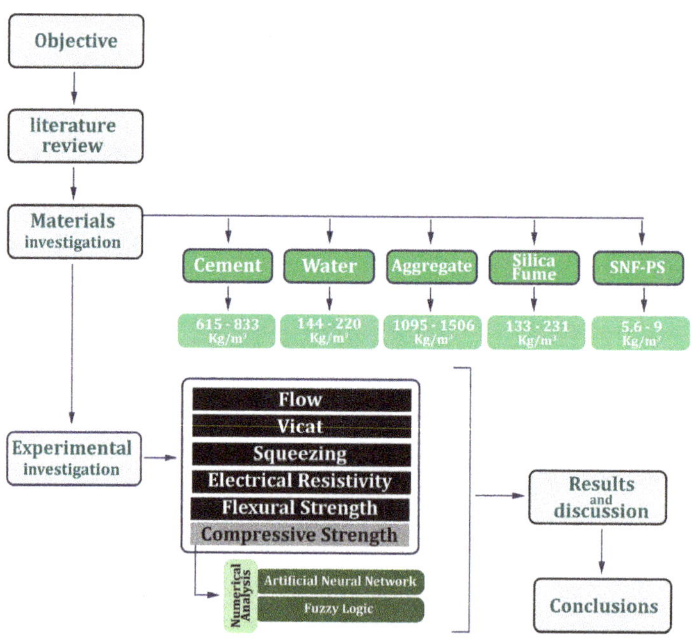

Figure 3. Methodology.

4. Materials and Test Methods

4.1. Materials

The constituents utilized in this investigation comprised (1) cement, (2) aggregates, (3) water, (4) superplasticizer, and (5) silica fume. Type II cement, sourced from the Hegmatan Cement Factory (Hamadan, Iran) and conforming to the ASTM C150 [56] standard, was utilized. The properties of the employed cement are delineated in Table 1. The silica fume (SF) was produced by the Ferroalloys Company (Tehran, Iran) according to the ASTM C1240 [57] standard. The specifications of the SF are presented in Table 2. River sand, which was sieved through #30 mesh (finer than 600 microns) in compliance with ASTM C778 [58], was utilized in the fabrication of the designated concrete. Table 3 delineates the characteristics of the sand employed. The superplasticizer, denoted as M200R and manufactured by Shimisakhteman Company (Tehran, Iran), conformed to the ASTM C494 [59] standard. The properties of the superplasticizer are listed in Table 4.

Table 1. Chemical and physical characteristics of the cement.

Chemical Properties		Physical Properties		
SiO_2	21.27	Compressive strength (MPa)	3 days	20.1
Al_2O_3	1.12		7 days	28.2
Fe_2O_3	4.03		28 days	40.3
CaO	62.95	Setting time (min)	Initial	154
MgO	1.55		Final	195
SO_3	2.26	Longitudinal expansion		1.5 mm-0.08%
Na_2O	0.49			
K_2O	0.65	Specific surface (cm^2/gr)		2910
C_3A	6.30			

Table 2. Chemical and physical characteristics of the silica fume.

Chemical Properties		Physical Properties	
SiO_2	89.26	Physical state	amorphous
Al_2O_3	4.95		
Fe_2O_3	1.8	Particle size (typical)	<1 μm
CaO	0.87		
MgO	1.1	Color	Light gray
Na_2O	0.5		
K_2O	0.66	Specific surface (cm^2/gr)	21,000

Table 3. Sand grading.

Sieve Size		ASTM C778 [58]	This Study
mm	No.		
1.18	16	100	100
0.6	30	100	96–100
0.425	40	67.5	65–75
0.3	50	28	20–30
0.15	100	3.5	0–4

Table 4. Specifications of superplasticizer.

Technical Features	
Generation	2
Physical State	Liquid
Color	Brown
Specific weight	1.2 ± 0.02 kg/lit
PH	8 ± 1
Chlorides (PPM)	500 max
Chemical Base	Modified sodium naphthalene sulfonate compounds

4.2. Mix Designs

The UHPC mix designs employed in this study are presented in Table 5. To fulfill the research objectives, the mix designs were categorized into two distinct groups, labeled A and B. Group A's focus was directed toward determining the optimal proportion of silica fume (SF), while group B's mixtures were analyzed to ascertain the influences of superplasticizer (SNF) and the water-to-binder (w/b) ratio. For SF, three varying ratios—20%, 25%, and 30% (relative to the weight of cement)—were examined. Concurrently, SNF-SP was incorporated at 0.7%, 0.8%, and 0.9% (in terms of binder weight). Furthermore, the w/b ratio was strategically varied at 0.18, 0.2, and 0.22 to evaluate its impact.

Table 5. UHPC mix designs.

Mix No.	Group	Design Code	Cement	Silica Fume		Water	w/b	S.P.	Aggregate	Dry Unit Weight (kg/m^3)	Final Water Absorption (%Wt.)
				(kg/m^3)	%						
N1	A	1-SF15WB20N08	696	104	15	160	0.2	6.4	1479	2083	3.42
N2		1-SF20WB20N08	667	133	20	160	0.2	6.4	1463	2074	3.24
N3		1-SF25WB20N08	640	160	25	160	0.2	6.4	1448	2064	2.99
N4		2-SF15WB20N08	783	117	15	180	0.2	7.2	1333	2067	3.47
N5		2-SF20WB20N08	750	150	20	180	0.2	7.2	1315	2051	3.15
N6		2-SF25WB20N08	720	180	25	180	0.2	7.2	1298	2042	2.82
N7		3-SF15WB20N08	870	131	15	200	0.2	8.0	1185	2045	3.21
N8		3-SF20WB20N08	833	167	20	200	0.2	8.0	1166	2024	3.26
N9		3-SF15WB20N08	800	200	25	200	0.2	8.0	1148	2015	2.88
N10	B	1-SF15WB18N08	696	104	15	144	0.2	6.4	1522	2114	3.02
N11		1-SF15WB22N08	696	104	15	176	0.2	6.4	1437	2065	3.64
N12		1-SF15WB20N07	696	104	15	160	0.2	5.6	1481	2097	3.51
N13		1-SF15WB20N09	696	104	15	160	0.2	7.2	1477	2090	3.31
N14		2-SF20WB18N08	750	150	20	162	0.2	7.2	1362	2081	3.03
N15		2-SF20WB22N08	750	150	20	198	0.2	7.2	1267	2026	3.23
N16		2-SF20WB20N07	750	150	20	180	0.2	6.3	1317	2054	3.23
N17		2-SF20WB20N09	750	150	20	180	0.2	8.1	1312	2051	3.1
N18		3-SF15WB18N08	870	131	15	180	0.2	8.0	1238	2053	3.33
N19		3-SF15WB22N08	870	131	15	220	0.2	8.0	1132	2006	3.29
N20		3-SF15WB20N07	870	131	15	200	0.2	7.0	1188	2028	3.37
N21		3-SF15WB20N09	870	131	15	200	0.2	9.0	1183	2026	3.09

Note: A = silica fume optimization; B = SNF and w/b ratio study; and 1, 2, 3 = binder content (800, 900, and 1000 kg/m^3).

The mix designs in this table are classified into three series—series 1, 2, and 3—which present total binder contents of 800, 900, and 1000 kg/m^3, respectively.

4.3. Test Methods

This study employed various tests to evaluate the properties of the fresh and hardened UHPC mixtures. Details regarding the specific tests conducted are presented in Table 6.

- The flowability of the fresh concrete was measured using a standard flow table (ASTM C230 [60]) in accordance with the ASTM C1437 [61] standard.
- The setting time was determined using the Vicat needle test (ASTM C191 [62]). The test procedure involved pouring the concrete mix into a mold and smoothing the surface. The Vicat needle was then introduced into the sample at predetermined intervals (starting at 30 min after mixing) to measure the depth of penetration. The initial setting time was recorded when the penetration depth fell below 25 mm, and the test continued until the needle penetration reached zero, indicating the final setting.
- The squeezing test, adapted from the work of Cardoso et al. [63], was employed to characterize the basic rheological properties of the fresh concrete. This test utilized a universal testing machine equipped with two grooved steel disks (diameter: 6 cm) positioned 21.5 cm apart within the clamps. An excess amount of concrete mix was placed between the disks, ensuring a uniform distribution. Excess material around the disks was removed with a spatula. Subsequently, the concrete was subjected to constant loading rates of 30 mm/min and 3000 mm/min, and the resulting load–displacement curve was recorded. A visual representation of the squeezing test setup is presented in Figure 4.
- Dry unit weight and water absorption were determined in accordance with ASTM C642 [64] on three specimens from each mixture.
- Electrical resistivity was measured at 28 days using the AASHTO TP119 [65] procedure on three samples per mix.
- Compressive strength was assessed at 7 and 28 days on 5 cm × 5 cm × 5 cm cube specimens following ASTM C109 [66].
- Flexural strength was determined at 28 days on 4 cm × 4 cm × 16 cm prism specimens, as per ASTM C348 [67].

Table 6. Details of tests.

Hardened Test			Fresh Test	
Test	**Standard**	**Curing**	**Test**	**Standard**
Compressive strength	ASTM C109 [66]	7, 28	Flow	ASTM C1437 [61]
Electrical Resistivity	AASHTO TP119 [65]	28	Vicat (Setting Time)	ASTM C191 [62]
Dry Unit Weight	ASTM C642 [64]	28	Squeezing	Cardoso et al. [63]
Water Absorption	ASTM C642 [64]	28	Cube	Prism
Flexural strength	ASTM C348 [67]	28		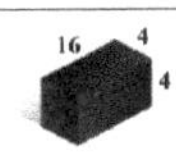

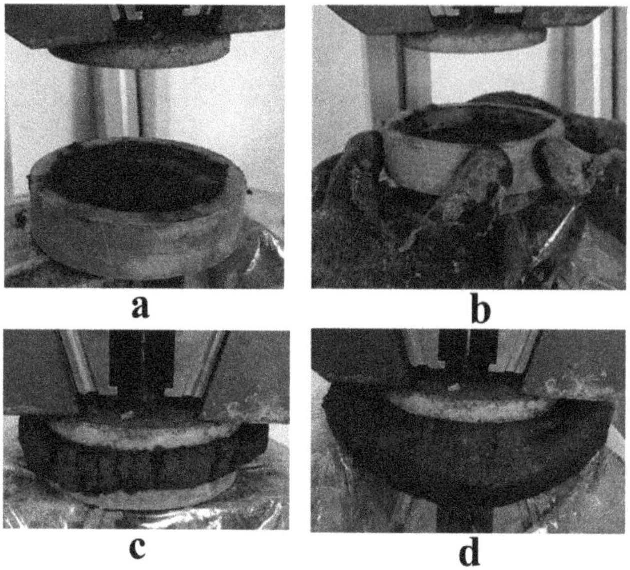

Figure 4. The squeezing test: (**a**) filling the ring mold, (**b**) removing the mold, (**c**) squeezing a high-viscosity mixture, and (**d**) squeezing a low-viscosity mixture.

5. Prediction Models

5.1. Artificial Neural Network (ANN)

Artificial neural networks (ANNs) are computational models inspired by the structure and function of biological neural networks. ANNs excel at modeling non-linear relationships between inputs and outputs, making them a valuable tool for predicting complex phenomena in concrete material science. In this study, an artificial neural network featuring a forward multi-layered perceptron structure was utilized. The backpropagation training algorithm employed the Levenberg–Marquardt (LM) algorithm optimization method, along with the scaled conjugate gradient (SCG) algorithm. The LM algorithm partitioned input and output vectors into training, validation, and test datasets. The SCG algorithm, by eliminating the need for second derivative calculations, ensured convergence to the quadratic function minimum, thus bypassing the time-intensive linear search process. The study's input variables included cement, silica fume, water, and naphthalene-based superplasticizer. Table 7 presents a report on the characteristics of these input and output variables.

Table 7. Characteristics of the independent and dependent variables.

Independent variables (kg/m^3)				
	Min.	**Max.**	**Averg.**	**SD.**
Cement (C)	640	870	763	75.3
Silica fume (SF)	104	200	137.24	26.58
Water (W)	144	220	180	19.44
Sulphonated naphthalene formaldehyde (SNF)	6	9	7.2	0.832
Dependent variables (MPa)				
	Min	**Max**	**Averg.**	**SD.**
Compressive strength—7 days	52.08	81.60	70.76	29.1
Compressive strength—28 days	115.07	145.44	132.05	9.13

To determine the optimal number of layers and neurons in the hidden layer of an artificial neural network, the learning and error method was utilized. The stop criterion for network training was selected as the mean squared error (MSE) (Equation (1)), representing the mean squared difference between the model's predicted value (network output) and the actual target value.

The R^2 value indicates the extent and strength of the relationship and dependency between the network's output value and the target, always ranging between 1 and −1 (Equation (2)). The closer the value is to 1, the stronger the relationship between the network's output and the target value. In this study, 10 neurons were chosen in the hidden layer to forecast the compressive strength at 7 and 28 days. Thus, the neural network's configuration in this study is 4–10–2: 4 inputs, 10 neurons in the hidden layer, and 2 outputs. Figure 5 depicts the overall architecture of the artificial neural network modeled in this study.

$$MSE = \frac{\sum_{j=0}^{P}\sum_{i=0}^{N}\left(d_{ij} - y_{ij}\right)^2}{N \times P}, \tag{1}$$

$$R^2 = 1 - \frac{\sum_{i=1}^{N}\left(y_i - \hat{y}_i\right)^2}{\sum_{i=1}^{N}\left(y_i - \overline{y}_i\right)^2}, \tag{2}$$

In Equations (1) and (2), the parameters are defined as follows: MSE (mean squared error) measures the average squared difference between the predicted values (y_{ij}) and the actual observed values (d_{ij}) across all samples, where N represents the total number of samples and P denotes the number of predictors or observations per sample. On the other hand, R^2 (coefficient of determination) evaluates the proportion of variance in the dependent variable (y_i) that is predictable from the independent variables. In the numerator, $\sum(y_i - \hat{y}_i)^2$ captures the sum of squared residuals, while the denominator $\sum(y_i - \bar{y}_i)^2$ accounts for the total variance, with $\hat{y}_i$ representing predicted values and $\bar{y}_i$ denoting the mean of observed values. Higher R^2 values indicate better model performance, while lower MSE corresponds to smaller prediction errors. Also, The R values indicate the extent and strength of the relationship and dependency between the network's output value and the target, always ranging between 1 and −1 (Equation (2)). The closer the value is to 1, the stronger the relationship between the network's output and the target value.

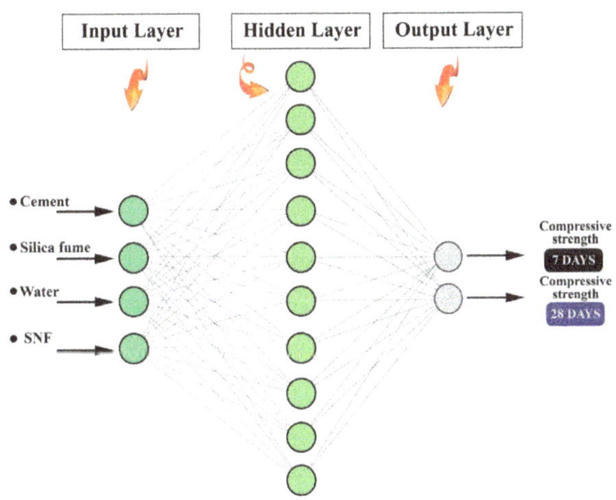

Figure 5. Architecture of ANN models for prediction of compressive strength (7 and 28 days).

5.2. Fuzzy Logic (FL)

Fuzzy logic (FL), as introduced in Section 2, provides a framework for reasoning with imprecise or uncertain information. Unlike traditional Boolean logic, FL allows for degrees of membership, enabling the representation of linguistic variables and the creation of rule-based systems that mimic human decision-making. In this study, we leveraged FL to predict the compressive strength of UHPC based on the mix design parameters. The FL model used four input variables to predict the compressive strength values (7 and 28 days) from experimental tests. Developed with the FL toolbox in MATLAB vR2024a, it utilized a rule-based if-then form for prediction. Since there are no mathematical equations in the FL model, the rule based on the elements expressed in the if-then form was used for prediction. Mamdani's inference system uses fuzzy sets as the result of the law, and the output of each law is nonlinear and fuzzy. The inputs for the system are categorized into four distinct groups: cement, silica fume, water, and NSF. These inputs were fuzzified into four gradations: low (L), medium (M), high (H), and very high (VH). The outputs are divided into two groups: 7-day compressive strength and 28-day compressive strength. Table 8 shows the details of the inputs and outputs of the FL model. Triangular membership functions, chosen for their simplicity, were constructed based on experience. The triangular shape can be considered the easiest shape to implement compared to others [22]. Figure 6 shows a schematic view of the FL model for this study.

Table 8. Range of input and output values and corresponding linguistic gradation.

Input	Fuzzy MF	Range
Cement (kg/m^3)	"L" (Low)	635–732
	"M" (Medium)	686–824
	"H" (High)	778–890
Silica fume (kg/m^3)	"L" (Low)	140–174
	"M" (Medium)	159–212
	"H" (High)	193–225
Water (kg/m^3)	"L" (Low)	140–174
	"M" (Medium)	159–212
	"H" (High)	193–225
NSF (kg/m^3)	"L" (Low)	4.5–7.1
	"M" (Medium)	6.6–8.4
	"H" (High)	7.8–9.5
Output	**Fuzzy MF**	**Range**
Compressive strength (MPa)—7 Days	"L" (Low)	45–60
	"M" (Medium)	57–68
	"H" (High)	65–77
	"V.H" (Very High)	74–90
Compressive strength (MPa)—28 Days	"L" (Low)	110–123
	"M" (Medium)	120–132
	"H" (High)	128–141
	"V.H" (Very High)	136–150

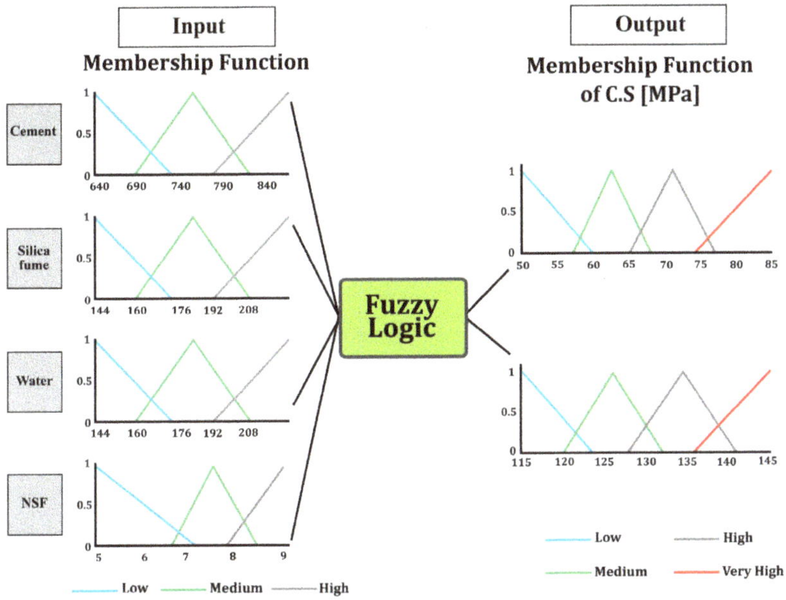

Figure 6. Displaying inputs and outputs in the FL model.

6. Results and Discussion

6.1. Fresh Tests

6.1.1. Slump Flow Test

Figure 7 shows the results of the slump flow test. The slump flow test indicated acceptable filling ability; the higher the slump flow, the greater the concrete's ability to fill the mold. Examining the mixtures of group A shows that the increase in SF substitution caused a decrease in slump flow. A decrease in workability with an increase in SF content has also been reported in other studies [68,69]. In series 1 mixtures, increasing SF from 15% to 20% caused a decrease in slump flow by almost 3%, and increasing SF to 25% resulted in a reduction of about 6%. Examining the mixtures of series 2 also shows a 2.92% and 5% decrease in slump flow when SF was increased to 20% and 25%, respectively. The series 3 mixtures also experienced 3.32% and 5.62% reductions in slump flow when SF was increased to 20% and 25%, respectively. SF, an exceptionally fine material (approximately 100 times smaller than cement particles) [70,71], effectively fills the voids between cement grains and larger aggregates when its dosage is increased. This phenomenon enhances the packing density of the concrete mix [70]. While this improvement translates to increased strength and durability, it can also lead to a stiffer mix. Furthermore, SF undergoes a chemical reaction with calcium hydroxide within the concrete matrix, resulting in the formation of additional calcium silicate hydrate gel (C-S-H). This reaction consumes free water in the mixture, potentially leading to a further reduction in slump flow [72–75].

Increasing the w/b ratio from 0.18 to 0.22 increased the slump flow. In this regard, the comparison of the SF15WB18N08 and SF15WB22N08 mixtures showed that in the mixtures of series 1, an increase in slump flow of about 11.9% was recorded; in the mixtures of series 2, an increase of about 13% was recorded; and in the mixtures of series 3, an increase of about 9.4% was recorded. The presence of water in a mixture acts to diminish interparticle friction, facilitating enhanced particle mobility [76]. This phenomenon translates to a reduction in cohesiveness and an increase in fluidity, consequently resulting in a higher slump flow value [76].

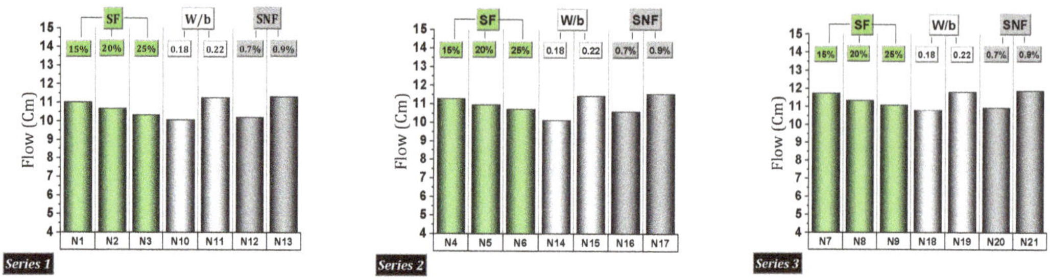

Figure 7. Slump flow test results for series 1, 2, and 3 mixtures.

Increasing SNF from 0.7 to 0.9 increased the slump flow. Therefore, slump flow was increased by 10.9%, 8.9%, and 8.6%, respectively, in the mixtures of series 1, 2, and 3 due to the increase in SNF. Superplasticizers, a class of surface-active agents, are proficient at dispersing cement particles. Their mechanism of action is twofold. First, due to their inherent electrostatic charge, superplasticizers adsorb onto the surface of cement particles, inducing electrostatic repulsion that hinders particle agglomeration [76–78]. Second, superplasticizers effectively reduce the surface tension of the water layer naturally surrounding cement particles. This phenomenon facilitates enhanced particle mobility by minimizing resistance to interparticle movement [76–78].

The main trends observed in the slump flow results indicate that increasing silica fume (SF) content reduces workability across all mix series, with greater SF levels (20–25%) leading to up to a 5–6% reduction in slump flow. Conversely, increasing the water-to-binder ratio from 0.18 to 0.22 improved slump flow by up to 13%. Additionally, increasing the superplasticizer dosage (SNF from 0.7% to 0.9%) led to an 8–11% rise in slump flow.

6.1.2. Vicat Needle Test (Setting Time)

As illustrated in Figure 8, increasing the dosage of SF from 15% to 20% led to a corresponding decrease in setting time. This effect was quantified by reductions of 4.5% and 19% in the initial and final setting times, respectively. Furthermore, increasing the SF dosage from 15% to 25% resulted in a more pronounced reduction in setting time, ranging from 13% to 33%. This phenomenon can be attributed to the acceleration of the hydration process caused by the presence of SF due to its fineness, chemical reactivity, and ability to enhance microstructural densification, all of which contribute to the faster setting of UHPC.

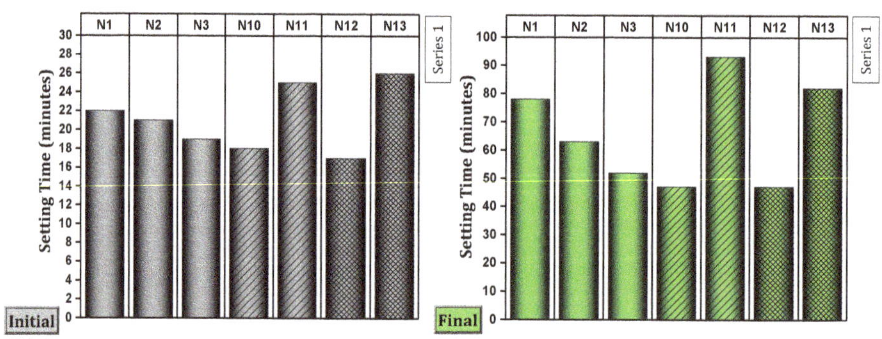

Figure 8. Vicat needle test (setting time) results for series 1 mixtures.

An increase in the w/b ratio from 0.18 to 0.22 demonstrably extended the setting times, as observed in the increased initial set time by 38% and final set time by more than

90% (Figure 8). This phenomenon can be attributed to the dilution effect of a higher water content. As the w/b ratio increases, the concentration of cement particles within the mix is reduced. This translates into a decrease in the probability of contact between particles, consequently slowing down the hydration reactions essential for setting. The hydration process is a chemical reaction between water and cement that forms the binding compounds responsible for concrete hardening. When the water content surpasses the binder content (including cement and supplementary cementitious materials like silica fume), the cement particles become more dispersed, hindering their interaction and subsequent hydration [79–81].

As observed, increasing the SNF dosage from 0.7% to 0.9% resulted in corresponding extensions of setting time, with increases of 52% and 74% in the initial set time and final set time, respectively (Figure 8). This phenomenon can be ascribed to the multi-faceted interplay between chemical and physical processes induced by the incorporation of superplasticizers within the concrete matrix. Expounding further, a higher dosage of superplasticizer promotes superior dispersion of cement particles, thereby enhancing the workability of the mix. However, this improved workability comes at the expense of a modified hydration process, ultimately leading to a prolonged setting time for the concrete [82]. Superplasticizers, particularly second-generation types like SNF, alter the hydration process by dispersing cement particles and delaying their interaction with water. This dispersion reduces the rate at which hydration products form, slowing the buildup of the initial structure needed for setting. Additionally, by improving workability without adding extra water, superplasticizers reduce early particle clustering, which further delays the formation of calcium silicate hydrate (C-S-H) gel, resulting in an extended setting time [83,84].

In contrast, increasing silica fume (SF) content significantly reduced both initial and final setting times. Raising the water-to-binder ratio (w/b) from 0.18 to 0.22 prolonged the setting time markedly by up to 90% for final setting. Additionally, higher superplasticizer (SNF) dosage also led to longer setting times, by up to 74%.

6.1.3. Squeezing Flow Test

According to prior research [63,85], the load–displacement curve schematic from the squeezing test comprises three stages, as depicted in Figure 9. Consequently, the assessment of various pastes in this study was conducted based on these three stages.

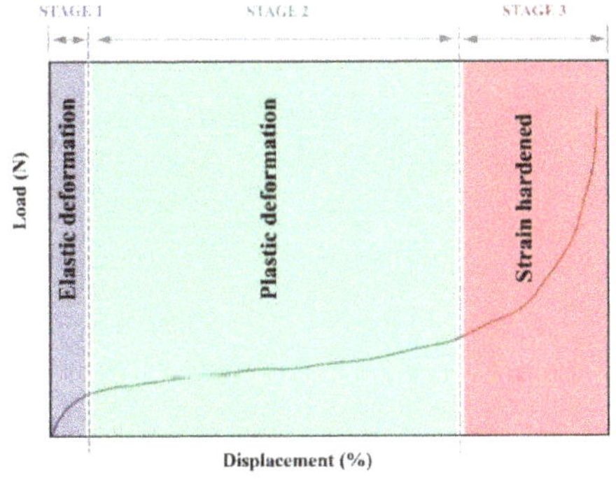

Figure 9. Different stages of the load–displacement curve under the squeezing test.

Examining the Effect of w/b and SF

Figure 10 shows the behavior of series 1 (Figure 10a) and series 2 (Figure 10b) pastes regarding the effect of the w/b ratio. According to Figure 10a, the analysis of the behavior of the pastes during the squeezing flow test reveals that all pastes exhibit three stages (elastic deformation, plastic deformation, and strain hardening). Pastes with higher w/b ratios are positioned lower than others, thus necessitating less compression load. Examining pastes with higher w/b ratios reveals a broader plastic stage. Conversely, pastes with lower w/b ratios (0.18) exhibit a shorter stage II due to drier conditions, advancing to stage III more quickly. Increasing the amount of water in the paste increases the likelihood of the dilatancy effect and results in a broader plastic phase [63,86–88]. Additionally, the higher water content in the paste reduces fluid viscosity, impacting the interaction between superplasticizer particles and plate gaps. Pastes with higher w/b ratios have a shorter I phase, while pastes with lower w/b ratios exhibit a more pronounced I phase. Additionally, increased water content in the paste reduces the likelihood of experiencing stage III.

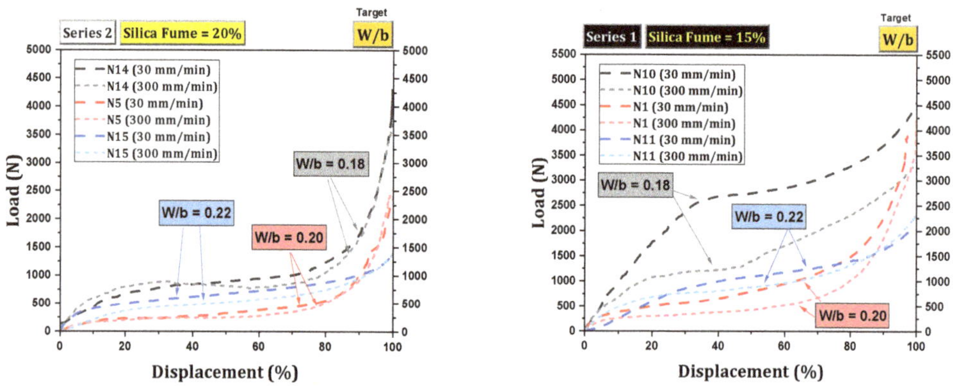

(**a**) (**b**)

Figure 10. Load–displacement curves for the squeeze flow test—checking the effect of w/b: (**a**) series 1; (**b**) series 2.

According to Figure 10b, in series 2, pastes with higher water content have a broader stage II and a shorter elastic stage. Pastes with a w/b ratio of 0.18 exhibit all three stages but with a shorter plastic stage. Pastes with higher water content exhibit partial strain hardening in a small portion of their plastic behavior.

A comparison between Figure 10a,b shows that the inclusion of SF in cement paste results in a more consistent behavior and enhances flowability, thereby lowering the need for compression force.

Examining the Effect of SNF and SF

Figure 11 illustrates the behavior of series 1 (Figure 11a) and 2 (Figure 11b) pastes concerning the impact of SNF dosage. Based on Figure 11a, the analysis of pastes shows that nearly all of them exhibit three stages. Lowering the SNF dosage appears to raise the necessary compression load. Pastes with a higher SNF dosage lead to greater displacement. The plastic stage (stage II) is broader in pastes with higher SNF doses. Conversely, pastes with a lower SNF dose exhibit a distinct stage I. An increase in SNF results in higher liquidity levels. The likelihood of encountering stage III decreases as the SNF dosage increases.

Based on Figure 11b, pastes with dosages below SNF undergo all three stages, but the elastic stage is not visible in pastes with dosages of 0.8% or 0.9%. Additionally, pastes with SNF dosage require more compression force and enter the hardening stage earlier than other pastes. Mixtures with higher SNF dosages show a broader plastic paste.

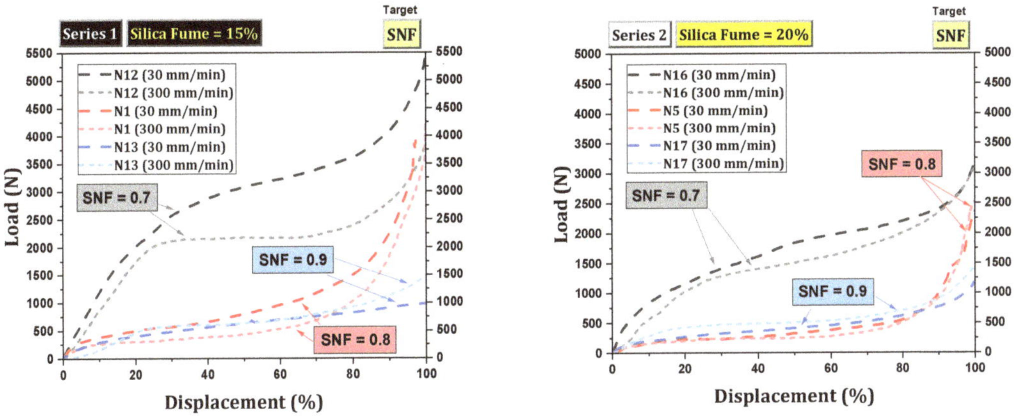

(**a**) (**b**)

Figure 11. Load–displacement curves for the squeeze flow test—checking the effect of SNF: (**a**) Series 1; (**b**) Series 2.

Comparing Figure 11a,b reveals that the rise in SF results in the relocation of the paste curves to a lower position. Essentially, the presence of SF in cement paste enhances fluidity, reducing the required compression force to reach the objective. In summary, the enhanced performance of series 2 (with 20% SF) stems from optimized microstructural packing, controlled water distribution, stronger early cohesion, and reduced internal friction, all of which contribute to more uniform flow behavior and lower compression requirements across varying SNF and w/b ratio conditions.

6.2. Hardened Tests

6.2.1. Electrical Resistivity

Figure 12 displays the results of electrical resistivity for various UHPC mixtures. Increasing SF to more than 15% in series 1 mixtures led to a rise in electrical resistance. Similarly, in series 2 and 3 mixtures, increasing SF enhanced electrical resistance. However, it is noteworthy that the increasing electrical resistance trend contradicts compressive resistance. Consequently, the authors undertook a more thorough investigation of this issue, which was previously unexplored. The conflict between compressive resistance and electrical resistivity, attributed to SF, may stem from two factors (to be elaborated on in the subsequent text): (1) SF filling C-S-H layers and (2) the inherent structure of SF.

The optimal dose of SF in UHPC is generally considered to be around 15–20%, according to previous studies. Exceeding this optimal dose results in a decrease in the expected C-S-H value of the mixture. In UHPC, some of the cement exists as crystalline particles, with only a small portion of real cement being hydrated to form C-S-H. Therefore, using SF at higher doses increases the distance between particles, posing a challenge for C-S-H formation and ultimately weakening resistance. Although using SF at a high dose reduces resistance in UHPC, it also increases electrical resistance. The C-S-H structure consists of layers with capillary paths between them that conduct and transmit a small electric current. These capillary paths facilitate 3D spatial communication. Increasing SF fills these capillary paths, reducing the distance between layers, disrupting the spatial communication, and decreasing compressive strength due to reduced layer adhesion (Figure 13). As a result, the concrete becomes more brittle, with a glass-like fracture pattern. Conversely, lower SF percentages allow for more flexible layers in the UHPC structure, preventing glass fractures. On the other hand, the filling between the layers increases the electrical resistance due to closed capillary paths and greatly increased density. However, this increase in density does not necessarily strengthen the structure.

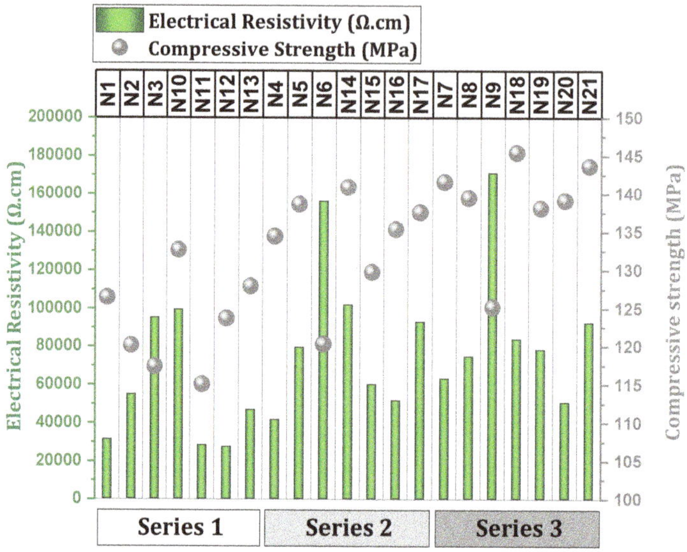

Figure 12. Results of electrical resistivity for series 1, 2, and 3 mixtures.

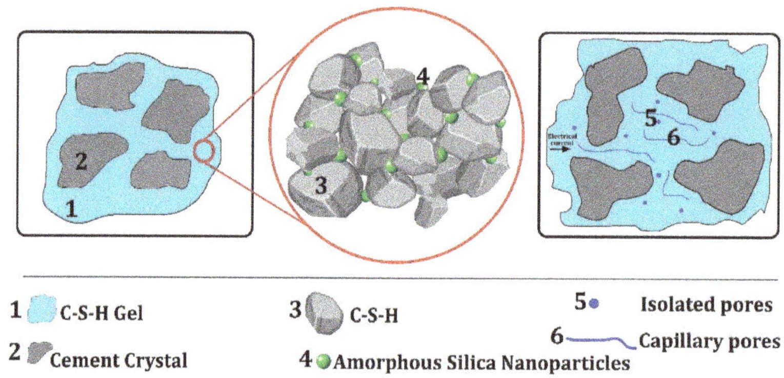

Figure 13. Mechanism of high-dose SF in UHPC.

Figure 12 reveals that increasing the w/b ratio from 0.18 to 0.22 (i.e., increasing water content in the mixture) leads to a decrease in electrical resistance (consistent with the compressive resistance trend). A higher w/b ratio results in more water in the mixture and increases the volume of pores in the concrete after hydration. These pores serve as small pathways for ion movement, aiding electricity flow and decreasing resistance [89–93].

The increase in SNF from 0.7% to 0.9% signifies a rise in density and an enhancement in electrical resistance (Figure 12), aligning with the trend of compressive strength. When superplasticizers are incorporated into concrete, they enhance workability by dispersing cement particles, thereby decreasing the thickness of the water layer surrounding them. This decrease boosts the packing density of the cement particles, resulting in a denser microstructure. Consequently, the route of electric current within the concrete becomes more circuitous, elevating electrical resistance. Furthermore, these additives can alter concrete's pore structure, diminishing pore network connectivity and further elevating electrical resistance [83,94,95].

6.2.2. Flexural Strength

Figure 14 displays the flexural strength outcomes of the mixtures in group A (SF effect) and group B (optimal SF-based, investigating w/b and SNF effects). In Figure 14a, it is observed that as the SF content increases, the flexural strength of series 1 decreases. The highest bending strength in series 1 was recorded in the mix with 15% SF (6.46 MPa). Examining series 2 mixtures reveals that raising SF from 15% to 20% leads to a 13% increase in flexural strength. In contrast, elevating SF from 15% to 25% caused a 23% decrease in flexural strength. In series 3, increasing SF from 15% to 20% resulted in a 9% reduction in flexural strength, and increasing SF from 15% to 25% resulted in an 11% reduction in flexural strength. The reduction in bending strength due to higher SF doses can be attributed to several factors [96–98]: (1) SF particles are extremely fine, leading to a high specific surface area. This results in increased water demand. Insufficient water results in a drier cement mix. (2) Higher SF content may reduce efficiency, making it harder to achieve proper density. (3) Excessive SF can cause uneven particle distribution in the cement paste, hindering the formation of calcium silicate hydrate (C-S-H) gel. (4) SF, by absorbing excess water, may deplete the water available for concrete hydration, disrupting the process. (5) Increasing SF dosage can weaken the interfacial transition zone (ITZ) between aggregate and cement paste. For series 1, 2, and 3, the best SF percentages for achieving maximum flexural strength are 15%, 20%, and 15%, respectively.

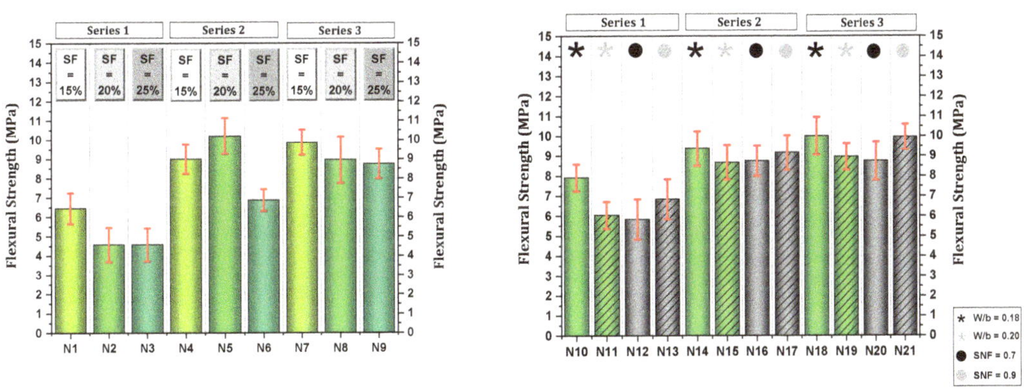

(**a**) (**b**)

Figure 14. Results of flexural strength (28 days): (**a**) effect of SF; (**b**) effect of w/b and SNF.

According to Figure 14b, the flexural strength of mixtures with increasing w/b ratios (from 0.18 to 0.22) shows a decreasing trend. Therefore, the rise in the w/b ratio in the mixtures of series 1, 2, and 3 led to 23%, 7.4%, and 10.4% reductions in flexural strength, respectively. A higher water-to-binder ratio results in more water in the mixture, potentially increasing porosity. Excess water also reduces the concentration of cement particles, decreasing the level of hydration. Consequently, fewer cement particles are accessible to create the robust bonds essential for achieving high flexural strength [99,100].

The increase in SNF from 0.7 to 0.9% enhanced bending strength (Figure 14b). Comparing SF15WB20N07 and SF15WB20N09 blends reveals a 17%, 4.8%, and 13% rise in flexural strength in series 1, 2, and 3, respectively, attributed to the higher SNF dosage. The improvement in flexural strength through SNF can be considered a result of the potential of SNF to create better alignment of the concrete microstructure [46,101]. In other words, the complete hydration of cement particles is ensured by reducing water content, improving particle dispersion, and providing optimal particle packing [46,101].

Figure 15 illustrates the impact of the w/b ratio and SNF on flexural strength. As shown in Figure 15a, a rise in the w/b ratio from 0.18 to 0.22 resulted in a decrease in flexural strength. Series 3 mixtures exhibited the highest flexural strength relative to the w/b ratio. In Figure 15b, it is shown that as the SNF ratio increased from 0.7 to 0.9, the flexural strength also increased. Series 3 mixtures exhibited the highest flexural strength based on the SNF ratio.

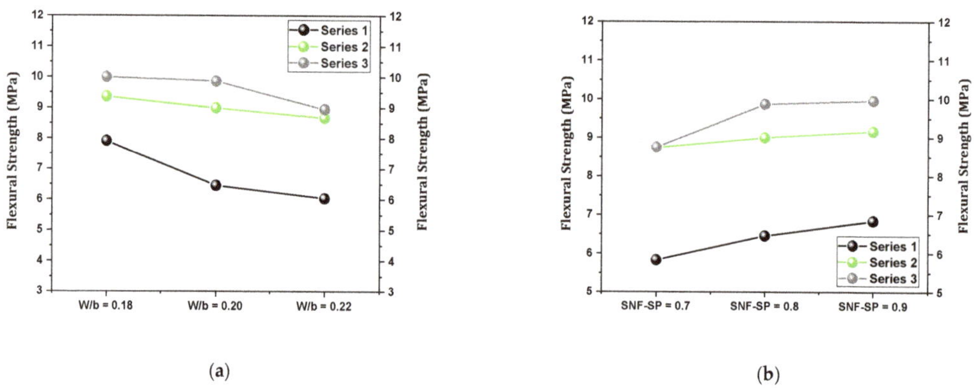

Figure 15. Flexural strength under the influence of (**a**) changes in the w/b ratio; (**b**) changes in the dosage of SNF.

6.2.3. Compressive Strength

Figure 16 displays the average compressive strength results at 7 and 28 days for group A mixtures (SF effect) and group B mixtures (optimized SF with varying w/b ratios and SNF effects). Based on Figure 16a, the compressive strength decreased at 7 and 28 days with higher SF replacement. For series 1 mixtures, increasing SF from 15% to 20% resulted in approximately 11% and 5% decreases in compressive strength at 7 and 28 days, respectively. The decrease in compressive strength with the increase in SF from 15% to 25% was also shown in series 1 mixtures, resulting in a 7% decrease in strength at 28 days. In series 2 mixtures, as the SF substitution rose from 15% to 20%, the compressive strength increased by 3% and 1% at 7 and 28 days, respectively. However, surpassing 20% SF replacement led to a decrease in strength. Elevating SF substitution from 15% to 25% resulted in a roughly 12% decrease in compressive strength at both 7 and 28 days. Series 3 mixtures showed a 6.5% strength reduction at 7 days and about 1.5% at 28 days when SF substitution increased from 15% to 20%. Furthermore, raising SF substitution from 15% to 25% in series 3 mixtures led to an 11% decrease in compressive strength. An excessive concentration of SF in concrete can adversely affect its compressive strength through multiple mechanisms. The incorporation of a high dosage of SF necessitates augmented water content to preserve workability, engendering a dichotomy [102–105]: (1) the addition of water elevates the w/b ratio, undermining the concrete's integrity, and (2) the omission of water compromises the mix's efficiency, culminating in inadequate compaction and increased porosity. Elevated levels of SF can induce heightened viscosity within the mix, impeding cohesive properties and leading to the entrapment of air voids [106]. Although SF is known to enhance the microstructural compactness of concrete, an overabundance may result in over-densification, thereby obstructing the hydration process and the structural evolution of the concrete, consequently diminishing its strength [107,108]. Based on the compressive strength (7 and 28 days) of mixtures in series 1, 2, and 3, the ideal SF dosages were determined to be 15%, 20%, and 15%, respectively.

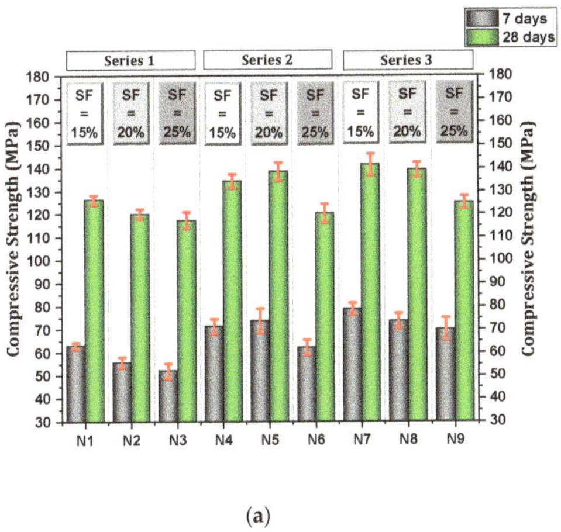

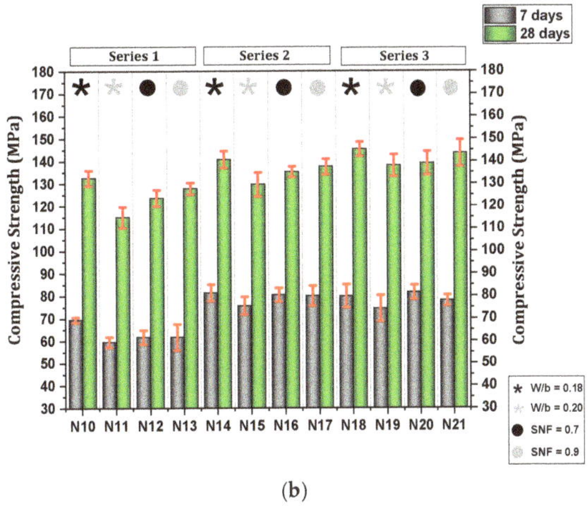

Figure 16. Results of compressive strength (7 and 28 days): (**a**) effect of SF; (**b**) effect of w/b ratio and SNF.

Figure 16b depicts the compressive strength of the mixtures at 7 and 28 days, considering the effects of the w/b ratio and SNF. The efficacy of SF in enhancing compressive strength is demonstrably linked to the w/b ratio. Generally, a lower w/b ratio translates into higher strength. However, improper adjustment of the w/b ratio when incorporating SF can compromise its effectiveness, leading to reduced compressive strength [109]. Increasing the w/b ratio from 0.18 to 0.22 resulted in a decrease in compressive strength. Comparing SF15WB18N08 and SF15WB22N08 mixtures revealed a reduction in compressive strength of approximately 13% for series 1, around 8% for series 2, and about 5% for series 3. Cement plays a crucial role in reacting with water to initiate a hydration reaction. This process results in the formation of a robust network of C-S-H gel, essential for concrete strength. A higher w/b ratio indicates more water relative to the binder (cement and other pozzolans). Excess water dilutes the cement paste, reducing the material for the vital C-S-H gel. With more water, not all cement particles may fully hydrate, leading to unreacted cement and weaker bonds in the concrete structure [100].

Increasing the SNF dosage from 0.7% to 0.9% also increased compressive strength. Comparing SF15WB20N07 and SF15WB20N09 mixtures reveals a 1.5 to 3.5% increase in compressive strength due to higher SNF substitution. SNF acts in the cement mixture by dispersing and separating cement particles and reducing internal friction. Also, SNF improves the contact between cement particles and water, which provides the basis for more complete hydration and results in a strong structure. On the other hand, the SNF used can reduce segregation (where coarse aggregates are separated from the cement paste) by creating more uniform areas in the concrete [94,110–113].

In contrast, increasing SF from 15% to higher levels generally led to reduced compressive strength at both 7 and 28 days, particularly beyond the optimal dosage. While small additions (up to 20%) may improve strength slightly in some series, excessive SF (25%) consistently results in strength losses. The optimal SF content for maximum strength was found to be 15% for series 1 and 3 and 20% for series 2. Raising the w/b ratio from 0.18 to 0.22 caused a notable decrease in compressive strength across all series. Furthermore, increasing SNF from 0.7% to 0.9% resulted in modest strength gains (1.5–3.5%).

Figure 17 illustrates the impact of the w/b ratio and SNF on compressive strength. As shown in Figure 17a, as the w/b ratio rises from 0.18 to 0.22, compressive strength decreases. Series 3 mixtures exhibit the highest bending strength relative to the w/b ratio. According to Figure 17b, as the SNF ratio increases from 0.7 to 0.9, compressive strength rises. Series 3 mixtures show the highest compressive strength concerning the SNF ratio.

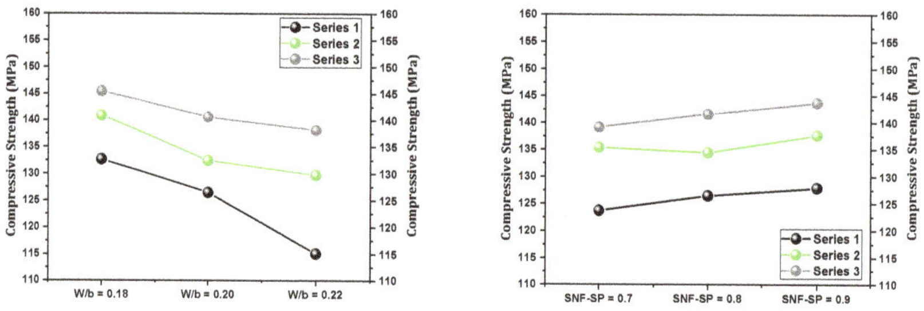

(a) (b)

Figure 17. Compressive strength under the influence of (**a**) changes in the w/b ratio; (**b**) changes in the dosage of SNF.

6.3. Analysis and Prediction of the Results with Artificial Neural Network (ANN) and Fuzzy Logic (FL)

Compressive strength at 7 and 28 days was assessed using ANN and FL models. After training the ANN model with various algorithms (LM and SCG) and different types of neuron arrangements in the hidden layer, the best neuron arrangement in the hidden layer was determined for each algorithm. The LM algorithm with 10 neurons in the hidden layer showed the highest correlation coefficient. Figure 18 illustrates the compressive strength outcomes at 7 days (Figure 18a) and 28 days (Figure 18b), obtained through actual and predicted results by the ANN and FL algorithms. The prediction error relative to the real results is depicted in Figure 19 for the 7-day (Figure 19a) and 28-day (Figure 19b) compressive strength.

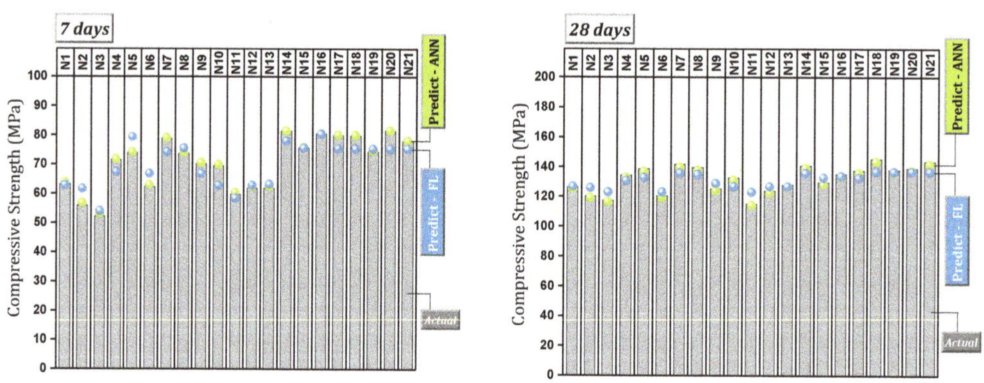

(a) (b)

Figure 18. Comparison of compressive strength results from prediction models (ANN and FL) and actual testing: (**a**) 7 days; (**b**) 28 days.

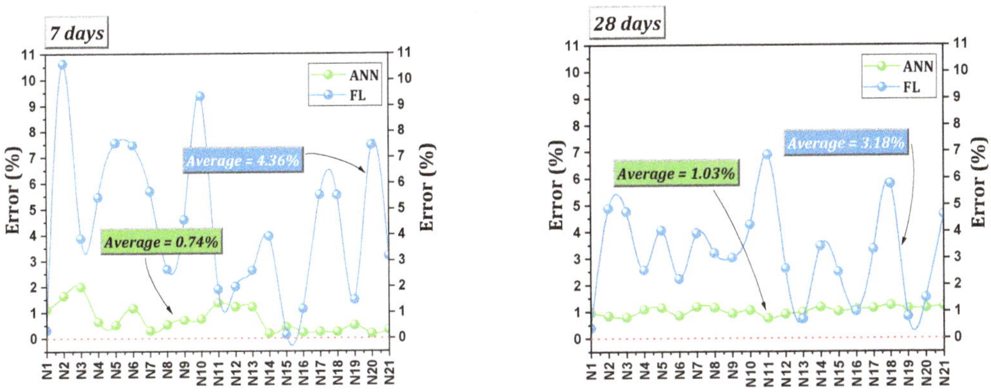

(a) (b)

Figure 19. Prediction error by ANN and FL models compared to reality: (**a**) 7 days; (**b**) 28 days.

The FL model had an average error rate of 4.36% for predicting 7-day compressive strength and 3.18% for predicting 28-day compressive strength. On the other hand, the average error in the ANN model was lower, suggesting that the results derived from the ANN model are more closely aligned with reality. Specifically, the ANN model achieved 0.74% in 7-day compressive strength prediction and 1.03% in 28-day compressive strength prediction (Figure 19). Figure 20 depicts the performance of the ANN model across training, validation, testing, and all datasets. It demonstrates the strong predictive capability of the ANN model in aligning with the experimental results, as indicated by the regression coefficient (>0.95).

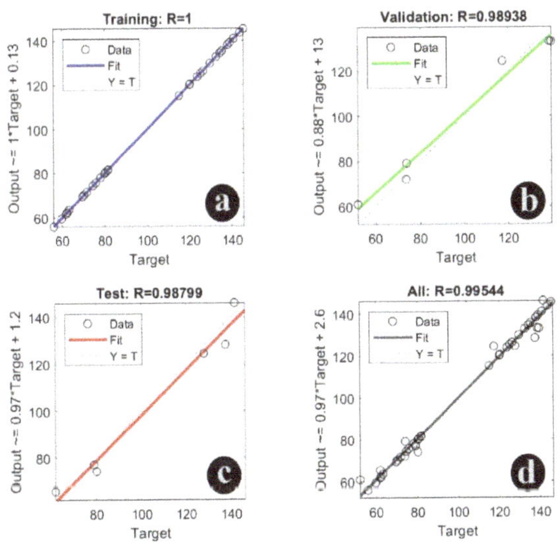

Figure 20. Performance of the artificial neural network (ANN) model based on (**a**) training, (**b**) validation, (**c**) test, and (**d**) all datasets.

Despite the satisfactory performance of both prediction models utilized in this study, the perceptron neural network model of the LM algorithm with a 2–10–4 architecture demonstrates superior performance compared to the FL model.

7. Conclusions

This study evaluated the fresh and mechanical performance of ultra-high-performance concrete (UHPC) mixtures incorporating varying replacement levels of silica fume (SF), sodium naphthalene formaldehyde (SNF) as a second-generation superplasticizer, and different water-to-binder (w/b) ratios. The key findings are summarized as follows:

Fresh Properties

- Increasing SF content from 15% to 25% led to reduced slump flow and shorter setting times, mainly due to reduced free water and accelerated hydration. Conversely, increasing the w/b ratio improved slump flow and extended setting time by enhancing particle mobility.
- A higher SNF dosage (from 0.7% to 0.9%) significantly improved both slump flow and setting time, demonstrating effective dispersion of cement particles and reduced agglomeration.
- Squeeze flow testing revealed three distinct paste behaviors:
- Low workability mixes with high stiffness and rapid transition to strain hardening;
- Highly flowable mixes exhibiting extended plastic phases and higher dilatancy;
- Transitional mixes highly sensitive to variables like flow rate and composition, affecting displacement trends.

Mechanical characteristics

- Increasing the SF dose led to the closure of capillary paths and enhanced density, resulting in higher electrical resistance. The rise in the w/b ratio provided a conducive substrate for ion movement, leading to reduced electrical resistance. Higher SNF substitution enhanced cement particle packing density and raised electrical resistance.
- Increasing SF dosage from 15% to 25% reduced compressive and flexural strength due to uneven cement particle distribution, hydration disruption, and ITZ weakening. A higher w/b ratio also decreased strength by limiting the material for C-S-H gel formation. However, raising the SNF dosage from 0.7% to 0.9% improved UHPC strength by promoting complete hydration and reducing internal friction.
- The ANN model with 4-10-2 architecture presented an acceptable performance, according to the regression coefficient of >0.98. Also, the good performance of the FL model should not be overlooked. The error rate of the fuzzy logic model resulted in a range of 3.18–4.36%, while the error rate recorded for the ANN model was much lower (0.74–1.03%).

Benefits and Limitations of Second-Generation Superplasticizers

- Naphthalene-based superplasticizers have the potential to improve workability without compromising strength while also improving particle dispersion for a denser, more homogeneous microstructure. However, their limitations include potential slump loss over time due to rapid adsorption on cement particles, reduced effectiveness in low-temperature conditions, and incompatibility with some cement types, which can lead to delayed setting or strength gain issues. Additionally, they may not provide the same level of flow retention as newer polycarboxylate ether (PCE)-based alternatives, making them less ideal for highly complex UHPC applications requiring prolonged workability. Nevertheless, using this type of economic superplasticizer for relatively expensive concrete mixtures like UHPC may present a partial solution for further industrial application of this material.

Recommended Mix Design

Among the tested mixes, the most balanced performance was observed for:

SF20WB20N09: 20% SF, w/b = 0.20, SNF = 0.9%.
This mixture exhibited high flexural and compressive strength, suitable setting time, and adequate slump flow, making it the recommended formulation for UHPC in practical applications.

Industrial perspective

- The study focused exclusively on second-generation SNF-based superplasticizers, and the conclusions remain within this scope. The results demonstrate that optimized SNF dosages (0.9%) can produce UHPC mixes with satisfactory fresh behavior and mechanical strength, suggesting the continued viability of SNF in cost-sensitive or prefabrication settings where flowability and early strength are critical. While third-generation superplasticizers offer enhanced performance, SNF is still a cost-effective and functional alternative, especially for applications not requiring extreme workability.

Author Contributions: This study is the result of the joint work of the authors. Conceptualization, A.S. and M.M.M.; methodology, A.S. and M.M.M.; software, A.S. and M.M.M.; validation, S.T., A.S. and M.M.M., S.E.R. and M.K.; formal analysis, A.S. and M.M.M.; investigation, S.T., J.B., A.S. and M.M.M.; resources, S.T., J.B., A.S. and M.M.M.; data curation, J.B., A.S. and M.M.M.; writing—original draft preparation, J.B., A.S. and M.M.M.; writing—review and editing, A.S., M.M.M., S.E.R. and M.K.; visualization, M.M.M.; supervision, A.S. and M.M.M.; project administration, A.S. and M.M.M. All authors have read and agreed to the published version of the manuscript.

Funding: This research received no external funding.

Institutional Review Board Statement: Not applicable.

Informed Consent Statement: Not applicable.

Data Availability Statement: The original contributions presented in this study are included in the article. Further inquiries can be directed to the corresponding authors.

Conflicts of Interest: Author Sam Rigby was employed by the company Arup Resilience, Security & Risk. The remaining authors declare that the research was conducted in the absence of any commercial or financial relationships that could be construed as a potential conflict of interest.

References

1. Li, J.; Wu, Z.; Shi, C.; Yuan, Q.; Zhang, Z. Durability of Ultra-High Performance Concrete—A Review. *Constr. Build. Mater.* **2020**, *255*, 119296. [CrossRef]
2. Du, J.; Meng, W.; Khayat, K.H.; Bao, Y.; Guo, P.; Lyu, Z.; Abu-obeidah, A.; Nassif, H.; Wang, H. New Development of Ultra-High-Performance Concrete (UHPC). *Compos. Part B Eng.* **2021**, *224*, 109220. [CrossRef]
3. Mohammad Nezhad Ayandeh, M.H.; Ghodousian, O.; Mohammad Nezhad, H.; Mohtasham Moein, M.; Saradar, A.; Karakouzian, M. Steel Slag and Zeolite as Sustainable Pozzolans for UHPC: An Experimental Study of Binary and Ternary Pozzolan Mixtures under Various Curing Conditions. *Innov. Infrastruct. Solut.* **2024**, *9*, 265. [CrossRef]
4. Shao, Y.; Nguyen, W.; Bandelt, M.J.; Ostertag, C.P.; Billington, S.L. Seismic Performance of High-Performance Fiber-Reinforced Cement-Based Composite Structural Members: A Review. *J. Struct. Eng.* **2022**, *148*, 03122004. [CrossRef]
5. Yang, C.; Okumus, P. Ultrahigh-Performance Concrete for Posttensioned Precast Bridge Piers for Seismic Resilience. *J. Struct. Eng.* **2017**, *143*, 03122004. [CrossRef]
6. Hung, C.-C.; El-Tawil, S.; Chao, S.-H. A Review of Developments and Challenges for UHPC in Structural Engineering: Behavior, Analysis, and Design. *J. Struct. Eng.* **2021**, *147*, 03121001. [CrossRef]
7. Cardoso, F.A.; John, V.M.; Pileggi, R.G.; Banfill, P.F.G. Characterisation of Rendering Mortars by Squeeze-Flow and Rotational Rheometry. *Cem. Concr. Res.* **2014**, *57*, 79–87. [CrossRef]
8. Cardoso, F.A.; Grandes, F.A.; Sakano, V.K.; Rego, A.C.A.; Lofrano, F.C.; John, V.M.; Pileggi, R.G. Experimental Developments of the Squeeze Flow Test for Mortars. In *Rheology and Processing of Construction Materials*; RILEM Bookseries; Springer: Cham, Switzerland, 2020; Volume 23.

9. Soltani, A.; Tarighat, A.; Rostami, R.; Tavakoli, D.; Moradi, A. Investigation of Mechanical Properties of Concrete with Clinoptilolite and Silica Fume Using Taguchi Method. *Innov. Infrastruct. Solut.* **2024**, *9*, 77. [CrossRef]
10. Ullah, R.; Qiang, Y.; Ahmad, J.; Vatin, N.I.; El-Shorbagy, M.A. Ultra-High-Performance Concrete (UHPC): A State-of-the-Art Review. *Materials* **2022**, *15*, 4131. [CrossRef]
11. Heidari, A.; Tavakoli, D. Performance of Ceramic Tile Powder as a Pozzolanic Material in Concrete. *Int. J. Adv. Mater. Sci.* **2012**, *3*, 1–11.
12. Kujawa, W.; Olewnik-Kruszkowska, E.; Nowaczyk, J. Concrete Strengthening by Introducing Polymer-Based Additives into the Cement Matrix-a Mini Review. *Materials* **2021**, *14*, 6071. [CrossRef] [PubMed]
13. Lei, D.; Yu, L.; Wang, S.; Li, Y.; Jia, H.; Wu, Z.; Bao, J.; Liu, J.; Xi, X.; Su, L. A State-of-the-Art on Electromagnetic and Mechanical Properties of Electromagnetic Waves Absorbing Cementitious Composites. *Cem. Concr. Compos.* **2025**, *157*, 105889. [CrossRef]
14. Li, Y.; Dong, C.; Wang, S.; Lei, D.; Yin, B.; Cui, Y.; Wang, Y.; Li, R. Micro-Macro Regulating Heterogeneous Interface Engineering in 3D N-Doped Carbon Fiber/MXene/TiO_2 Nano-Aerogel for Boosting Electromagnetic Wave Absorption. *Nano Res.* **2025**, *18*, 94907169. [CrossRef]
15. Lei, D.; Liu, C.; Wang, S.; Zhang, P.; Li, Y.; Yang, D.; Jin, Y.; Liu, Z.; Dong, C. Multiple Synergistic Effects of Structural Coupling and Dielectric-Magnetic Loss in Promoting Microwave Absorption of Bark-Derived Absorbers. *Adv. Compos. Hybrid. Mater.* **2025**, *8*, 158. [CrossRef]
16. Lei, D.; Jia, H.; Yu, L.; Li, Y.; Wu, Z.; Wang, B.; Zhou, A.; Qin, L.; Sun, J.; Wang, W.; et al. Recent Developments in Low-Carbon Engineered Cementitious Composites (ECC). *J. Build. Eng.* **2025**, *100*, 111734. [CrossRef]
17. Li, Y.; Liu, Z.; Lu, Y.; Yang, M.; Zhang, P.; Lei, D.; Liu, C.; Wang, S.; Dong, C. Surface Energy Induced Microstructural Engineering of Bio-Derived N-Doped Carbon Fibers Anchored by CoNi Nanoparticles for Superior Microwave Absorption. *Adv. Compos. Hybrid. Mater.* **2025**, *8*, 176. [CrossRef]
18. al-Swaidani, A.M.; Khwies, W.T.; al-Baly, M.; Lala, T. Development of Multiple Linear Regression, Artificial Neural Networks and Fuzzy Logic Models to Predict the Efficiency Factor and Durability Indicator of Nano Natural Pozzolana as Cement Additive. *J. Build. Eng.* **2022**, *52*, 104475. [CrossRef]
19. Rostami, J.; Rasekhi Sahneh, A.; Sedighardekani, R.; Latifinowsoud, M.; Rostami, R.; Kaltaei, A.; Sanaei Ataabadi, H.; Bahrami, N.; Mahmoudy, S.A.; Khandel, O. Influence of Phase Change Material and Nano Silica Aerogel Aggregates on the Characteristics of Cementitious Composite: An Experimental and Predictive Study. *J. Build. Eng.* **2024**, *82*, 108148. [CrossRef]
20. Heyran Najafi, M.R.; Saradar, A.; Mohtasham Moein, M.; Karakouzian, M. Investigation Mechanical Characteristics and Permeability of Concrete with Pozzolanic Materials: A Sustainable Approach. *Multiscale Multidiscip. Model. Exp. Des.* **2024**, *7*, 5051–5078. [CrossRef]
21. Moein, M.M.; Moein, A.M.; Saradar, A.; Rigby, S.E.; Tazari, H.; Karakouzian, M. Mechanical Properties of Portland Cement Concrete Mixed with Different Doses of Recycled Brick Powder and Steel Fiber. *Heliyon* **2025**, *11*, e41900. [CrossRef]
22. Tayfur, G.; Erdem, T.K.; Kırca, Ö. Strength Prediction of High-Strength Concrete by Fuzzy Logic and Artificial Neural Networks. *J. Mater. Civ. Eng.* **2014**, *26*, 04014079. [CrossRef]
23. Yuan, Y.; Yang, M.; Shang, X.; Xiong, Y.; Zhang, Y. Predicting the Compressive Strength of UHPC with Coarse Aggregates in the Context of Machine Learning. *Case Stud. Constr. Mater.* **2023**, *19*, e02627. [CrossRef]
24. Behforouz, B.; Tavakoli, D.; Gharghani, M.; Ashour, A. Bond Strength of the Interface between Concrete Substrate and Overlay Concrete Containing Fly Ash Exposed to High Temperature. *Structures* **2023**, *49*, 183–197. [CrossRef]
25. Song, H.; Ahmad, A.; Farooq, F.; Ostrowski, K.A.; Maślak, M.; Czarnecki, S.; Aslam, F. Predicting the Compressive Strength of Concrete with Fly Ash Admixture Using Machine Learning Algorithms. *Constr. Build. Mater.* **2021**, *308*, 125021. [CrossRef]
26. Sadrmomtazi, A.; Sobhani, J.; Mirgozar, M.A. Modeling Compressive Strength of EPS Lightweight Concrete Using Regression, Neural Network and ANFIS. *Constr. Build. Mater.* **2013**, *42*, 205–216. [CrossRef]
27. Mohtasham Moein, M.; Rahmati, K.; Mohtasham Moein, A.; Saradar, A.; Rigby, S.E.; Akhavan Tabassi, A. Employing Neural Networks, Fuzzy Logic, and Weibull Analysis for the Evaluation of Recycled Brick Powder in Concrete Compositions. *Buildings* **2024**, *14*, 4062. [CrossRef]
28. Mohtasham Moein, M.; Rahmati, K.; Mohtasham Moein, A.; Rigby, S.E.; Saradar, A.; Karakouzian, M. Utilizing Construction and Demolition Waste in Concrete as a Sustainable Cement Substitute: A Comprehensive Study on Behavior Under Short-Term Dynamic and Static Loads via Laboratory and Numerical Analysis. *J. Build. Eng.* **2024**, *97*, 110778. [CrossRef]
29. Sobhani, J.; Ejtemaei, M.; Sadrmomtazi, A.; Mirgozar, M.A. Modeling flexural strength of EPS lightweight concrete using regression, neural network and ANFIS. *Int. J. Optim. Civil Eng.* **2019**, *9*, 313–329.
30. Wang, S.; Xia, P.; Chen, K.; Gong, F.; Wang, H.; Wang, Q.; Zhao, Y.; Jin, W. Prediction and Optimization Model of Sustainable Concrete Properties Using Machine Learning, Deep Learning and Swarm Intelligence: A Review. *J. Build. Eng.* **2023**, *80*, 108065. [CrossRef]
31. Zadeh, L.A. Fuzzy Sets. *Inf. Control* **1965**, *8*, 338–353. [CrossRef]

32. Başyigit, C.; Akkurt, I.; Kilincarslan, S.; Beycioglu, A. Prediction of Compressive Strength of Heavyweight Concrete by ANN and FL Models. *Neural Comput. Appl.* **2010**, *19*, 507–513. [CrossRef]
33. Soltanifar, M.; Sharafi, H.; Hosseinzadeh Lotfi, F.; Pedrycz, W.; Allahviranloo, T. Introduction to Fuzzy Logic. In *Studies in Systems, Decision and Control*; Springer: Cham, Switzerland, 2023; Volume 471.
34. Zadeh, L.A. Is There a Need for Fuzzy Logic? *Inf. Sci.* **2008**, *178*, 2751–2779. [CrossRef]
35. Mirgozar Langaroudi, M.A.; Mohtasham Moein, M.; Saradar, A.; Karakouzian, M. Investigation of the Mechanical Properties and Durability of Fiber-Reinforced Geopolymer Mortars Containing Metakaolin and Glass Powder. *Infrastruct.* **2025**, *10*, 25. [CrossRef]
36. Saridemir, M. Predicting the Compressive Strength of Mortars Containing Metakaolin by Artificial Neural Networks and Fuzzy Logic. *Adv. Eng. Softw.* **2009**, *40*, 920–927. [CrossRef]
37. Topçu, I.B.; Saridemir, M. Prediction of Compressive Strength of Concrete Containing Fly Ash Using Artificial Neural Networks and Fuzzy Logic. *Comput. Mater. Sci.* **2008**, *41*, 305–311. [CrossRef]
38. Zhang, Q.; Shu, X.; Yang, Y.; Wang, X.; Liu, J.; Ran, Q. Preferential Adsorption of Superplasticizer on Cement/Silica Fume and Its Effect on Rheological Properties of UHPC. *Constr. Build. Mater.* **2022**, *359*, 129519. [CrossRef]
39. Li, P.P.; Yu, Q.L.; Brouwers, H.J.H. Effect of PCE-Type Superplasticizer on Early-Age Behaviour of Ultra-High Performance Concrete (UHPC). *Constr. Build. Mater.* **2017**, *153*, 740–750. [CrossRef]
40. Wang, R.; Gao, X.; Huang, H.; Han, G. Influence of Rheological Properties of Cement Mortar on Steel Fiber Distribution in UHPC. *Constr. Build. Mater.* **2017**, *144*, 65–73. [CrossRef]
41. Teng, L.; Jin, M.; Du, J.; Khayat, K.H. Synergetic Effect of Viscosity Modifying Admixtures and Polycarboxylate Ether Superplasticizer on Key Characteristics of Thixotropic UHPC for Bonded Bridge Deck Overlay Rehabilitation. *Case Stud. Constr. Mater.* **2024**, *20*, e02739. [CrossRef]
42. Yu, R.; Spiesz, P.; Brouwers, H.J.H. Effect of Nano-Silica on the Hydration and Microstructure Development of Ultra-High Performance Concrete (UHPC) with a Low Binder Amount. *Constr. Build. Mater.* **2014**, *65*, 140–150. [CrossRef]
43. Van Tuan, N.; Ye, G.; Van Breugel, K.; Fraaij, A.L.A.; Bui, D.D. The Study of Using Rice Husk Ash to Produce Ultra High Performance Concrete. *Constr. Build. Mater.* **2011**, *25*, 2030–2035. [CrossRef]
44. Ahmadi, S.M.; Honarbakhsh, A.; Zhiani, R.; Tavakoli, D. Effects of KCC-1/Ag Nanoparticles on the Mechanical Properties of Concrete. *Int. J. Eng.* **2022**, *35*, 1388–1397. [CrossRef]
45. Heidari, A.; Tavakoli, S.; Tavakoli, D. Reusing Waste Ceramic and Waste Sanitary Ware in Concrete as Pozzolans with Nano-Silica and Metakaolin. *Int. J. Sustain. Constr. Eng. Technol.* **2019**, *10*, 55–67. [CrossRef]
46. Murugesan, A.; Umapathi, N.; Mohamed Ismail, A.A.; Srinivasan, D. Compatibility Matrix of Superplasticizers in Ultra-High-Performance Concrete for Material Sustainability. *Innov. Infrastruct. Solut.* **2023**, *8*, 260. [CrossRef]
47. Ramkumar, K.B.; Kannan Rajkumar, P.R.; Noor Ahmmad, S.; Jegan, M. A Review on Performance of Self-Compacting Concrete—Use of Mineral Admixtures and Steel Fibres with Artificial Neural Network Application. *Constr. Build. Mater.* **2020**, *261*, 120215. [CrossRef]
48. Balf, F.R.; Kordkheili, H.M.; Kordkheili, A.M. A New Method for Predicting the Ingredients of Self-Compacting Concrete (SCC) Including Fly Ash (FA) Using Data Envelopment Analysis (DEA). *Arab. J. Sci. Eng.* **2021**, *46*, 4439–4460. [CrossRef]
49. Ly, H.-B.; Nguyen, T.-A.; Thi Mai, H.-V.; Tran, V.Q. Development of Deep Neural Network Model to Predict the Compressive Strength of Rubber Concrete. *Constr. Build. Mater.* **2021**, *301*, 124081. [CrossRef]
50. Liu, J.C.; Zhang, Z. Neural Network Models to Predict Explosive Spalling of PP Fiber Reinforced Concrete under Heating. *J. Build. Eng.* **2020**, *32*, 101472. [CrossRef]
51. Biswas, R.; Kumar, M.; Singh, R.K.; Alzara, M.; El Sayed, S.B.A.; Abdelmongy, M.; Yosri, A.M.; Yousef, S.E.A.S. A Novel Integrated Approach of RUNge Kutta Optimizer and ANN for Estimating Compressive Strength of Self-Compacting Concrete. *Case Stud. Constr. Mater.* **2023**, *18*, e02163. [CrossRef]
52. Qayyum Khan, A.; Ahmad Awan, H.; Rasul, M.; Ahmad Siddiqi, Z.; Pimanmas, A. Optimized Artificial Neural Network Model for Accurate Prediction of Compressive Strength of Normal and High Strength Concrete. *Clean. Mater.* **2023**, *10*, 100211. [CrossRef]
53. Shafaie, V.; Ghodousian, O.; Ghodousian, A.; Cucuzza, R.; Movahedi Rad, M. Integrating Push-out Test Validation and Fuzzy Logic for Bond Strength Study of Fiber-Reinforced Self-Compacting Concrete. *Constr. Build. Mater.* **2024**, *425*, 136062. [CrossRef]
54. Abbas, Y.M.; Iqbal Khan, M. Prediction of Compressive Stress–Strain Behavior of Hybrid Steel–Polyvinyl-Alcohol Fiber Reinforced Concrete Response by Fuzzy-Logic Approach. *Constr. Build. Mater.* **2023**, *379*, 131212. [CrossRef]
55. Demir, F. A New Way of Prediction Elastic Modulus of Normal and High Strength Concrete-Fuzzy Logic. *Cem. Concr. Res.* **2005**, *35*, 1531–1538. [CrossRef]
56. *ASTM C150/C150M-19a*; Standard Specification for Portland Cement. ASTM International: West Conshohocken, PA, USA, 2019.
57. *ASTM C1240*; Standard Specification for Silica Fume Used in Cementitious Mixtures. ASTM International: West Conshohocken, PA, USA, 2020.
58. *ASTM C778*; Standard Specification for Standard Sand. ASTM International: West Conshohocken, PA, USA, 2017.

59. *ASTM C494*; Standard Specification for Chemical Admixtures for Concrete. ASTM International: West Conshohocken, PA, USA, 2019.
60. *ASTM C230/230M-21*; Standard Specification for Flow Table for Use in Tests of Hydraulic Cement. ASTM International: West Conshohocken, PA, USA, 2021.
61. *ASTM C1437*; Standard Test Method for Flow of Hydraulic Cement Mortar. ASTM International: West Conshohocken, PA, USA, 2013.
62. *ASTM C191-08*; Standard Test Methods for Time of Setting of Hydraulic Cement by Vicat Needle. ASTM International: West Conshohocken, PA, USA, 2008.
63. Cardoso, F.A.; John, V.M.; Pileggi, R.G. Rheological Behavior of Mortars under Different Squeezing Rates. *Cem. Concr. Res.* **2009**, *39*, 748–753. [CrossRef]
64. *ASTM-C642-13*; Standard Test Method for Density, Absorption, and Voids in Hardened Concrete. ASTM International: West Conshohocken, PA, USA, 2013.
65. *AASHTO TP 119*; Standard Method of Test for Electrical Resistivity of a Concrete Cylinder Tested in a Uniaxial Resistance Test. American Association of State Highway and Transportation Officials: Washington, DC, USA, 2015.
66. *ASTM C109/C109M-20b*; Standard Test Method for Compressive Strength of Hydraulic Cement Mortars (Using 2-in. or [50 Mm] Cube Specimens). ASTM International: West Conshohocken, PA, USA, 2020.
67. *ASTM C348-21*; Standard Test Method for Flexural Strength of Hydraulic-Cement Mortars. ASTM International: West Conshohocken, PA, USA, 2021.
68. Ali, K.; Qureshi, M.I.; Saleem, S.; Khan, S.U. Effect of Waste Electronic Plastic and Silica Fume on Mechanical Properties and Thermal Performance of Concrete. *Constr. Build. Mater.* **2021**, *285*, 122952. [CrossRef]
69. Sharma, M.; Behera, P.; Saha, S.; Mohanty, T.; Saha, P. Effect of Silica Fume and Red Mud on Mechanical Properties of Ferrochrome Ash Based Concrete. *Mater. Today Proc.* **2022**, *60*, 55–61. [CrossRef]
70. Fidjestol, P. *ACI 234R-06: Guide for the Use of Silica Fume in Concrete*; American Concrete Institute: Mumbai, MH, USA, 2006; Volume 96.
71. Saradar, A.; Rezakhani, Y.; Rahmati, K.; Johari Majd, F.; Mohtasham Moein, M.; Karakouzian, M. Investigating the Properties and Microstructure of High-Performance Cement Composites with Nano-Silica, Silica Fume, and Ultra-Fine TiO_2. *Innov. Infrastruct. Solut.* **2024**, *9*, 84. [CrossRef]
72. Gleize, P.J.P.; Müller, A.; Roman, H.R. Microstructural Investigation of a Silica Fume-Cement-Lime Mortar. *Cem. Concr. Compos.* **2003**, *25*, 171–175. [CrossRef]
73. Jo, B.W.; Chakraborty, S.; Lee, S.T.; Lee, Y.S. Durability Study of Silica Fume-Mortar Exposed to the Combined Sulfate and Chloride-Rich Solution. *KSCE J. Civ. Eng.* **2019**, *23*, 356–366. [CrossRef]
74. Bahrami, N.; Zohrabi, M.; Mahmoudy, S.A.; Akbari, M. Optimum Recycled Concrete Aggregate and Micro-Silica Content in Self-Compacting Concrete: Rheological, Mechanical and Microstructural Properties. *J. Build. Eng.* **2020**, *31*, 101361. [CrossRef]
75. Sadrmomtazi, A.; Langeroudi, M.A.M.; Fasihi, A.; Haghi, A.K. An Investigation on Effect of Using PP Fibers and Different Cementitious Materials on Mechanical Properties of EPS Concrete. In Proceedings of the 3rd International Conference on Concrete and Development, Tehrant, Iran, 27–29 April 2009.
76. Huynh, P.T.; Bui, P.T.; Ogawa, Y.; Kawai, K. Effect of Water-to-Binder Ratio on Cementing Efficiency Factor of Fly Ash Regarding Compressive Strength of Concrete. In *ICSCEA 2019: Proceedings of the International Conference on Sustainable Civil Engineering and Architecture*; Lecture Notes in Civil Engineering; Springer: Singapore, 2020; Volume 80.
77. Melo, K.A.; Repette, W.L. Optimization of Superplasticizer Content in Self-Compacting Concrete. In *Measuring, Monitoring and Modeling Concrete Properties*; Springer: Dordrecht, The Netherlands, 2007.
78. Kawai, T.; Okada, T. *Effect of Superplasticizer and Viscosity-Increasing Admixture on Properties of Lightweight Aggregate Concrete*; ACI Special Publication; American Concrete Institute: Farmington Hills, MI, USA, 1989; Volume SP-119.
79. Zhang, G.; Wang, S.; Wang, B.; Zhao, Y.; Kang, M.; Wang, P. Properties of Pervious Concrete with Steel Slag as Aggregates and Different Mineral Admixtures as Binders. *Constr. Build. Mater.* **2020**, *257*, 119543. [CrossRef]
80. Wang, G.; Kong, Y.; Sun, T.; Shui, Z. Effect of Water-Binder Ratio and Fly Ash on the Homogeneity of Concrete. *Constr. Build. Mater.* **2013**, *38*, 1129–1134. [CrossRef]
81. Verma, M.; Dev, N. Effect of Liquid to Binder Ratio and Curing Temperature on the Engineering Properties of the Geopolymer Concrete. *Silicon* **2022**, *14*, 1743–1757. [CrossRef]
82. Xun, W.; Wu, C.; Leng, X.; Li, J.; Xin, D.; Li, Y. Effect of Functional Superplasticizers on Concrete Strength and Pore Structure. *Appl. Sci.* **2020**, *10*, 3496. [CrossRef]
83. Yousuf, F.; Wei, X.; Tao, J. Evaluation of the Influence of a Superplasticizer on the Hydration of Varying Composition Cements by the Electrical Resistivity Measurement Method. *Constr. Build. Mater.* **2017**, *144*, 25–34. [CrossRef]
84. Ma, B.; Ma, M.; Shen, X.; Li, X.; Wu, X. Compatibility between a Polycarboxylate Superplasticizer and the Belite-Rich Sulfoaluminate Cement: Setting Time and the Hydration Properties. *Constr. Build. Mater.* **2014**, *51*, 47–54. [CrossRef]
85. Min, B.H.; Erwin, L.; Jennings, H.M. Rheological Behaviour of Fresh Cement Paste as Measured by Squeeze Flow. *J. Mater. Sci.* **1994**, *29*, 1374–1381. [CrossRef]

86. Lai, M.H.; Binhowimal, S.A.M.; Hanzic, L.; Wang, Q.; Ho, J.C.M. Dilatancy Mitigation of Cement Powder Paste by Pozzolanic and Inert Fillers. *Struct. Concr.* **2020**, *21*, 1164–1180. [CrossRef]
87. Lai, M.H.; Lao, W.C.; Tang, W.K.; Hanzic, L.; Wang, Q.; Ho, J.C.M. Dilatancy Swerve in Superplasticized Cement Powder Paste. *Constr. Build. Mater.* **2023**, *362*, 129524. [CrossRef]
88. Daukšys, M.; Skripkiunas, G. Investigation of Dilatancy Mechanism of Portland Cement Paste. *Constr. Build. Mater.* **2015**, *83*, 53–61. [CrossRef]
89. Medeiros, R.A.; Lima, M.G. Electrical Resistivity of Unsaturated Concrete Using Different Types of Cement. *Constr. Build. Mater.* **2016**, *107*, 11–16. [CrossRef]
90. Faraj, R.H.; Ahmed, H.U.; Rafiq, S.; Sor, N.H.; Ibrahim, D.F.; Qaidi, S.M.A. Performance of Self-Compacting Mortars Modified with Nanoparticles: A Systematic Review and Modeling. *Clean. Mater.* **2022**, *4*, 100086. [CrossRef]
91. Cosoli, G.; Mobili, A.; Tittarelli, F.; Revel, G.M.; Chiariotti, P. Electrical Resistivity and Electrical Impedance Measurement in Mortar and Concrete Elements: A Systematic Review. *Appl. Sci.* **2020**, *10*, 9152. [CrossRef]
92. Pourahmadi, H.; Arabani, S.; Sadrmomtazi, A.; Langaroudi, M.A.M.; Khoshkbijari, R.K.; Amooie, M. Durability of Self-Compacting Lightweight Aggregate Concretes (LWSCC) as Repair Overlays ARTICLE INFO ABSTRACT. *J. Rehabil. Civ. Eng.* **2017**, *5*, 101–113.
93. Sadrinejad, I.; Madandoust, R.; Ranjbar, M.M. The Mechanical and Durability Properties of Concrete Containing Hybrid Synthetic Fibers. *Constr. Build. Mater.* **2018**, *178*, 72–82. [CrossRef]
94. AlHassan, M.K.; Nasser, M.S.; Onaizi, S.A.; Hussein, I.A.; Hassan, M.K. Production of High-Performance Concrete through the Addition of New Generation of Superplasticizers: A Comprehensive Review on the Electrokinetics and Rheological Behaviors. *Emergent Mater.* **2023**, *7*, 403–441. [CrossRef]
95. Liao, Y.; Wei, X. Penetration Resistance and Electrical Resistivity of Cement Paste with Superplasticizer. *Mater. Struct.* **2014**, *47*, 563–570. [CrossRef]
96. Toutanji, H.A.; El-Korchi, T. The Influence of Silica Fume on the Compressive Strength of Cement Paste and Mortar. *Cem. Concr. Res.* **1995**, *25*, 1591–1602. [CrossRef]
97. Shafieyzadeh, M. Prediction of Flexural Strength of Concretes Containing Silica Fume and Styrene-Butadiene Rubber (SBR) with an Empirical Model. *J. Inst. Eng. (India) Ser. A* **2015**, *96*, 349–355. [CrossRef]
98. Beshkari, M.; Amani, B.; Rahmati, K.; Mohtasham Moein, M.; Saradar, A.; Karakouzian, M. Synergistic Effects of Pozzolan and Carbon Fibers on the Performance of Self-Consolidating Concrete under Plastic Shrinkage and Dynamic Loading. *Innov. Infrastruct. Solut.* **2024**, *9*, 160. [CrossRef]
99. He, T.; Xiang, W.; Zhang, J.; Hu, C.; Zhang, G.; Kou, B. Influence of Water-Binder Ratio on the Mechanical Strength and Microstructure of Arch Shell Interface Transition Zone. *Buildings* **2022**, *12*, 491. [CrossRef]
100. Yang, Y.; Gao, X.; Deng, H.; Yu, P.; Yao, Y. Effects of Water/Binder Ratio on the Properties of Engineered Cementitious Composites. *J. Wuhan. Univ. Technol. Mater. Sci. Ed.* **2010**, *25*, 298–302. [CrossRef]
101. Gagné, R.; Boisvert, A.; Pigeon, M. Effect of Superplasticizer Dosage on Mechanical Properties, Permeability, and Freeze-Thaw Durability of High-Strength Concretes with and without Silica Fume. *ACI Mater. J.* **1996**, *93*, 111–120. [CrossRef]
102. Adebanjo, A.U.; Shafiq, N.; Razak, S.N.A.; Kumar, V.; Farhan, S.A.; Singh, P.; Abubakar, A.S. Design and Modeling the Compressive Strength of High-Performance Concrete with Silica Fume: A Soft Computing Approach. *Soft Comput.* **2023**, *28*, 6059–6083. [CrossRef]
103. Eren, Ö.; Çelik, T. Effect of Silica Fume and Steel Fibers on Some Properties of High-Strength Concrete. *Constr. Build. Mater.* **1997**, *11*, 373–382. [CrossRef]
104. Nabighods, K.; Saradar, A.; Mohtasham Moein, M.; Mirgozar Langaroudi, M.A.; Byzyka, J.; Karakouzian, M. Evaluation of Self-Compacting Concrete Containing Pozzolan (Zeolite, Metakaolin & Silica Fume) and Polypropylene Fiber against Sulfate Attacks with Different PH: An Experimental Study. *Innov. Infrastruct. Solut.* **2023**, *9*, 1. [CrossRef]
105. Mousavinejad, S.H.G.; Saradar, A.; Jabbari, M.; Moein, M.M. Evaluation of Fresh and Hardened Properties of Self-Compacting Concrete Containing Different Percentages of Waste Tiles. *J. Build. Pathol. Rehabil.* **2023**, *8*, 81. [CrossRef]
106. Xiaofeng, C.; Shanglong, G.; Darwin, D.; McCabe, S.L. Role of Silica Fume in Compressive Strength of Cement Paste, Mortar, and Concrete. *ACI Mater. J.* **1992**, *89*, 375–387. [CrossRef]
107. Yogendran, V.; Langan, B.W.; Haque, M.N.; Ward, M.A. Silica Fume in High-Strength Concrete. *ACI Mater. J.* **1987**, *84*, 124–129. [CrossRef]
108. Zareei, S.A.; Ameri, F.; Shoaei, P.; Bahrami, N. Recycled Ceramic Waste High Strength Concrete Containing Wollastonite Particles and Micro-Silica: A Comprehensive Experimental Study. *Constr. Build. Mater.* **2019**, *201*, 11–32. [CrossRef]
109. Burhan, L.; Ghafor, K.; Mohammed, A. Modeling the Effect of Silica Fume on the Compressive, Tensile Strengths and Durability of NSC and HSC in Various Strength Ranges. *J. Build. Pathol. Rehabil.* **2019**, *4*, 19. [CrossRef]
110. Govin, A.; Bartholin, M.C.; Schmidt, W.; Grosseau, P. Combination of Superplasticizers with Hydroxypropyl Guar, Effect on Cement-Paste Properties. *Constr. Build. Mater.* **2019**, *215*, 595–604. [CrossRef]

111. Matias, D.; De Brito, J.; Rosa, A.; Pedro, D. Mechanical Properties of Concrete Produced with Recycled Coarse Aggregates—Influence of the Use of Superplasticizers. *Constr. Build. Mater.* **2013**, *44*, 101–109. [CrossRef]
112. Shi, C.; He, T.S.; Zhang, G.; Wang, X.; Hu, Y. Effects of Superplasticizers on Carbonation Resistance of Concrete. *Constr. Build. Mater.* **2016**, *108*, 48–55. [CrossRef]
113. Malhotra, V.M. Effect of Repeated Dosages of Superplasticizers on Slump, Strength and Freeze-Thaw Resistance of Concrete. *Matériaux Constr.* **1981**, *14*, 79–89. [CrossRef]

applied sciences

MDPI

Article

Concrete Crack Width Measurement Using a Laser Beam and Image Processing Algorithms

Mthabisi Adriano Nyathi, Jiping Bai * and Ian David Wilson

Faculty of Computing, Engineering and Science, University of South Wales, Pontypridd CF37 1DL, UK
* Correspondence: jiping.bai@southwales.ac.uk

Abstract: The presence of concrete cracks in structures indicates possible structural deterioration, but it is quite difficult to measure crack width accurately. While much research has been conducted on crack detection using image processing, there is a gap in the accurate quantification of crack width in millimeters. Current methods either measure in pixels or require the attachment of scales or markers onto a measured surface, which can pose safety hazards in hard-to-reach areas. This paper addresses these issues by proposing a novel image-based method for measuring concrete crack width in millimeters using a laser beam and image processing. The proposed method was validated in the laboratory by capturing images of concrete cracks with two cameras of different resolutions, each attached with lasers. The lasers projected a circular laser beam onto the surface of the concrete under inspection. The images were then pre-processed, segmented, and skeletonized for crack width measurement in pixels. The relationship between the laser diameter and camera distance from the surface was used to convert the measured crack width from pixels to millimeters. The method was performed with high accuracy, as indicated by the absolute error. The largest absolute error was 0.57 mm, while the smallest absolute error was 0.02 mm. The proposed method allows real-world interpretation of results in millimeters. As a result, measured crack widths can easily be compared to allowable limits in international standards, which are typically expressed in metric or SI units. The proposed method can also promote safer inspections in areas of low accessibility by attaching the laser to devices such as drones.

Keywords: crack detection; crack width measurement; laser beam; image processing

1. Introduction

Concrete structures are increasingly experiencing deterioration due to various factors such as aging, increased traffic loads, and extreme weather conditions [1,2]. While concrete structures experience various kinds of damage, cracks are typically one of the most common indicators of the presence of damage. The presence of cracks does not always indicate imminent collapse; however, they often lead to reduced structural integrity. Cracks weaken structures by leading to issues such as corrosion of internal rebar and reduced loading capacity [3]. It is important to detect and measure cracks in a timely manner in order to infer decisions on necessary repairs and maintenance [4]. Failure to do so can lead to catastrophic failures with a high cost to human life and the economy [5]. It is worth noting that it is common for concrete to crack, and not all cracks are a cause of concern. The width of cracks is typically a good indicator of the severity of the cracks.

The conventional method for detecting and measuring concrete cracks has been through visual inspections performed by trained professionals on-site. However, this method has limitations such as a potential for human error, high costs of operation, disruptions to traffic, and access limitations [2,6]. Crack widths are typically recorded manually during the visual inspection using crack-width comparator gauges [7], crack scales, or specialized bridge inspection vehicles.

Extensive research has been conducted on image-based crack detection methods. The images of the concrete surfaces are captured using a variety of image acquisition devices. The most common devices include DSLR cameras [6,8–10], unmanned aerial vehicles [6,8,11,12], and smartphones [9,13,14]. Other techniques include infrared cameras, 3D laser microscopes [15], and stereovision [16]. Early works looked at the use of image-processing techniques that used threshold-based methods [17]. Threshold-based methods rely on pixel intensity information for the detection of cracks. Fujita and Hamamoto [18] proposed a method for the detection of cracks on noisy concrete surfaces which applied locally adaptive thresholding to an image that had undergone pre-processing and probabilistic relaxation. Kim et al. [19] detected cracks using a hybrid image processing method that used the image binarization technique by Sauvola [20]. The technique was initially designed for detecting text in noisy backgrounds. The detected cracks were then skeletonized, allowing crack width to be measured as the distance between the two extreme ends of the crack skeleton. Ioli et al. [21] also used binarization and skeletonization based on medial axis transform and the Canny edge detection algorithm for crack detection and measurement. Yang et al. [4] measured crack widths in pixels by using a medial axis transform for the skeletonization of cracks. Thus, representing the cracks as single-pixel wide crack representations. The use of morphological operations prior to skeletonization is a widely adopted technique in image-based concrete crack detection. Common morphological operations include thinning, edge detection, closing, opening, erosion, and dilation. This step is often applied to improve the accuracy of the results [4,19]. Other skeletonization algorithms have been proposed, such as the Hilditchs approach [22] and the 3D medial axis thinning algorithm [23,24]. However, these methods tend to be computationally demanding due to their iterative nature [4].

The majority of existing studies in the literature measure concrete crack width in pixels and do not provide a direct understanding of the extent of damage in the real world. Recent studies have explored methods for measuring concrete crack width in millimeters from images. Tomczak et al. [25] measured the change in crack width in millimeters by keeping the distance between the concrete surface and the image scanner of known resolution constant. The estimated standard uncertainty of average crack width was found to be ±0.008 mm. Ito et al. [26] attached a transparent scale to the surface of a concrete block being measured and used image processing techniques such as thresholding and sub-pixel interpolation for crack width measurement in millimeters. Peng et al. [6] were able to measure the width of cracks on images in millimeters by utilizing a rangefinder and three laser points projected onto the measured surface, which were used to calculate the corrected angle of the measured plane, $cos\theta$. Peng et al. [6] then used the corrected angle of the measured plane in Equation (1) to find the ratio of actual width and pixel width (J).

$$J = \frac{(L - f) \cdot d}{f \cdot D} cos\theta \text{ (mm/pixel)} \quad (1)$$

where L is distance, f is the focal length, d is the longest dimension of the image sensor, and D is the number of pixels along the long side of the image sensor.

Li and Zhao [13] measured the maximum crack width using the Euclidean distance transform and the least squares principle. The maximum crack width was measured in millimeters by using the relation between pixel distance and real distance, acquired using a laser range finder, camera field of view, and camera distance to the surface. Kim et al. [19] and Kim and Cho [27] converted the crack widths measured in pixels to millimeters by using the camera pinhole model [28], as shown in Equation (2).

$$W_r = D_p W_p = \frac{10\ D_w W_p}{P_c L_f} \quad (2)$$

where W_r is the real crack width in millimeters, D_p is the camera resolution, W_p is the crack width in pixels, D_w is the working distance in mm, P_c is the pixels per centimeter of the camera and L_f is the focal length of the camera in mm.

Jeong et al. [29] used Otsu image processing to measure crack width in pixels. The crack width was converted to millimeters using a proposed equation shown in Equation (3).

$$CW = GSD \times N_m = \left(\frac{L - f}{f} \times \frac{S_S}{S_R} \right) \times N_m \tag{3}$$

where CW is the crack width in millimeters, GSD is the ground sample distance, N_m is the number of measured pixels, L is the working distance, f is the focal length, S_S is the horizontal dimensions of the sensor, S_R is the horizontal or vertical resolution values in pixels of the sensor.

Previous methods for measuring concrete crack width are effective, but they are not without limitations. The need for knowledge of key parameters such as the camera properties such as resolution and focal length has a direct impact on accuracy. The use of incorrect parameters can lead to inaccurate measurements. Other methods face limitations, such as the requirement for markers or scales to be attached to the measured surface. These methods pose significant safety hazards in areas of low accessibility.

The aim of this paper is to overcome these limitations by developing a novel concrete crack width measurement method. This is achieved by proposing an affordable and highly flexible approach that utilizes laser beam and image processing algorithms. The crack width is measured in millimeters without the need to know any camera parameters or the need to attach markers or scales to the structure. The effectiveness of this approach is demonstrated through a laboratory experiment and images captured in the real world. The advantages of this method include its ease of application, promotion of safer inspections in hazardous cases, and the ability to measure crack width in millimeters. This method will allow inspectors to interpret results and compare them to allowable limits in international guidelines and codes typically stated in SI or metric units.

2. Materials and Methods

2.1. Overview of Proposed Method

The proposed crack width measuring technique uses a laser beam and image processing algorithms to accurately measure concrete crack widths in millimeters. An overview of the proposed method is illustrated in the flow chart shown in Figure 1. The steps involved in the crack width measurement process, such as image acquisition, crack segmentation, morphological operations, and skeletonization, are presented in the subsequent sections.

2.2. Determining the Pixel to Millimeter Conversion Factor Using the Laser Beam

The laser beam is a key factor in the ability of the proposed method to accurately measure concrete crack width in millimeters. A circular laser beam was chosen because it is easier to detect and measure its diameter more accurately using image processing algorithms. Prior to crack width measurement, the relationship between the laser diameter and the distance to the measuring plane was first established. Knowledge of this relationship is essential because the diameter of the laser beam projected onto the measuring plane changes with distance. Understanding the behavior of this change in diameter allowed the calculation of a conversion factor, α_c. The conversion factor converts concrete crack widths from pixels to millimeters. An indoor experiment was conducted to establish this relationship. A camera with an attached laser was placed at varying distances from the measured plane, ranging from 50 mm to 1800 mm. The distance and diameter of the laser beam were recorded for each instance and plotted, as shown in Figure 2. The relationship between the diameter of the laser beam and the distance to the measured plane was found to be directly proportional. The relationship is defined by Equation (4).

$$\varnothing_{real} = 0.055D + 2.644 \text{ (mm)} \tag{4}$$

where $\varnothing_{real}$ is the actual diameter of the laser in millimeters, and D is the distance to the measuring plane in millimeters.

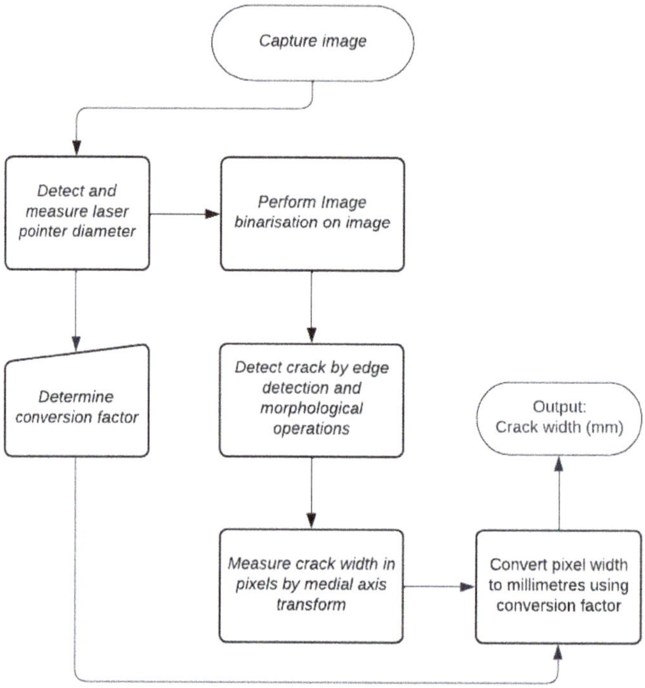

Figure 1. Flowchart showing the crack width measurement process.

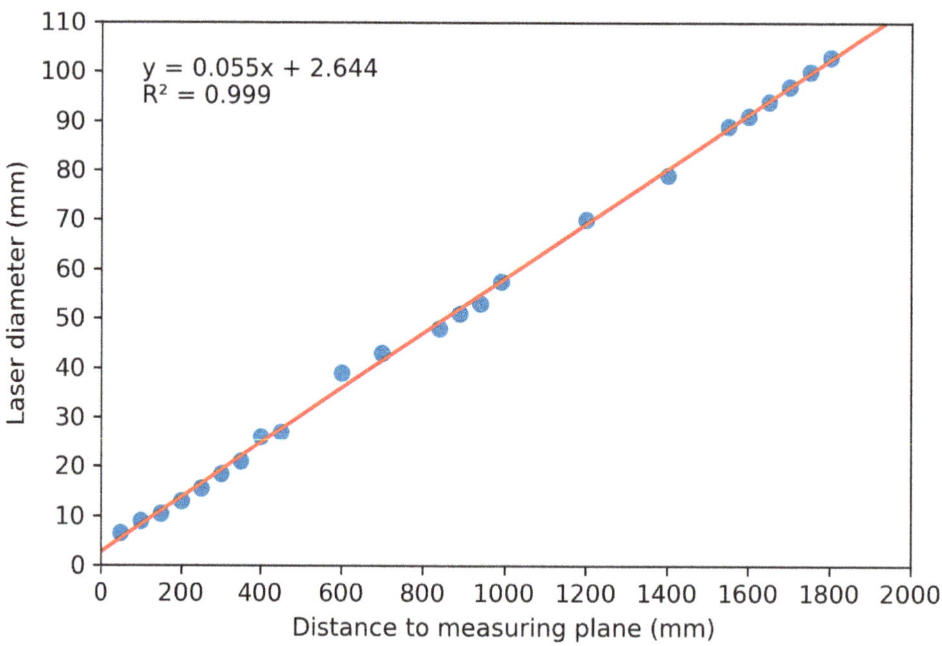

Figure 2. Relationship between distance to measuring plane and laser beam diameter.

Using this relationship, in conjunction with the diameter of the laser beam measured in pixels, the conversion factor, α_c, is defined by Equation (5).

$$\alpha_c = \frac{\varnothing_{real}}{\varnothing_{px}} \text{ (mm/px)} \tag{5}$$

where α_c is the conversion factor, $\varnothing_{real}$ is the actual diameter of the laser in millimeters and $\varnothing_{px}$ is the diameter of the laser in the image measured in pixels.

2.3. Image Acquisition

In order to develop and validate the proposed crack width measurement method, reinforced concrete beams were subjected to four-point bending loading until failure in the laboratory. The width of the cracks on the beams was manually measured in millimeters. Images of the cracks were captured using two different image acquisition devices. The first device was an iPhone 11 Pro Max smartphone equipped with a 12 mega-pixel f/1.8 aperture wide lens. The second device was a Trust webcam with a resolution of 640 × 480 pixels. The reason for choosing these two devices was their difference in image resolution, with the iPhone producing high-quality images while the Trust webcam produced low-quality images. The distance at which some of the images were captured is shown in Table 1, along with details of the image acquisition devices. This measurement and acquisition process was repeated on cracks in buildings for validation purposes.

Table 1. Some of the images of cracks were captured using two different devices.

Images	Distance to Measuring Plane (mm)	Acquisition Device
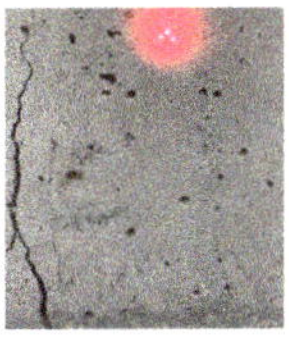	250	iPhone Pro Max 11
	460	Trust Webcam
	1150	iPhone Pro Max 11

2.4. Image Processing

Image processing algorithms were applied to the captured images using Python. Libraries such as NumPy, and Matplotlib were used, but the main library used for the image processing was the Open Computer Vision (OpenCV) Library [30].

Before the crack width could be measured, it was necessary to pre-process the images. Image processing was carried out in two stages: (1) detecting the laser beam in the image and measuring its pixel diameter and (2) further image processing to segment cracks and measure their respective crack widths. The steps taken in carrying out these operations are briefly explained below:

- Cropping: The images were cropped based on the region of interest, which was the area with the crack pattern that was to be measured. This was conducted to remove unnecessary information from the image that might affect the accuracy of measurements or lead to increased computation costs.
- Laser beam detection and diameter measurement: In order to detect the presence of the laser beam in the images captured in the laboratory and in the field, a color filtering technique was used to isolate the color red in the image. The projected laser beam was red in color, as shown in Figure 3a. The color filtering approach allowed for red pixels in the image to be identified, as shown in Figure 3b. Hough Circle Transform, a technique commonly used in computer vision for the task of circle detection, was then applied to the filtered image. Hough Circle Transform is defined by Equation (6) and detects circles by searching for points (x, y) in the image that satisfy Equation (6) for a given set of values of $(x_0,\ y_0)$ and r. Figure 3c shows the detected circle and the center of the circle.

$$(x - x_0)^2 + (y - y_0)^2 = r^2 \tag{6}$$

  The radius of the detected circle can then be used to determine the diameter of the circle in pixels, $\varnothing_{px}$, which is inputted into Equation (5), enabling the conversion factor, α_c, to be calculated. This specific approach was chosen because the Hough Circle Transform is a well-established method for detecting circular shapes in images, and it has been widely used in various applications [8,31–33]. Additionally, filtering out the color red was the most effective approach in eliminating the chance of other objects that might resemble circles from being falsely detected.
- BGR to grayscale conversion: Images are made up of thousands of pixels, which are typically represented by three color components; red (R), green (G), and blue (B), commonly referred to as RGB. Each pixel has an R, G, and B value between 0 and 255. However, OpenCV reads images in BGR format. Converting the images from BGR to Grayscale simplifies the representation of the pixels into a single value between 0 (black) and 255 (white). The BGR images are converted to grayscale by Equation (7).

$$Gray\_image = 0.114\ B + 0.587\ G + 0.299\ R \tag{7}$$

- Edge detection and morphological operations: Some of the images containing text on the surface needed removal of the text, as it could easily be mistaken for a crack. The image's noise was reduced through image filters to improve the visibility of the cracks. These filters were chosen because they are effective in removing noise while preserving the edges of the cracks. The specific parameter values for the filters were chosen through trial and error to achieve the best results. The canny edge detection algorithm [34] was used to detect the cracks in the images by identifying gradient changes in the image intensity. After successful edge detection, morphological opening and closing operations were performed on the image to remove noise and fill small holes in the image. The initial kernel sizes used for each image differed and were chosen at random and fine-tuned until satisfactory results were achieved.

(a) (b) (c)

Figure 3. (a) Original image; (b) color filtered image (c) grayscale image with laser detected and center marked.

2.5. Crack Width Calculation

The images resulting from the image processing step are then further processed by skeletonization of the crack. Skeletonization is conducted using a method known as medial axis transform, which converts an object shape into a single-pixel-wide representation, thus clearly outlining the crack's topology. The maximum crack width is measured in pixels as the maximum distance from the center of the medial axis to the two edges of the crack on either side of the medial axis, as shown in Figure 4.

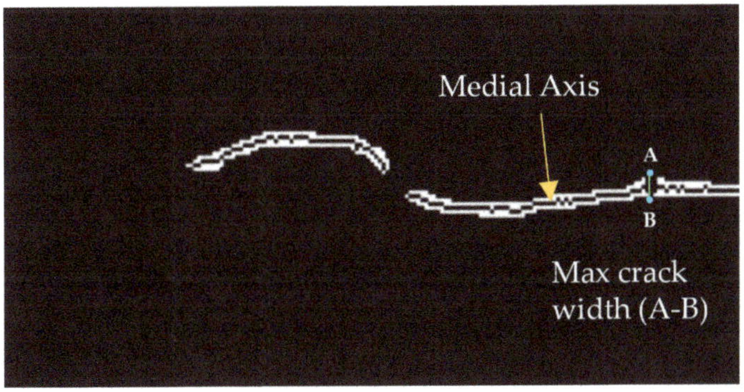

Figure 4. Calculation of maximum crack width based on medial axis.

The medial axis is calculated by using the distance transform a technique that assigns to each pixel in the image the minimum distance to the nearest non-zero pixel. The crack width in pixels is converted to millimeters by multiplying it by the conversion factor derived in Equation (5), as shown in Equation (8).

$$C_w \text{ (mm)} = Cw_p \times \alpha_c = Cw_p \times \frac{\varnothing_{real}}{\varnothing_{px}} \tag{8}$$

where C_w is the crack width in millimeters, Cw_p is the crack width in pixels and α_c is the conversion factor in (mm/pixels).

Table 2 shows a comparison of the necessary parameters required by different methods to calculate crack width in millimeters. It can be seen that the proposed method only depends on one parameter, which is the distance to the measuring plane. The distance to the measuring plane is unaffected by lens size and zooming, unlike focal length, which is used in other studies. Therefore, this promotes flexibility and higher accuracy as this minimizes accumulative error due to multiple parameters, which may not also be readily known. The proposed method allows the quick use of any available camera device to capture images with cracks.

Table 2. Comparison of image-based crack width measurement methods.

References	Method	User Parameters
[6]	$CW \text{ (mm)} = Pixel\ width \times \frac{(L-f) \cdot d}{f\ D} \cos\theta$	$[L, f, d, D, \theta]$ [1]
[19,27]	$W_r(\text{mm}) = D_p W_p = \frac{10\ D_w W_p}{P_c L_f}$	$[D_w, P_c, L_f]$ [2]
[29]	$CW \text{ (mm)} = \left(\frac{L-f}{f} \times \frac{S_S}{S_R}\right) \times N_m$	$[L, f, S_s, S_R]$ [3]
Proposed method in this paper	$Crack\ width\ \text{(mm)} = Cw_p \times \alpha_c$	Distance to measuring plane (used to calculate α_c)

[1] Where L = distance, f = focal length, d = longer dimension of image sensor, D = number of pixels along long side of image sensor, θ = corrected angle of measuring plane. [2] Where D_w = Working distance, P_c = pixels per centimeter of camera, L_f = focal length. [3] Where L = Working distance, f = focal length, S_s = horizontal or vertical dimensions of sensor, S_R = horizontal or vertical resolution of sensor.

3. Results and Discussion

This section presents and discusses the results obtained from the proposed method. The results of the crack segmentations, which were achieved through image processing techniques, are presented first. Their impact on the accuracy of crack width measurement and the overall performance of the method are then discussed. Finally, the results of the concrete crack width measurement are presented, discussed, and compared to other approaches in the literature.

3.1. Crack Segmentation

Figure 5 shows the image processing steps that were applied to the images to create a crack mask through segmentation. The crack masks, seen after the morphological closing stage, were used for the maximum crack width measurement.

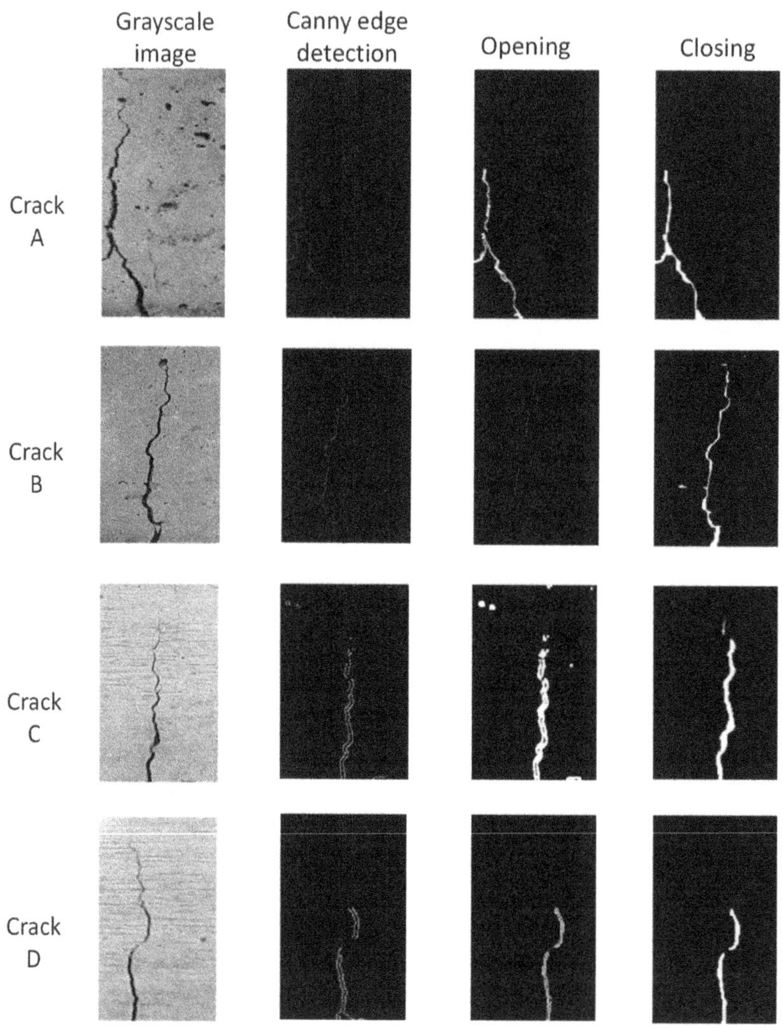

Figure 5. Segmentation of cracks using image processing algorithms.

The proposed implementation of the image processing algorithms achieved reasonably accurate crack segmentation. The initial parameters used in the image processing algorithms were manually specified and fine-tuned until satisfactory results were achieved. The reliance on the user to specify the initial parameters means the efficiency can vary depending on the experience of the user specifying the initial parameters. Kim et al. [19] also highlighted how this could affect the overall accuracy of methods relying on manual parameter selection.

Although this study does not focus on the measurement of crack lengths, it can be seen from Figure 5 that the lengths of cracks A, C, and D are shorter than before segmentation. Crack A and crack D were the most affected, followed by crack C. Crack B retained its length even after segmentation. This is a common problem. Kim et al. [19] found there to be a conflict between obtaining high accuracy for both crack width and crack length and, as a result, proposed a hybrid method [19]. There are several factors that generally lead to this when using image processing algorithms for segmentation, such as image quality, lighting conditions, and distance from the surface [5,35]. These factors also have a direct effect on the accuracy of crack width measurement.

3.2. Maximum Crack Width Measurement

The proposed method was validated on a total of eight images with cracks, four of which were captured indoors in a laboratory, while the other four were captured outdoors on a sunny day. Table 3 presents the results of the maximum crack width measurements obtained from the images.

Table 3. Measured maximum crack widths.

Crack ID	Distance to Surface (mm)	α_c (mm/Pixels)	Pixel Width (Pixels)	Converted Width (mm)	Actual Width (mm)	Absolute Error (mm)	Relative Error (RE) (%)
Indoors A	250	0.391	10.8	4.23	4.19	0.04	0.95
B	250	0.205	21.0	4.39	4.03	0.26	6.56
C	460	1.071	3.162	3.39	3.54	0.15	4.30
D	460	1.071	1.00	1.07	1.14	0.07	6.02
Outdoors E	330	0.138	6.00	0.83	0.85	0.02	2.56
F	1150	0.357	7.00	2.50	2.00	0.50	25.0
G	1350	0.507	11.0	5.57	5.00	0.57	11.45
H	950	0.351	18.86	6.62	7.10	0.48	6.82

3.2.1. Accuracy

It can be observed from Table 3 that the method was capable of measuring crack widths as small as 0.85 mm, as observed with crack E. The overall performance of the method by calculating the mean absolute error (MAE) by taking a summation of the absolute errors and dividing it by the number of measured cracks. The MAE was found to be 0.26 mm; this means that, on average, our method will tend to measure crack widths to a value ±0.26 mm away from the true value. Park et al. [2] also used lasers to measure maximum crack width and achieved results with errors of less than 1.5 mm. While Yang et al. [4] had a relative error of between −13.27% and 24.01%. The results are within a reasonable error margin compared to previous studies.

The smallest absolute error was observed for crack E, with a converted maximum crack width of 0.83 mm, which was an underestimation of 0.02 mm. Crack G, measured to be 5.57 mm instead of 5.00 mm, had the largest observed absolute error of 0.57 mm, which was an overestimation.

3.2.2. Effect of Distance Away from Measuring Plane

Figure 6a shows the relative error (RE) is sensitive to the distance away from the measuring surface. Crack F, captured at a distance of 1150 mm from the surface, had an RE of 25% which was higher than that of crack G, 13.5%, captured at 1350 mm. The larger error of crack F is most likely due to image processing error in the segmentation stage, more than it is due to the distance away from the surface. Figure 6b shows that the absolute error is affected by the distance to the measuring plane. The R^2 value of 0.827 shows that the absolute error is highly dependent on the distance to the measuring plane. The high level of variability can be explained by other factors that can affect the accuracy of the proposed method, such as the camera quality and parameter selection in segmentation processing. If a camera of low resolution is used, there will be an increase in errors with an increase in the distance away from the surface increases, especially for smaller cracks. This is in line with the study by Kim et al. [23] that found when measuring smaller crack widths, larger errors can be expected.

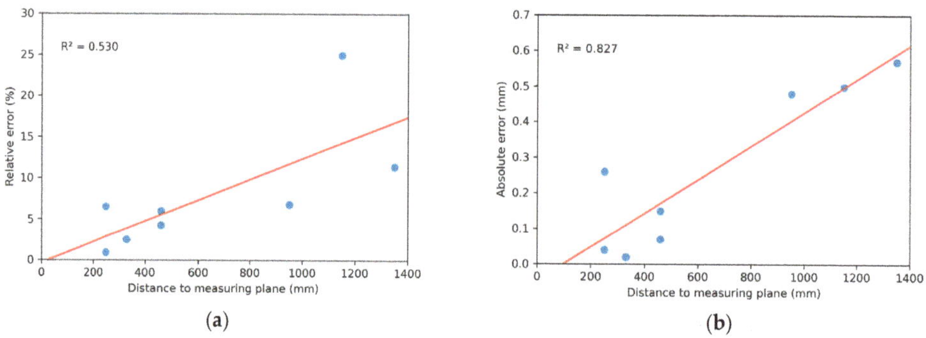

Figure 6. (**a**) Effect of distance to measuring plane on the relative error; (**b**) Effect of distance to measuring plane on the absolute error.

3.2.3. Performance in Outdoor Scenarios

To evaluate the method's suitability for use in outdoor scenarios, images for crack E to crack H were captured outside on a sunny day, as shown in Figure 7. The yellow bounding boxes in the images show how the images were cropped to focus on the critical regions that need measuring. Cropping or resizing an image does not affect the accuracy of the laser measurement method if the distance from the camera to the measurement surface is known. The proposed method in this study only requires knowledge of the camera-to-surface distance and uses image processing techniques to calculate other necessary parameters. This makes it usable with any camera without the need to know its specifications, unlike other methods [6,19] that require additional information such as focal length and the number of pixels. It can be seen from Table 3 that the proposed method achieves results of low absolute error and low mean relative error even for outdoor scenarios. The MAE for cracks measured indoors was 0.13 mm, while the MAE for cracks measured outdoors was 0.39 mm. Maximum crack widths measured on images captured outdoors had higher MAE mainly because the images captured outdoors were captured from long distances away from the surface being measured. The absolute error increases as the distance away from the measuring plane increases, as shown in Figure 6b.

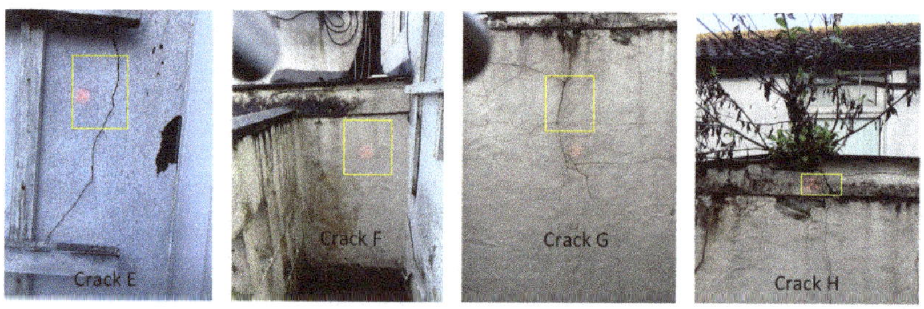

Figure 7. Application of the laser method in outdoor scenarios.

The conversion factor developed using the laser beam is effective in converting the results of the maximum crack width from pixels to millimeters. We achieved this by attaching a laser to the image acquisition devices. Kim et al. [36] measured crack width in millimeters by using planer markers which needed to be physically attached to the surface being measured. Good results with a 1–2% relative error were achieved. However, the attachment of markers to structures with low accessibility poses a safety hazard. The proposed method does not require any physical attachment of markers onto the surface of the structure being measured and thus promoting safety. The laser is easily attachable to any image acquisition device; attaching it to an unmanned aerial vehicle will promote safer inspections of structures.

4. Conclusions and Recommendations

This paper presented a novel method for measuring crack width in millimeters using image processing algorithms coupled with a unique laser beam technique. The main contribution of this method is the use of the laser beam projected onto the measuring surface to allow for the conversion of the crack width measured in pixels to millimeters. The following conclusions can be drawn from the study:

- The relationship between the distance to the measurement plane and the diameter of the laser beam is well established.
- The conversion factor, α_c, is defined and can be obtained using the established relationship between the distance to the plane of the measurement and the laser beam diameter, in conjunction with the laser beam diameter in pixels resulting from image processing algorithms.
- A unique method for determining the width of concrete cracks in millimeters has been devised using a laser beam and image processing algorithms through the conversion factor.
- The developed method is innovative and produced highly accurate findings that were closely in agreement with the actual crack width.
- The outcomes obtained through this method are suitable for verifying compliance with the allowable limits established in international standards, which are commonly expressed in metric or SI units.

It is expected further work could maximize the effective use of the developed method, such as expanding the image database for refined deep learning for the binarization and segmentation process and acquiring images by attaching the laser to a drone with cameras. Moreover, it is highly desirable to automate the process of performing calculations and crack segmentations on acquired images.

Author Contributions: Conceptualization, M.A.N., J.B. and I.D.W.; methodology, M.A.N., J.B. and I.D.W.; software, M.A.N.; validation, J.B. and I.D.W.; investigation, M.A.N.; resources, J.B.; data curation, M.A.N.; writing—original draft preparation, M.A.N.; writing—review and editing, J.B. and I.D.W.; visualization, M.A.N.; supervision, J.B. and I.D.W.; project administration, J.B. and I.D.W.; funding acquisition, J.B. All authors have read and agreed to the published version of the manuscript.

Funding: This research received no external funding.

Institutional Review Board Statement: Not applicable.

Informed Consent Statement: Not applicable.

Data Availability Statement: Not applicable.

Conflicts of Interest: The authors declare no conflict of interest.

References

1. Choudhary, G.K.; Dey, S. Crack Detection in Concrete Surfaces Using Image Processing, Fuzzy Logic, and Neural Networks. In Proceedings of the 2012 IEEE 5th International Conference on Advanced Computational Intelligence, ICACI 2012, Nanjing, China, 18–20 October 2012; pp. 404–411. [CrossRef]
2. Park, S.E.; Eem, S.H.; Jeon, H. Concrete Crack Detection and Quantification Using Deep Learning and Structured Light. *Constr. Build. Mater.* **2020**, *252*, 119096. [CrossRef]
3. Chen, J.; Xiong, F.; Zhu, Y.; Yan, H. A Crack Detection Method for Underwater Concrete Structures Using Sensing-Heating System with Porous Casing. *Measurement* **2020**, *168*, 108332. [CrossRef]
4. Yang, X.; Li, H.; Yu, Y.; Luo, X.; Huang, T.; Yang, X. Automatic Pixel-Level Crack Detection and Measurement Using Fully Convolutional Network. *Comput.-Aided Civ. Infrastruct. Eng.* **2018**, *33*, 1090–1109. [CrossRef]
5. Munawar, H.S.; Hammad, A.W.A.; Haddad, A.; Soares, C.A.P.; Waller, S.T. Image-Based Crack Detection Methods: A Review. *Infrastructures* **2021**, *6*, 115. [CrossRef]
6. Peng, X.; Zhong, X.; Zhao, C.; Chen, A.; Zhang, T. A UAV-Based Machine Vision Method for Bridge Crack Recognition and Width Quantification through Hybrid Feature Learning. *Constr. Build. Mater.* **2021**, *299*, 123896. [CrossRef]
7. Carrasco, M.; Araya-letelier, G.; Velázquez, R.; Visconti, P. Image-Based Automated Width Measurement of Surface Cracking. *Sensors* **2021**, *21*, 7534. [CrossRef]
8. Dobson, R.J.; Brooks, C.; Roussi, C.; Colling, T. Developing an Unpaved Road Assessment System for Practical Deployment with High-Resolution Optical Data Collection Using a Helicopter UAV. In Proceedings of the 2013 International Conference on Unmanned Aircraft Systems, ICUAS 2013-Conference Proceedings, Atlanta, GA, USA, 28–31 May 2013; pp. 235–243. [CrossRef]
9. Kim, B.; Cho, S. Automated Vision-Based Detection of Cracks on Concrete Surfaces Using a Deep Learning Technique. *Sensors* **2018**, *18*, 3452. [CrossRef]
10. Chaiyasarn, K.; Buatik, A.; Mohamad, H.; Zhou, M.; Kongsilp, S.; Poovarodom, N. Integrated Pixel-Level CNN-FCN Crack Detection via Photogrammetric 3D Texture Mapping of Concrete Structures. *Autom. Constr.* **2022**, *140*, 104388. [CrossRef]
11. Sankarasrinivasan, S.; Balasubramanian, E.; Karthik, K.; Chandrasekar, U.; Gupta, R. Health Monitoring of Civil Structures with Integrated UAV and Image Processing System. *Procedia Comput. Sci.* **2015**, *54*, 508–515. [CrossRef]
12. Yoon, J.; Shin, H.; Song, M.; Gil, H.; Lee, S. A Crack Width Measurement Method of UAV Images Using High-Resolution Algorithms. *Sustainability* **2022**, *15*, 478. [CrossRef]
13. Li, S.; Zhao, X. Automatic Crack Detection and Measurement of Concrete Structure Using Convolutional Encoder-Decoder Network. *IEEE Access* **2020**, *8*, 134602–134618. [CrossRef]
14. Zhao, W.; Liu, Y.; Zhang, J.; Shao, Y.; Shu, J. Automatic Pixel-Level Crack Detection and Evaluation of Concrete Structures Using Deep Learning. *Struct. Control Health Monit.* **2022**, *29*, e2981. [CrossRef]
15. Khosravani, M.R.; Silani, M.; Weinberg, K. Fracture Studies of Ultra-High Performance Concrete Using Dynamic Brazilian Tests. *Theor. Appl. Fract. Mech.* **2018**, *93*, 302–310. [CrossRef]
16. Kim, H.; Sim, S.H.; Spencer, B.F. Automated Concrete Crack Evaluation Using Stereo Vision with Two Different Focal Lengths. *Autom. Constr.* **2022**, *135*, 104136. [CrossRef]
17. Ai, D.; Jiang, G.; Lam, S.K.; He, P.; Li, C. Computer Vision Framework for Crack Detection of Civil Infrastructure—A Review. *Eng. Appl. Artif. Intell.* **2023**, *117*, 105478. [CrossRef]
18. Fujita, Y.; Hamamoto, Y. A Robust Automatic Crack Detection Method from Noisy Concrete Surfaces. *Mach. Vision Appl.* **2010**, *22*, 245–254. [CrossRef]
19. Kim, H.; Lee, J.; Ahn, E.; Cho, S.; Shin, M.; Sim, S.H. Concrete Crack Identification Using a UAV Incorporating Hybrid Image Processing. *Sensors* **2017**, *17*, 2052. [CrossRef]
20. Sauvola, J.; Pietikäinen, M. Adaptive Document Image Binarization. *Pattern Recognit.* **2000**, *33*, 225–236. [CrossRef]
21. Ioli, F.; Pinto, A.; Pinto, L. UAV PHOTOGRAMMETRY FOR METRIC EVALUATION OF CONCRETE BRIDGE CRACKS. *Int. Arch. Photogramm. Remote Sens. Spat. Inf. Sci.* **2022**, *XLIII-B2-2*, 1025–1032. [CrossRef]
22. Naccache, N.J.; Shinghal, R. An Investigation into the Skeletonization Approach of Hilditch. *Pattern Recognit.* **1984**, *17*, 279–284. [CrossRef]
23. Lee, T.C.; Kashyap, R.L.; Chu, C.N. Building Skeleton Models via 3-D Medial Surface Axis Thinning Algorithms. *CVGIP Graph. Model. Image Process.* **1994**, *56*, 462–478. [CrossRef]
24. Zhang, T.Y.; Suen, C.Y. A Fast Parallel Algorithm for Thinning Digital Patterns. *Commun. ACM* **1984**, *27*, 236–239. [CrossRef]
25. Tomczak, K.; Jakubowski, J.; Fiołek, P. Method for Assessment of Changes in the Width of Cracks in Cement Composites with Use of Computer Image Processing and Analysis. *Stud. Geotech. Mech.* **2017**, *39*, 73–80. [CrossRef]
26. Ito, A.; Aoki, Y.; Hashimoto, S. Accurate Extraction and Measurement of Fine Cracks from Concrete Block Surface Image. *IECON Proc. (Ind. Electron. Conf.)* **2002**, *3*, 2202–2207. [CrossRef]
27. Kim, B.; Cho, S. Image-Based Concrete Crack Assessment Using Mask and Region-Based Convolutional Neural Network. *Struct. Control Health Monit.* **2019**, *26*, e2381. [CrossRef]
28. Kim, H.; Ahn, E.; Cho, S.; Shin, M.; Sim, S.H. Comparative Analysis of Image Binarization Methods for Crack Identification in Concrete Structures. *Cem. Concr. Res.* **2017**, *99*, 53–61. [CrossRef]
29. Jeong, H.; Jeong, B.; Han, M.; Cho, D. Analysis of Fine Crack Images Using Image Processing Technique and High-Resolution Camera. *Appl. Sci.* **2021**, *11*, 9714. [CrossRef]

30. Culjak, I.; Abram, D.; Pribanic, T.; Dzapo, H.; Cifrek, M. A Brief Introduction to OpenCV. In Proceedings of the 2012 Proceedings of the 35th International Convention MIPRO, Opatija, Croatia, 21–25 May 2012.
31. Supriyanti, R.; Setiawan, B.; Widodo, H.B.; Murdyantoro, E. Detecting Pupil and Iris under Uncontrolled Illumination Using Fixed-Hough Circle Transform. *Int. J. Signal Process.* **2012**, *5*, 175–188.
32. Cornelia, A.; Setyawan, I. Ball Detection Algorithm for Robot Soccer Based on Contour and Gradient Hough Circle Transform. In Proceedings of the 2017 4th International Conference on Information Technology, Computer, and Electrical Engineering, ICITACEE 2017, Semarang, Indonesia, 18–19 October 2017; pp. 136–141. [CrossRef]
33. Liu, T.; Lei, Y.; Mao, Y. Computer Vision-Based Structural Displacement Monitoring and Modal Identification with Subpixel Localization Refinement. *Adv. Civ. Eng.* **2022**, *2022*, 5444101. [CrossRef]
34. Canny, J. A Computational Approach to Edge Detection. *IEEE Trans. Pattern Anal. Mach. Intell.* **1986**, *PAMI-8*, 679–698. [CrossRef]
35. Dorafshan, S.; Thomas, R.J.; Maguire, M. Benchmarking Image Processing Algorithms for Unmanned Aerial System-Assisted Crack Detection in Concrete Structures. *Infrastructures* **2019**, *4*, 19. [CrossRef]
36. Kim, I.H.; Jeon, H.; Baek, S.C.; Hong, W.H.; Jung, H.J. Application of Crack Identification Techniques for an Aging Concrete Bridge Inspection Using an Unmanned Aerial Vehicle. *Sensors* **2018**, *18*, 1881. [CrossRef] [PubMed]

applied sciences

MDPI

Article

Sustainability Enhancement and Evaluation of a Concrete Dam Using Recycling

Hasan Mostafaei [1], Niyousha Fallah Chamasemani [2], Mohammadreza Mashayekhi [3], Naser Safaeian Hamzehkolaei [4] and Paulo Santos [5,*]

1 School of Civil and Environmental Engineering, University of Technology, Ultimo, Sydney, NSW 2007, Australia; hasan.mostafaei@student.uts.edu.au
2 Department of Civil and Environmental Engineering, Politecnico di Milano, P.za L. da Vinci 32, 20133 Milan, Italy; niyousha.fallah@mail.polimi.it
3 Department of Civil Engineering, K. N. Toosi University of Technology, Tehran 15433-19967, Iran; m.mashayekhi@kntu.ac.ir
4 Department of Civil Engineering, Bozorgmehr University of Qaenat, Qaen 97619-86844, Iran; nsafaeian@buqaen.ac.ir
5 University of Coimbra, Department of Civil Engineering, ISISE, ARISE, 3030-788 Coimbra, Portugal
* Correspondence: pfsantos@dec.uc.pt

Abstract: Examining the life cycle of structures, such as concrete dams, holds paramount importance for engineers, as it facilitates a comprehensive assessment of overall sustainability, enabling the balancing of the benefits and costs associated with dam development. The recycling of materials emerges as a crucial factor in mitigating environmental impacts. This study employs the IMPACT 2002+ methodology to perform a life cycle assessment (LCA) of a concrete dam, covering the stages from construction to decommissioning. Additionally, carbon footprint analysis (CFA) and life cycle costing (LCC) are performed to pinpoint greenhouse gas (GHG) emission sources and access economic performance. This investigation spans three key-stages: (1) initial construction; (2) decommissioning; (3) hypothetical scenarios with recycling rates for demolished concrete and steel, evaluating how different recycling percentages influence both the environmental benefits and LCC outcomes. The results emphasize the significance of reducing air pollution, with climate change identified as the primary environmental concern compared to ecosystem and resource indicators. The findings show that the carbon footprint associated with the construction of 1 m width of the dam is estimated to be around 355 ton CO_2 eq. Furthermore, the total carbon emissions resulting from the demolition of the dam were identified to amount to 735 ton CO_2 eq/m. The recycling of the dam materials after demolition led to a notable reduction in pollution associated with the decommissioning process of the dam, compared to the dams' destruction without recycling.

Keywords: sustainability assessment; carbon footprint analysis (CFA); life cycle assessment (LCA); life cycle cost (LCC); concrete dam; recycling

1. Introduction

Dams hold a significant role in the water resources management within nations, serving as pivotal structures that contribute not only to flood crisis management, but also fulfill various functions such as water supply, irrigation, and power generation [1–3]. Beyond the socioeconomic ramifications, dam deterioration carries the potential for profound environmental repercussions, encompassing both the immediate consequences of failure and the long-term impacts arising from the dam's presence and operations [4,5]. These

https://doi.org/10.3390/app15052479

include alterations to downstream river ecosystems, habitat loss for aquatic species, and the disruption of animal migration patterns [6,7].

The construction industry, encompassing various activities from raw material extraction to demolition waste recycling, constitutes a significant portion of Europe's environmental footprint [8–11]. It represents half of the consumption of natural raw materials, accounts for 40% of total energy use (as the primary consumer), generates 46% of waste, and contributes to 36% of all greenhouse gas emissions. Within this sector, mineral materials, concrete, and other cement-based materials play a significant role in environmental impacts. The vast global production of concrete, reaching 25 billion tons annually, leads to the substantial utilization of natural resources and energy, primarily for cement and reinforcement steel production, as well as for transportation, construction, and demolition. This production also emits large amounts of greenhouse gases, primarily carbon dioxide (CO_2) [12], primarily originating from cement production, along with emissions of SO_2 contributing to acidification, mainly from transportation. Additionally, it generates a considerable amount of construction and demolition waste, mostly being inert and non-dangerous [13].

The recycling process of materials from the structures' decommissioning phase has been evaluated, with a focus on concrete and steel. The concrete is crushed and processed to create recycled concrete aggregate (RCA), which can be reused in new concrete mixes for construction projects, reducing the demand for virgin aggregates. The steel is separated and melted down for reuse in various steel products, supporting circular economy principles by minimizing waste and conserving natural resources. These recycled materials not only help in reducing landfill waste, but also lower the environmental impact associated with the extraction and processing of raw materials. RCA, in particular, can be used in a variety of applications, including road base, foundations, and non-structural concrete, although further research is needed to fully evaluate its performance in high-stress structural applications such as dam construction. This recycling process contributes to the sustainability of large-scale projects by enabling the recovery and repurposing of valuable materials, thus reducing both the carbon footprint and life cycle costs of future construction [14]. Research on RCA has demonstrated that while RCA reduces landfill use and conserves virgin materials, it often results in reduced material strength. As a result, additional cement or raw materials may be required to meet performance standards, which introduces economic allocation complexities [15].

However, replacing primary aggregates with RCA does not inherently ensure improved environmental performance across the entire life cycle of a concrete structure. A comprehensive evaluation of sustainability should encompass social and economic dimensions alongside environmental considerations. Therefore, it is imperative to employ scientifically grounded approaches to assess all three aspects of sustainability. For environmental evaluation, the widely accepted and standardized method of life cycle Assessment (LCA) is commonly utilized, enabling the examination of environmental impacts across processes and product life cycles.

Dams play a significant role in GHG emissions due to the extensive use of concrete in their construction, which involves the production of cement, a process known to emit substantial amounts of CO_2 during manufacturing [16]. The environmental implications spanning a dam's complete life cycle could be categorized into construction, maintenance, operation, and decommissioning stages. In pursuit of optimal sustainability in dam construction, LCA offers a valuable framework for exploring the trade-offs inherent in different alternatives [17].

LCA stands as a standardized technique to assess the ecological ramifications of a system or product over its complete life cycle [18]. This comprehensive analysis encompasses factors such as emissions, energy usage, and waste production. LCA has the capacity to gauge both the overall influence of a construction project throughout its existence and the specific effects associated with individual materials employed in its construction. Within the spectrum of environmental aspects subject to LCA scrutiny, the practice of carbon footprint analysis (CFA) emerges, a methodology designed to quantify the cumulative direct and indirect emissions of greenhouse gases (GHGs) stemming from an individual, entity, or product within a defined timeframe. While CO_2 represents the most prevalent GHG under consideration, the analysis also encompasses gases like methane (CH_4) and nitrous oxide (N_2O). Conversely, life cycle costing (LCC) serves as a method employed to appraise the comprehensive economic expenditure linked to a service or product over its life cycle [19].

In the realm of construction, LCA and CFA have been employed to assess the ecological effects stemming from different materials, construction techniques, and the sustainability of projects, as well as pinpoint avenues for enhancing environmental performance [20]. Implementing these approaches on a significant and influential construction holds paramount importance, as it facilitates a balanced assessment of the trade-offs between the advantages and drawbacks linked to its environmental impacts and resource utilization [21]. Researchers have delved into different dimensions of dam analysis within LCA, encompassing aspects like the utilization of dredged sediments [22], the decommissioning of dams [23], the decay of biomass within reservoirs, and the functioning of hydroelectric plants [23–25].

Zhang et al. [26] proposed an LCA method for carbon assessment in pumped storage hydropower projects, addressing a gap in estimating emissions at the design stage. By calculating diesel, electricity, and material use, the method helps identify emission influences of design factors, supporting greener pumped storage hydropower evaluations and improved emission estimates. In tandem with distinct building materials, Liu et al. [27] utilized an LCA approach to assess the environmental repercussions of two types of dams, including conventional concrete and rock-filled concrete. The findings indicate that rock-filled concrete led to a reduction of 25% in transportation, 51% in construction, 72% in CO_2 emissions during material production, and 15.6% in maintenance and operation.

Similarly, Zhang et al. [28] utilized a hybrid LCA method to juxtapose the carbon footprints of two varieties of hydropower schemes: a concrete gravity dam and an earth-core rockfill dam. The outcomes reveal that earth-core rockfill dams exhibit superior environmental responsibility throughout their life cycle, contributing to a roughly 24.7% decrease in CO_2 emissions in comparison to concrete gravity dams. Nonetheless, the current investigations focus solely on existing factors, neglecting the potential influence of future considerations such as recycling, which could significantly impact the maintenance and operation of structures.

Mostafaei et al. [16] investigated the LCA of the pre-existing Pine Flat dam, a concrete gravity structure, utilizing the ReCiPe 2016 methodology. They evaluated the influence of two strategies—seismic retrofitting and non-retrofitting—on the dam's life cycle. Special attention was given to the impact of retrofitting on environmental indicators like carbon footprint and human health. Their findings underscore the positive environmental effects of dam retrofitting, showcasing a reduction in impact indicators compared to the scenario of dam disposal.

Sadok et al. investigated the use of calcined dredged sediments as a sustainable cement substitute [22]. Their findings show enhanced eco-strength efficiency and reduced embodied carbon in mortars with up to 25% sediment replacement. Microscopic analyses indicated improved C-A-S-H and C-S-H gel formation. Their cost assessments reveal significant savings, with sediment treatment up to 86% cheaper than cement.

The Bluestone dam, located in West Virginia, serves as a case study for this research. The process of evaluating its life cycle is carried out through the utilization of SimaPro v9.5 software along with the IMPACT 2002+ methodology, encompassing stages ranging from construction to decommissioning. Consequently, the LCA adopts a cradle-to-grave approach encompassing both the construction and decommissioning stages, while a cradle-to-gate approach is applied exclusively to the construction stage. Moreover, the influences of recycling for demolished concrete and steel in the final stage on the CFA and LCC results are evaluated. Notably, this investigation exclusively concentrates on evaluating the dam's construction aspects, excluding assessments pertaining to the hydropower plant or other interconnected elements.

2. Methodology

2.1. Design of Concrete Gravity Dams

Concrete gravity dams are designed to resist external forces primarily through their weight, ensuring stability under various loading conditions. This fundamental design approach involves analyzing the forces acting on the dam, computing the stresses induced in the structure, and ensuring that safety factors against failure are met. The stability of these dams is governed by well-established engineering principles, including the equilibrium of forces and moments.

These dams rely on their own weight to resist external forces such as hydrostatic pressure, silt pressure, uplift pressure, and seismic loads. Key considerations in dam design include evaluating stability against sliding and overturning, assessing the effects of uplift pressure on structural integrity, and ensuring adequate safety factors under normal, flood, and seismic conditions. Additionally, material properties, foundation conditions, and load combinations must be carefully analyzed to optimize the dam's performance. International design standards, such as IS 6512:2019 [29] and other regulatory codes, provide guidelines for assessing these factors, ensuring that dams are designed to withstand extreme environmental conditions and long-term operational stresses.

2.2. LCA Method

Life cycle assessment (LCA) is a systematic and holistic methodology employed to assess the environmental impact of a product or process throughout its entire life cycle. This includes the extraction of raw materials, manufacturing, transportation, use, and end-of-life disposal [10,30]. The primary goal of LCA is to quantify the environmental burdens associated with a particular system and identify opportunities for improvement in terms of resource use, energy efficiency, and emissions. The LCA process typically involves four main stages: goal and scope definition, life cycle inventory, life cycle impact assessment, and interpretation. During the goal and scope definition stage, the objectives and boundaries of the assessment are established. Life cycle inventory involves compiling an exhaustive list of inputs and outputs associated with the system under study. The life cycle impact assessment phase evaluates the potential environmental impacts of these inputs and outputs, considering factors such as climate change, resource depletion, and human health [31,32].

Interpretation is a critical step where the results of the LCA are analyzed and communicated. This involves considering trade-offs and identifying areas where changes can be made to reduce environmental impacts [33]. LCA is a versatile tool used across various industries to inform decision-making processes, guide sustainable product development, and support environmentally conscious practices [34]. In recent years, life cycle thinking has gained prominence as organizations and consumers increasingly recognize the importance of considering the entire life cycle of products and services to achieve more sustainable and environmentally friendly outcomes. LCA provides a comprehensive framework for making informed choices that balance economic, environmental, and social considerations.

Impact 2002+ is a widely used methodology in the field of LCA, a systematic approach to evaluating the environmental impacts of a product, process, or activity throughout its entire lifecycle [35]. Impact 2002+ encompasses a comprehensive set of impact categories and indicators that allow for a holistic analysis of environmental consequences. By assessing factors such as resource depletion, greenhouse gas emissions, air and water pollution, and other ecological considerations, Impact 2002+ provides valuable insights into the potential environmental burdens associated with various alternatives [35]. Its incorporation into LCA studies aids in making informed decisions and formulating strategies for minimizing the overall environmental footprint of a given system, thereby contributing to sustainable and environmentally responsible practices across industries and sectors.

Hence, this research employs the Impact 2002+ approach to acquire and assess the ecological repercussions of the dam. The environmental impacts associated with dam construction, considering materials such as concrete, fuel, wood, blasting material, steel, and rebars, are evaluated in this study. Using LCA, each material's environmental footprint is assessed across its entire life cycle from extraction and processing to transportation, construction, and disposal. Figure 1 shows the system boundary for dam construction.

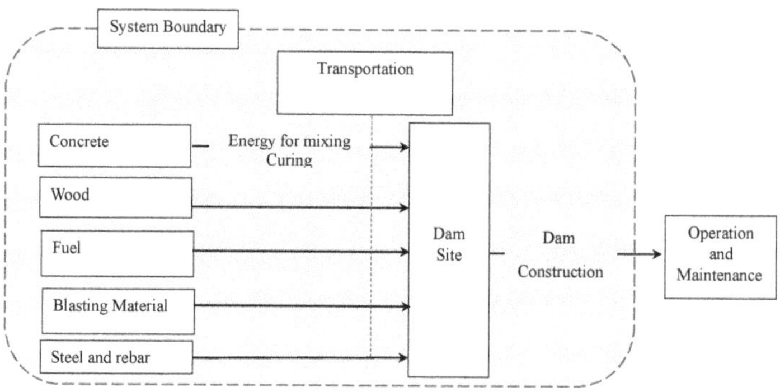

Figure 1. System boundary for the dam construction stage.

2.3. CFA Method

Carbon footprint analysis (CFA) is a comprehensive assessment that quantifies the total amount of greenhouse gas emissions, measured in carbon dioxide equivalents, associated with a particular activity, product, or process throughout its life cycle [36–40]. It is believed that the exact origin of CFA has emerged from the concept of Ecological Footprint since the 1990s. This analysis aims to provide insights into the environmental impact and sustainability of a given entity by accounting for emissions resulting from resource extraction, manufacturing, transportation, usage, and disposal. By examining the carbon footprint, one can identify key contributors to greenhouse gas emissions and explore opportunities to reduce or offset these impacts [41]. This methodology is crucial in addressing climate change concerns and promoting sustainable practices across various industries, guiding individuals, businesses, and policymakers toward more environmentally responsible choices. In the present work, the outcomes of the LCA conducted using SimaPro are utilized to pinpoint primary emission sources, thereby facilitating an evaluation of the carbon footprint attributed to the dam's construction [16].

2.4. LCC Method

The concept of life cycle cost (LCC) originated within the US Department of Defense in the 1960s. It presents a comprehensive approach for assessing the total expenses associated with a project or asset across its entire lifespan, employing economic principles [42,43]. The primary objective of LCC is to identify the most economically efficient choices for a material or system and ensure its financial viability over the long term [44]. Recent efforts have focused on adapting LCC methodologies for the construction sector, encompassing investigations that employ LCC to appraise property and construction alternatives and quantify disposal expenditures. In the context of a construction endeavor, LCC considers costs related to design, construction, maintenance, and operation.

3. Model Description

In this study, the Bluestone Dam is selected for LCA. This dam, located on the New River in West Virginia (USA), is a significant structure. Originally constructed in 1930, the dam serves the dual purpose of flood control and hydropower generation [45]. A schematic representation of one of the dam models developed using CADAM [46] is shown in Figure 2.

Figure 2. A schematic view of the dam modeled using CADAM v1.4.12 software.

A spectral acceleration of 0.36 g corresponds to a return period of 2475 years, where g represents gravitational acceleration. Additionally, the hydrologic analysis suggests that, in the event of an impending dam failure, the reservoir water level could rise to 52 m [47]. In this study, it is assumed that the reservoir is fully filled, with a water depth of 48.15 m, and all analyses are conducted under this condition.

To conduct a comprehensive evaluation encompassing LCA, CFA, and LCC for the dam, it becomes essential to encompass various phases of its existence, ranging from initial construction to eventual demolition and recycling. The creation of an exhaustive life cycle inventory (LCI) database with minimal uncertainties stands as a notable challenge faced by scholars in these domains, acting as a limiting factor in the execution of these assessments.

Table 1 presents the relevant material properties and construction details essential for the dam's lifecycle, considering 1 m width of the dam [48]. This tabulated information furnishes intricate insights into material weights and on-site and off-site transportation considerations across all phases. Wood is commonly used as a material for constructing temporary formwork and scaffolding during the concrete pouring phase of dam construction. It is important to underscore that, in the phase of demolition and recycling, an assumption has been introduced positing a 20% concrete recycling rate, where a portion of the demolished dam's concrete finds renewed utility in other construction projects [49].

Table 1. LCI database for the construction stage.

Material or Operation	Value (ton)	Transportation (km)
Concrete	2142	30
Steel and rebar	0.35	550
Blasting	4.14	620
Fuel oil	66.3	150
Wood	27	150
Earth and rock excavation	192	25

Table 2 forms the foundational framework for LCC computation, encompassing not only the expenses linked to materials and construction procedures, but also factoring in their consequential impacts, such as the financial implications stemming from environmental emissions.

Table 2. Price of each operation and construction materials.

Construction Material or Operation	Unit Cost ($)
Concrete (USD/ton)	45
Wood (USD/ton)	345
Fuel oil (USD/ton)	407
Steel and Rebar (USD/ton)	1050
Earth and rock excavation (USD/ton)	200
Transportation (USD/ton.km)	0.25
Blasting materials (USD/ton)	599
Landfill (USD/ton)	23
Recycling concrete (USD/ton)	55

4. Results and Discussion

In this part, the outcomes of the LCA, CFA, and LCC investigations are displayed for both phases of the dam's lifespan: building and eventual decommissioning with recycling. Following this, a comparative examination of the findings from these assessments is carried out for two separate scenarios representing different phases of the dam's life cycle.

4.1. Construction Stage

4.1.1. Environmental Impacts

Figure 3 demonstrates the carbon footprint distribution across different components and processes involved in the production of concrete. Cement is shown to be the most significant contributor, accounting for nearly 90% of the total emissions, aligning with the typical findings in the literature [16] that highlight cement's high environmental impact due to energy-intensive production. Aggregates, both fine and coarse, contribute a smaller but notable amount, reflecting the environmental cost of extraction and transportation. Admixture production and transportation have measurable impacts, suggesting that logistics management and minimizing admixture quantities can further reduce emissions. Concrete batching, although a relatively minor component, also adds to the footprint, emphasizing the need for efficient batching processes.

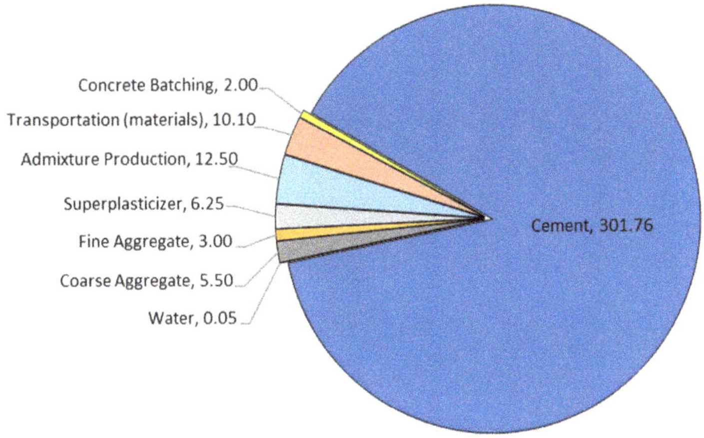

Figure 3. Carbon footprint distribution across different components and processes involved in the production of concrete (kg CO_2 eq).

The analysis' outcomes, as presented in Table 3, outline the environmental impacts throughout the entire construction phase of 1 m width of the dam. Moreover, Figure 4 presents the distribution of environmental impacts during the construction phase of the dam based on the IMPACT 2002+ method. The bar chart delineates the influence of six components—concrete, transportation, diesel fuel, wood, blasting materials, and steel and rebar—on the overall environmental impacts. The results reveal that the total carbon dioxide equivalent (CO_2 eq) emissions amount to approximately 355 ton CO_2 eq/m, with the primary contributors being concrete production and fuel usage. Concrete, as one of the most extensively used materials in dam construction, accounts for over 305 ton CO_2 eq/m, indicating its substantial role in greenhouse gas emissions. Fuel use, largely associated with transportation and machinery, contributes an additional 33 ton CO_2 eq/m. Additionally, aquatic ecotoxicity is primarily driven by the use of blasting materials, which also contribute heavily to terrestrial impacts and land occupation. These findings point to the need for improved practices in material selection and handling to mitigate ecological damage.

Table 3. The environmental impacts throughout the entire construction of 1 m width of the dam.

Impact Category	Total
Carcinogens (ton C_2H_3Cl eq)	2.405
Non-carcinogens (ton C_2H_3Cl eq)	5.059
Respiratory inorganics (ton PM2.5 eq)	0.269
Ionizing radiation (Bq C-14 eq)	2,384,000
Ozone layer depletion (ton CFC-11 eq)	0.00006
Respiratory organics (ton C_2H_4 eq)	0.126
Aquatic ecotoxicity (ton TEG water)	58,727
Terrestrial ecotoxicity (ton TEG soil)	16,074
Terrestrial acid (ton SO_2 eq)	6.253
Land occupation (m^2org.arable)	62,759
Aquatic acidification (ton SO_2 eq)	1.348
Aquatic eutrophication (ton PO_4 P-lim)	0.042
Global warming (ton CO_2 eq)	355
Non-renewable energy (MJ primary)	5,700,000
Mineral extraction (MJ surplus)	5215

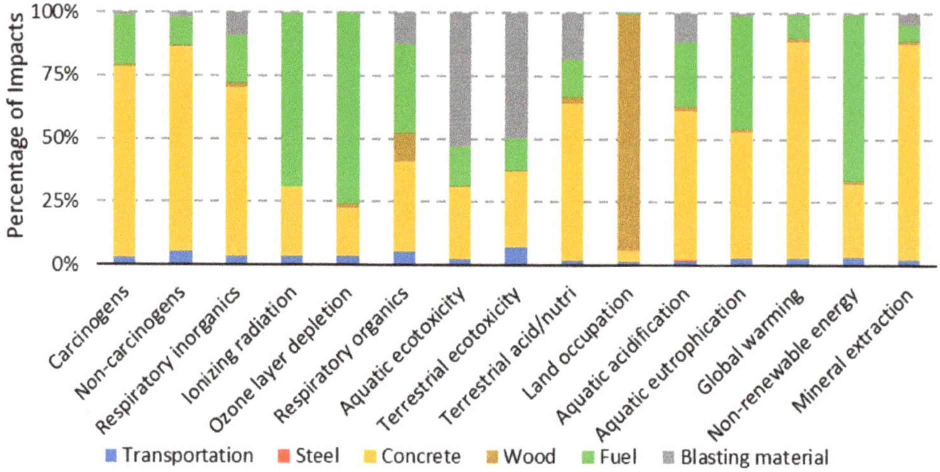

Figure 4. Distribution of environmental impacts during the construction phase of the dam based on the IMPACT 2002+ method.

Energy consumption in the project is another key factor, with over 5,700,000 MJ/m of non-renewable energy being consumed, mostly attributed to fuel use and concrete production. Concrete significantly contributes to carcinogens, non-carcinogens, aquatic eutrophication, global warming, and material extraction. Furthermore, the use of wood is associated with deforestation, leading to reduced clean air production and contributing to environmental degradation, particularly affecting land occupation.

Table 4 outlines the LCA outcomes for dam construction across four specific categories—human health, ecosystem quality, climate change, and resources—utilizing the IMPACT 2002+ methodology. Furthermore, Figure 5 represents the impact of six key components—concrete, transportation, diesel fuel, wood, blasting materials, and steel and rebar—on these aforementioned categories. Concrete production significantly impacts human health, with the total effect estimated at 0.143 disability-adjusted life years (DALY). This impact arises primarily from emissions during cement production, including particulate matter and pollutants such as nitrogen oxides and sulfur dioxide, which contribute to respiratory and cardiovascular issues.

Table 4. LCA results of construction of 1 m width of the dam using the IMPACT 2002+ method.

Damage Category	Total
Human health (DALY)	0.210
Ecosystem quality (PDF m^2 r)	205,000
Climate change (kg CO_2 eq)	355
Resources (MJ primary)	5,700,000

Additionally, the extraction and processing of raw materials, as well as the energy-intensive nature of concrete manufacturing, release greenhouse gases and airborne toxins that further harm air quality and human health. For concrete batching, the DALY primarily stems from air pollution during cement production, including emissions of particulate matter (PM2.5), nitrogen oxides (NOx), and sulfur dioxide (SO_2). Moreover, the results indicate that, in terms of human health (DALY), the most significant contributors are fuel and concrete, together accounting for a large portion of the impact. Fuel contributes the highest, followed closely by concrete, indicating that energy consumption and emissions from these materials and processes are critical factors affecting human health. Other elements like blasting material and transportation also play a role, but are less impactful in comparison.

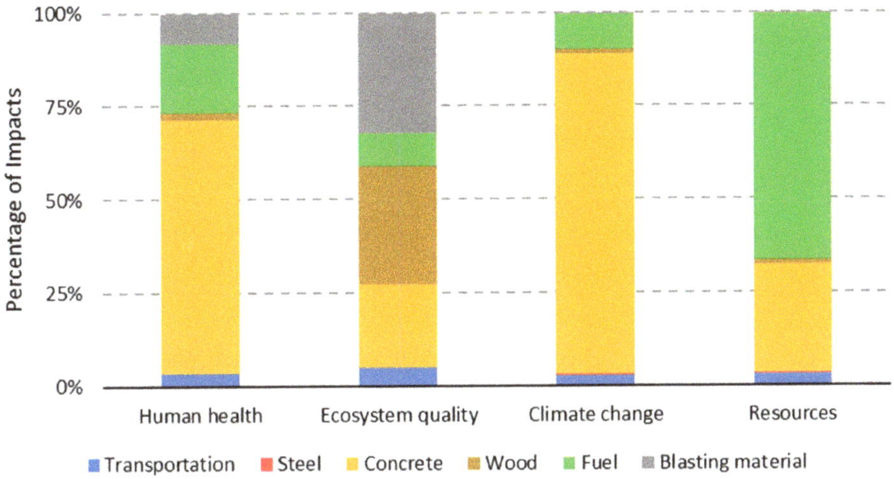

Figure 5. The distribution of the dam's components on the environmental impacts.

For ecosystem quality (PDF m^2 yr), the most striking result is the overwhelming contribution of blasting materials, which account for the majority of the damage in this category. This suggests that activities related to blasting are highly disruptive to ecosystems, possibly due to habitat destruction. Wood also has a notable impact, likely related to deforestation or land use changes, while other materials like concrete have a much smaller effect on ecosystems. In the climate change (ton CO_2 eq) category, concrete is by far the largest contributor, responsible for the vast majority of carbon emissions. This aligns with the energy-intensive nature of concrete production, particularly in cement manufacturing. Fuel also plays a significant role in emissions, reinforcing the need to address fuel usage to mitigate climate change's impacts. Transportation contributes a smaller but still noteworthy amount of CO_2 emissions. Finally, in the resources (MJ primary) category, fuel consumption dominates resource use, indicating a high demand for energy in the processes involved. Concrete follows as the next largest consumer of resources, suggesting that both material production and energy consumption are driving factors in resource depletion.

Figure 6 illustrates the impact of pollution across four categories—resource consumption, ecosystem pollution, climate change, and endangerment to human health—according to the IMPACT 2002+ methodology. The graph provides a comprehensive representation of pollution's environmental effects by normalizing and weighting the created pollution based on its importance factors, measured in MPt (megapoints). Notably, the graph reveals that pollution generated for climate change and human health holds the top two positions, while pollution related to resources and the ecosystem ranks third and fourth. Concrete consistently emerges as the component with the highest impact in all categories, particularly in climate change, where it scores nearly 0.04 MPt, surpassing the other components such as blasting material.

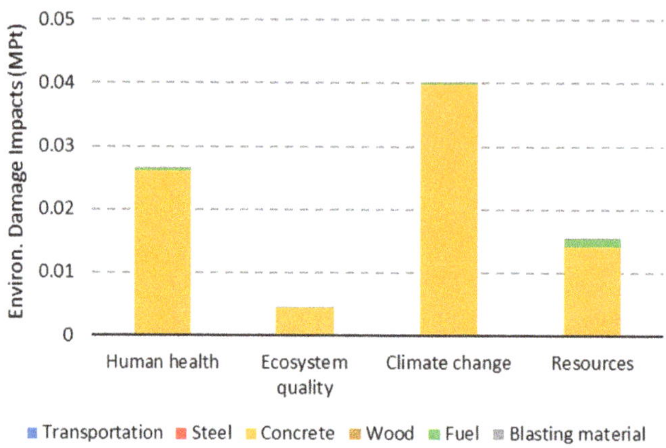

Figure 6. Total weighted and normalized environmental impacts using IMPACTS 2002+ on the construction of 1 m width of the dam.

4.1.2. LCC Results of the Construction Stage

The results of the LCC analysis reveal that the estimated economic expenses for the dam's construction phase are anticipated to reach approximately USD 0.195 million. In Figure 7, a graphical depiction illustrates the contributions of different construction materials and operations to these costs. Concrete emerges as the most significant factor, making an impactful contribution of USD 0.0959 million. Excavation follows closely as the second most substantial contributor at USD 0.0384 million, succeeded by oil at USD 0.0270 million.

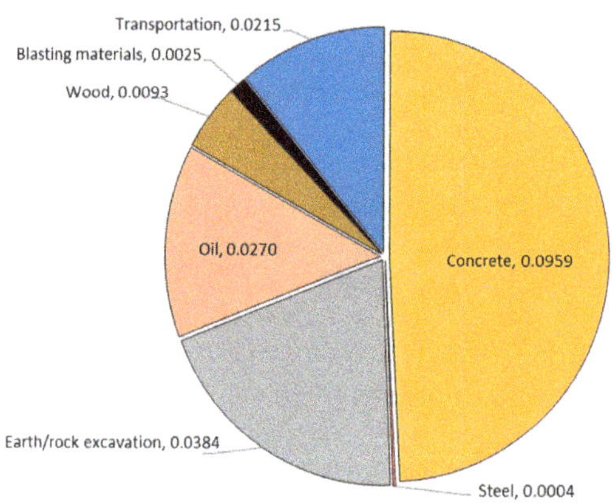

Figure 7. The results of the LCC analysis for the dam's construction phase (million USD).

4.2. *The Effects of Demolition*

By conducting CFA and LCC analyses, the total carbon emission and cost of demolishing the 1 m width of the dam were determined to be 735 ton CO_2 eq/m and USD 0.138 million, respectively. Figure 8 illustrates the contributions of the parameters to the CFA and LCC results of dam demolition. The analysis reveals that the most influential parameters affecting carbon emissions during dam demolition are landfill (51%) and blasting material (46%), with transportation accounting for 3%, influenced by landfill location. Similarly, in the LCC results, blasting material is the most significant parameter, contributing to approximately 82% of the total cost.

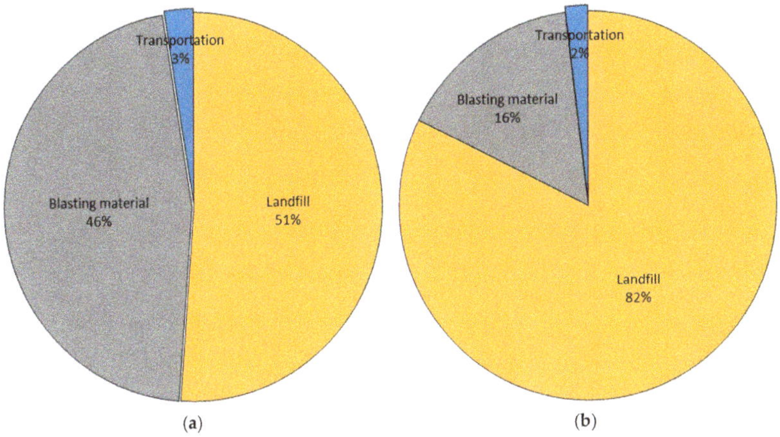

Figure 8. The contribution of the parameters on the (**a**) carbon emissions due to the destruction of the dam and (**b**) the total price of the destruction of the dam.

4.3. *The Effects of Recycling*

This section assesses the economic and environmental implications linked to the recycling of dam materials. The effects of recycling were considered within the framework of the LCA by specifically examining the environmental impacts associated with the use of recycled materials, such as RCA, in concrete production. This involved quantifying the reduction in energy consumption, emissions, and resource depletion resulting from the substitution of virgin aggregates with recycled materials. Additionally, the end-of-life phase of the recycled materials was analyzed to assess their potential for reuse or recycling, thus contributing to the overall sustainability of the concrete lifecycle.

4.3.1. CFA Results

Figure 9 depicts the carbon footprint generated during the decommissioning of the dam, taking into account the recycling of used concrete and steel. The carbon footprint attributed to the decommissioning process without recycling was calculated using CFA, amounting to approximately 735 ton CO_2 eq/m. The results emphasize the significant impact of material recycling on the carbon footprint, showcasing a substantial 95% reduction in greenhouse gas emissions. This underscores the potential of incorporating sustainable waste management practices to alleviate the environmental impacts.

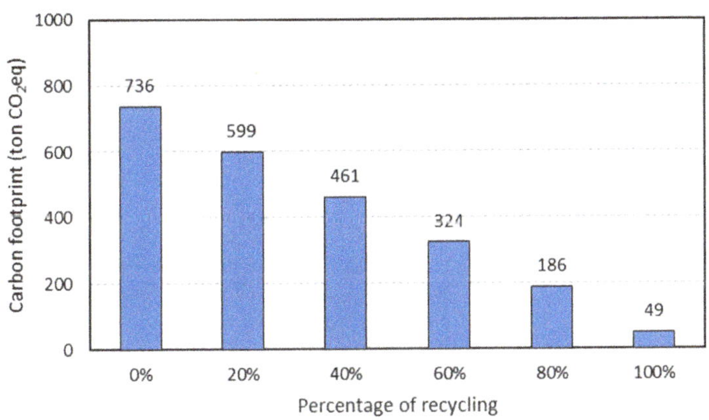

Figure 9. The effect of recycling steel and concrete on carbon footprint.

4.3.2. LCC Results of Recycling

The findings indicate that the total cost for the demolition of the dam amount to approximately USD 91.6 million. In Figure 10, the impact of recycling on the total cost of dam demolition is illustrated. Notably, the inclusion of recycling concrete and steel results in an increase in the overall cost of dam destruction. For instance, if all the utilized steel and concrete are recycled, the total cost of the dam's demolition reaches USD 0.142 million. These results emphasize the importance of factoring in both economic and environmental considerations related to waste disposal during the initial design and planning stages of construction projects. This proactive approach is crucial for mitigating adverse impacts and ensuring cost-effectiveness.

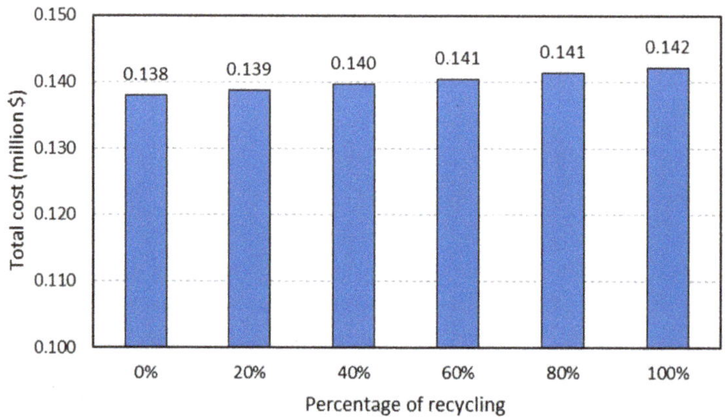

Figure 10. The effect of recycling steel and concrete on the overall cost of dam destruction.

4.4. The Effects of Compressive Strength of the Concrete

Concrete is a major contributor to the environmental impacts of dam construction, with cement being the most critical factor due to its high carbon footprint. The compressive strength of concrete plays a crucial role in determining the overall sustainability of the structure. Increasing compressive strength can reduce the required volume of concrete in the dam, thereby lowering material consumption. However, this comes at the cost of increased cement content in the mix, which has a significant environmental impact due to its energy-intensive production process.

To explore this trade-off, the effects of different concrete compressive strengths—20 MPa, 30 MPa, 40 MPa, and 50 MPa—are investigated. For each case, the dam is designed to meet the required structural stability while minimizing material use.

By considering the design criteria, the required volume of concrete for a 1 m width of the dam varies with the compressive strength of the concrete. For compressive strengths of 20 MPa, 30 MPa, 40 MPa, and 50 MPa, the corresponding concrete volumes are 994.3 m^3, 895.4 m^3, 890.5 m^3, and 887.2 m^3, respectively. Figure 11 presents the carbon footprint associated with the construction of 1 m width of dam using different compressive strengths. The results indicate that although lower compressive strength requires a larger volume of concrete, the total carbon footprint remains lower due to the reduced cement content in the mix. In other words, increasing compressive strength leads to a decrease in concrete volume, but results in a higher carbon footprint, primarily due to the increased cement consumption, which is the dominant contributor to greenhouse gas emissions in concrete production.

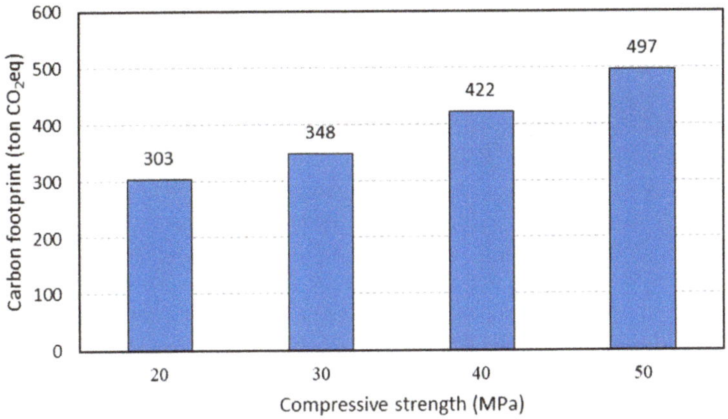

Figure 11. The effect of increasing the compressive strength of concrete on carbon footprint per 1 m width of the dam.

5. Conclusions

This paper conducts assessments of the LCA, CFA, and LCC for a concrete dam across construction and decommissioning scenarios. While other studies [16,28] have investigated the environmental impacts and costs of dams, this study focuses on the sustainability of dams through material recycling and dam design considerations. It further examines the impact of integrating recycling practices for demolished concrete and steel on the carbon footprint and LCC results by considering different end-of-life scenarios, including material recovery, reuse, and recycling processes. These scenarios assess the environmental benefits of reducing raw material extraction and landfill use, along with the financial implications of recycling logistics and potential savings from reusing materials. In essence, the evaluation of these stages for the dam can be summarized as follows:

1. Cement is the largest contributor to the carbon footprint of concrete due to its energy-intensive production. Aggregates and admixtures contribute smaller but meaningful emissions, while transportation and batching also add to the total footprint.
2. The findings highlight the significant environmental impact of concrete production and fuel usage, which are the primary contributors to carbon emissions in construction of 1 m width of the dam, with total CO_2 eq emissions reaching 355 ton/m. Concrete alone accounts for over 305 ton CO_2 eq/m. The project's non-renewable energy consumption exceeds 5,700,000 MJ/m, largely driven by concrete and fuel.
3. Climate change and human health rank as the top environmental concerns, with concrete consistently emerging as the dominant factor in these categories. These findings reinforce the importance of addressing material production and energy consumption to mitigate environmental impacts, particularly in large-scale construction projects such as dam construction.
4. The total estimated cost for the dam's construction phase is approximately USD 0.195 million, with concrete being the most significant contributor, accounting for USD 0.0959 million. Excavation is the second-largest cost at USD 0.0384 million, followed by oil usage at USD 0.0270 million. The overall cost for the decommissioning of the dam is estimated at around USD 0.138 million. Integrating the recycling of concrete and steel leads to an augmentation in the overall cost of dam demolition. To illustrate, if all the employed steel and concrete are recycled, the total cost for the dam's demolition escalates to USD 0.142 million.

5. The carbon footprint resulting from the decommissioning process, excluding recycling, was assessed using CFA and amounted to around 735 ton CO_2 eq/m. The outcomes underscore the considerable influence of material recycling on the carbon footprint, demonstrating a notable 95% reduction in greenhouse gas emissions. Recycling concrete as RCA not only reduces waste and conserves natural resources, but also significantly lowers the carbon footprint by decreasing the demand for virgin aggregate extraction. Recycled steel, similarly, can be repurposed in various construction applications, promoting a circular economy approach. These practices contribute to environmental and economic benefits by minimizing landfill usage, reducing greenhouse gas emissions, and aligning with sustainable development goals.
6. While lower compressive strength requires more concrete, its total carbon footprint remains lower due to its reduced cement content. Conversely, increasing compressive strength reduces concrete volume, but significantly raises carbon emissions, as cement consumption is the primary contributor to greenhouse gas emissions.

While this study provides valuable insights into the environmental and economic impacts of recycling practices in the context of dam construction and decommissioning, there are some limitations that should be acknowledged. Firstly, the reliance on the assumption of a 20% recycling rate of concrete, based on global averages, may not reflect more optimistic or region-specific recycling practices. Additionally, this study heavily depends on data obtained from previous investigations due to the lack of detailed construction and demolition records for the concrete dam. This reliance on secondary data may introduce uncertainties in the results.

For future research, investigating the operational phases of dams could yield significant insights into their long-term environmental and economic impacts. Specifically, assessing the carbon footprint, life cycle costs, and life cycle impacts associated with dam operations—such as maintenance practices, water management, and emissions from reservoirs—can provide a clearer understanding of how to optimize these phases for sustainability. Additionally, exploring the effects of using RCA and sustainable material innovations in dam construction offers a promising pathway to reduce environmental impacts. RCA's role in concrete dam design could impact LCC, LCA, and CFA by reducing reliance on virgin materials and mitigating waste. Furthermore, research focused on optimization approaches that minimize environmental impacts—considering factors such as resource consumption, emissions, and ecosystem effects—can lead to the development of more sustainable dam designs.

Author Contributions: Conceptualization, H.M., N.F.C. and P.S.; methodology, H.M., N.F.C. and P.S.; formal analysis, H.M. and N.F.C.; investigation, H.M. and N.F.C.; resources, H.M. and N.F.C.; writing—original draft preparation, H.M. and N.F.C.; writing—review and editing, H.M., M.M., N.S.H. and P.S.; visualization, H.M. and N.F.C.; supervision, H.M. and P.S. All authors have read and agreed to the published version of the manuscript.

Funding: This research received no external funding.

Informed Consent Statement: Not applicable.

Data Availability Statement: The data presented in this study are available on request from the corresponding author. The data are not publicly available due to confidentiality issues.

Conflicts of Interest: The authors declare no conflicts of interest.

Abbreviations

The following abbreviations are used in this manuscript:

CFA	Carbon footprint analysis
GHGs	Greenhouse gases
LCA	Life cycle assessment
LCC	Life cycle costs
LCI	Life cycle inventory
RCA	Recycled concrete aggregate

References

1. Mostafaei, H. Modal Identification Techniques for Concrete Dams: A Comprehensive Review and Application. *Sci* **2024**, *6*, 40. [CrossRef]
2. Schmitt, R.J.P.; Rosa, L. Dams for hydropower and irrigation: Trends, challenges, and alternatives. *Renew. Sustain. Energy Rev.* **2024**, *199*, 114439. [CrossRef]
3. Angelakis, A.N.; Baba, A.; Valipour, M.; Dietrich, J.; Fallah-Mehdipour, E.; Krasilnikoff, J.; Bilgic, E.; Passchier, C.; Tzanakakis, V.A.; Kumar, R. Water Dams: From ancient to present times and into the future. *Water* **2024**, *16*, 1889. [CrossRef]
4. Mostafaei, H.; Bahmani, H. Sustainable High-Performance Concrete Using Zeolite Powder: Mechanical and Carbon Footprint Analyses. *Buildings* **2024**, *14*, 3660. [CrossRef]
5. Ali, H.O. Advancing integrated water resources management approaches for the sustainability of small dams in arid and semi-arid regions: A case study of the Um Dafoug dam in South Darfur State, Sudan. *Int. J. Water Resour. Environ. Eng.* **2025**, *17*, 1–16.
6. Bahmani, H.; Mostafaei, H.; Santos, P.; Fallah Chamasemani, N. Enhancing the mechanical properties of Ultra-High-Performance Concrete (UHPC) through silica sand replacement with steel slag. *Buildings* **2024**, *14*, 3520. [CrossRef]
7. Udara Willhelm Abeydeera, L.H.; Wadu Mesthrige, J.; Samarasinghalage, T.I. Global research on carbon emissions: A scientometric review. *Sustainability* **2019**, *11*, 3972. [CrossRef]
8. Schützenhofer, S.; Kovacic, I.; Rechberger, H.; Mack, S. Improvement of environmental sustainability and circular economy through construction waste management for material reuse. *Sustainability* **2022**, *14*, 11087. [CrossRef]
9. Santos, P.; Cervantes, G.C.; Zaragoza-Benzal, A.; Byrne, A.; Karaca, F.; Ferrández, D.; Salles, A.; Bragança, L. Circular Material Usage Strategies and Principles in Buildings: A Review. *Buildings* **2024**, *14*, 281. [CrossRef]
10. Al Omar, S.; Abdelhadi, A. Comparative Life-Cycle Assessment of Steel and GFRP Rebars for Procurement Sustainability in the Construction Industry. *Sustainability* **2024**, *16*, 3899. [CrossRef]
11. Saffari, R.; Habibagahi, G.; Nikooee, E.; Niazi, A. Biological stabilization of a swelling fine-grained soil: The role of microstructural changes in the shear behavior. *Iran. J. Sci. Technol. Trans. Civ. Eng.* **2017**, *41*, 405–414. [CrossRef]
12. Shabani, K.; Bahmani, M.; Fatehi, H.; Chang, I. Improvement of the geotechnical engineering properties of dune sand using a plant-based biopolymer named serish. *Geomech. Eng.* **2022**, *29*, 535–548.
13. Banias, G.F.; Karkanias, C.; Batsioula, M.; Melas, L.D.; Malamakis, A.E.; Geroliolios, D.; Skoutida, S.; Spiliotis, X. Environmental assessment of alternative strategies for the management of construction and demolition waste: A life cycle approach. *Sustainability* **2022**, *14*, 9674. [CrossRef]
14. Zhang, L.W.; Sojobi, A.O.; Kodur, V.K.R.; Liew, K.M. Effective utilization and recycling of mixed recycled aggregates for a greener environment. *J. Clean. Prod.* **2019**, *236*, 117600. [CrossRef]
15. Yu, Y.; Pacheco-Torgal, F.; Zhao, X.-Y.; Wang, X.-L. Cleaner production of the precast concrete industry: Comparative life cycle analysis of concrete using recycled aggregates from crushed precast rejects. *Eur. J. Environ. Civ. Eng.* **2024**, *28*, 1014–1038. [CrossRef]
16. Mostafaei, H.; Rostampour, M.A.; Chamasemani, N.F.; Wu, C. An In-Depth Exploration of Carbon Footprint Analysis in the Construction Sector with Emphasis on the Dam Industry. In *Carbon Footprint Assessments: Case Studies & Best Practices*; Springer: Berlin/Heidelberg, Germany, 2024; pp. 45–80.
17. Song, C.; Gardner, K.H.; Klein, S.J.W.; Souza, S.P.; Mo, W. Cradle-to-grave greenhouse gas emissions from dams in the United States of America. *Renew. Sustain. Energy Rev.* **2018**, *90*, 945–956. [CrossRef]
18. Zhang, C.; Hu, M.; van der Meide, M.; Di Maio, F.; Yang, X.; Gao, X.; Li, K.; Zhao, H.; Li, C. Life cycle assessment of material footprint in recycling: A case of concrete recycling. *Waste Manag.* **2023**, *155*, 311–319. [CrossRef] [PubMed]
19. Suwanit, W.; Gheewala, S.H. Life cycle assessment of mini-hydropower plants in Thailand. *Int. J. Life Cycle Assess.* **2011**, *16*, 849–858. [CrossRef]

20. Chhun, B.; Bonnet, S.; Gheewala, S.H. Life cycle assessment of the Kamchay hydropower plant in Cambodia. *J. Sustain. Energy Environ.* **2021**, *12*, 23–33.
21. Pang, M.; Zhang, L.; Wang, C.; Liu, G. Environmental life cycle assessment of a small hydropower plant in China. *Int. J. Life Cycle Assess.* **2015**, *20*, 796–806. [CrossRef]
22. Sadok, R.H.; Belaribi, N.B.; Mazouzi, R.; Sadok, F.H. Life cycle assessment of cementitious materials based on calcined sediments from Chorfa II dam for low carbon binders as sustainable building materials. *Sci. Total Environ.* **2022**, *826*, 154077. [CrossRef] [PubMed]
23. Pacca, S. Impacts from decommissioning of hydroelectric dams: A life cycle perspective. *Clim. Change* **2007**, *84*, 281–294. [CrossRef]
24. Rosa, L.P.; Dos Santos, M.A.; Matvienko, B.; Dos Santos, E.O.; Sikar, E. Greenhouse gas emissions from hydroelectric reservoirs in tropical regions. *Clim. Change* **2004**, *66*, 9–21. [CrossRef]
25. Fearnside, P.M. Greenhouse gas emissions from hydroelectric dams in tropical forests. In *Alternative Energy and Shale Gas Encyclopedia*; John Wiley & Sons, Inc.: Hoboken, NJ, USA, 2016; pp. 426–436.
26. Zhang, S.; Zhang, S.; Wu, Z.; Wang, X.; Jiang, Z.; Wang, C.; Zhao, G. IFC-enabled LCA for carbon assessment in pumped storage hydropower (PSH) with concrete face rockfill dams. *Autom. Constr.* **2023**, *156*, 105121. [CrossRef]
27. Liu, C.; Ahn, C.R.; An, X.; Lee, S. Life-cycle assessment of concrete dam construction: Comparison of environmental impact of rock-filled and conventional concrete. *J. Constr. Eng. Manag.* **2013**, *139*, A4013009. [CrossRef]
28. Zhang, S.; Pang, B.; Zhang, Z. Carbon footprint analysis of two different types of hydropower schemes: Comparing earth-rockfill dams and concrete gravity dams using hybrid life cycle assessment. *J. Clean. Prod.* **2015**, *103*, 854–862. [CrossRef]
29. *IS 6512*; Criteria for Design of Solid Gravity Dams (Second Revision). Bureau of Indian Standards (BIS): New Delhi, India, 2019.
30. Sharma, S. Fostering Green Product Design and Innovation for a Sustainable Future. In *Waste Management and Life Cycle Assessment for Sustainable Business Practice*; IGI Global: Hershey, PA, USA, 2024; pp. 86–110.
31. Sakib, M.N.; Kabir, G.; Ali, S.M. A life cycle analysis approach to evaluate sustainable strategies in the furniture manufacturing industry. *Sci. Total Environ.* **2024**, *907*, 167611. [CrossRef]
32. Akintayo, B.D.; Babatunde, O.M.; Olanrewaju, O.A. Comparative Analysis of Cement Production Methods Using a Life Cycle Assessment and a Multicriteria Decision-Making Approach. *Sustainability* **2024**, *16*, 484. [CrossRef]
33. Yuguda, T.K.; Li, Y.; Xiong, W.; Zhang, W. Life cycle assessment of options for retrofitting an existing dam to generate hydroelectricity. *Int. J. Life Cycle Assess.* **2020**, *25*, 57–72. [CrossRef]
34. Curran, M.A. *Life-Cycle Assessment: Principles and Practice*; National Risk Management Research Laboratory, Office of Research and Development, U.S. Environmental Protection Agency: Cincinnati, OH, USA, 2006.
35. Barbhuiya, S.; Das, B.B. Life Cycle Assessment of construction materials: Methodologies, applications and future directions for sustainable decision-making. *Case Stud. Constr. Mater.* **2023**, *19*, e02326. [CrossRef]
36. Momotaz, H.; Rahman, M.M.; Karim, M.R.; Zhuge, Y.; Ma, X.; Levett, P. Comparative study on properties of kerb concrete made from recycled materials and related carbon footprint. *J. Build. Eng.* **2023**, *72*, 106484. [CrossRef]
37. Wang, C.; Chen, B.; Vo, T.L.; Rezania, M. Mechanical anisotropy, rheology and carbon footprint of 3D printable concrete: A review. *J. Build. Eng.* **2023**, *76*, 107309. [CrossRef]
38. Mostafaei, H.; Kelishadi, M.; Bahmani, H.; Wu, C.; Ghiassi, B. Development of sustainable HPC using rubber powder and waste wire: Carbon footprint analysis, mechanical and microstructural properties. *Eur. J. Environ. Civ. Eng.* **2025**, *29*, 399–420. [CrossRef]
39. Bahmani, H.; Mostofinejad, D. Sustainable high-performance concrete with sugar factory lime waste-activated slag and fiber reinforcement: Mechanical properties and environmental impact. *Constr. Build. Mater.* **2025**, *464*, 140202. [CrossRef]
40. Bahmani, H.; Mostofinejad, D. A novel development of ultra-high-performance concrete with calcium oxide-activated materials and fibers: Engineering properties and sustainability evaluation. *Mag. Concr. Res.* **2024**, *76*, 882–902. [CrossRef]
41. Cordoba, G.; Irassar, E.F. Carbon footprint of reinforced concrete columns with and without supplementary cementitious materials. *Int. J. Life Cycle Assess.* **2023**, *28*, 800–812. [CrossRef]
42. Hassan, H.F.; Al-Shamsi, K.; Al-Jabri, K. Effect of Steel Slag on the Permanent Deformation and Life Cycle Cost of Asphalt Concrete Pavements. *Int. J. Pavement Res. Technol.* **2023**, *17*, 1513–1530. [CrossRef]
43. Njotea, B.A.; Okonkwo, V.O.; Mezie, E.O.; Nwankwoeke, C.H. Investigation of the Economic and Life Cycle Cost Benefits of Concrete Pavement over Asphalt Pavement: Case Study of Isieke Road, Ebonyi State. *Saudi J. Civ. Eng.* **2023**, *7*, 29–39.
44. Kromoser, B.; Butler, M.; Hunger, M.; Kimm, M.; Kopf, F.; Mechtcherine, V.; Pressmair, N.; Traverso, M. Article of RILEM TC 292-MCC: Life cycle assessment (LCA) of non-metallic reinforcement for reinforcing concrete: Manufacturing, durability, dismantling, recycling and reuse: A review. *Mater. Struct.* **2023**, *56*, 126. [CrossRef]
45. Ellingwood, B.; Tekie, P.B. Fragility analysis of concrete gravity dams. *J. Infrastruct. Syst.* **2001**, *7*, 41–48. [CrossRef]
46. Leclerc, M.; Léger, P.; Tinawi, R. Computer aided stability analysis of gravity dams—CADAM. *Adv. Eng. Softw.* **2003**, *34*, 403–420. [CrossRef]

47. Tekie, P.B.; Ellingwood, B.R. Seismic fragility assessment of concrete gravity dams. *Earthq. Eng. Struct. Dyn.* **2003**, *32*, 2221–2240. [CrossRef]
48. Baghlani, A.; Sattari, M.; Makiabadi, M.H. Application of genetic programming in shape optimization of concrete gravity dams by metaheuristics. *Cogent Eng.* **2014**, *1*, 982348. [CrossRef]
49. Gauthier, D. *European Cement Research Academy Technical Report*; Report, no. 1001-0742; European Cement Research Academy: Duesseldorf, Germany, 2015.

applied sciences

MDPI

Article

Evaluating 3D-Printed Polylactic Acid (PLA)-Reinforced Materials: Mechanical Performance and Chemical Stability in Concrete Mediums

Hanna Csótár [1], Szabolcs Szalai [1], Dmytro Kurhan [2], Mykola Sysyn [3] and Szabolcs Fischer [1,*]

[1] Central Campus Győr, Széchenyi István University, H-9026 Győr, Hungary; csotar.hanna@sze.hu (H.C.); szalaisz@sze.hu (S.S.)

[2] Department of Transport Infrastructure, Ukrainian State University of Science and Technologies, UA-49005 Dnipro, Ukraine; d.m.kurhan@ust.edu.ua

[3] Department of Planning and Design of Railway Infrastructure, Technical University Dresden, D-01069 Dresden, Germany; mykola.sysyn@tu-dresden.de

[*] Correspondence: fischersz@sze.hu; Tel.: +36-(96)-503-400

Abstract: The optimization and evaluation of 3D-printed polylactic acid (PLA) materials for reinforcing concrete elements present a promising avenue for advancing sustainable construction methods. This study addresses the challenges associated with PLA's dual nature—biodegradable yet mechanically limited for long-term applications—while leveraging its potential to enhance concrete reinforcement. The research identifies gaps in understanding PLA's mechanical and chemical behavior in alkaline environments, particularly its interactions with concrete matrices. To bridge this gap, four distinct PLA variants (high-impact PLA, engineering PLA, electrical ESD PLA, and gypsum PLA) and ABS (acrylonitrile butadiene styrene) were subjected to dissolution tests in NaOH solutions (pH 12 and 12.55) and mechanical evaluation under three-point bending using digital image correlation (DIC) technology. Test specimens were prepared using optimized 3D printing strategies to ensure structural consistency and were embedded in concrete beams to analyze their reinforcement potential. Force–displacement data and GOM ARAMIS measurements revealed significant differences in mechanical responses, with peak loads ranging from 0.812 kN (high-impact PLA) to 1.021 kN (electrical ESD PLA). Notably, electrical ESD PLA exhibited post-failure load-bearing capacity, highlighting its reinforcement capability. Chemical dissolution tests revealed material-specific degradation patterns, with high-impact and Gypsum PLA showing accelerated surface changes and precipitation phenomena. Observations indicated white crystalline precipitates, likely lime (calcium hydroxide—$Ca(OH)_2$), residue from the dissolution tests (sodium hydroxide—NaOH), or material-derived residues formed on and near PLA elements, suggesting potential chemical interactions. These findings underline the critical role of material selection and optimization in achieving effective PLA–concrete integration. While PLA's environmental sustainability aligns with industry goals, its structural reliability under long-term exposure remains a challenge. The study concludes that electrical ESD PLA demonstrates the highest potential for application in reinforced concrete, provided its chemical stability is managed, as its peak value (1.021 kN) showed 25.7% higher load-bearing capacity than high-impact PLA (0.812 kN) and did not lose any of its structural stability in the dissolution tests. This work advances the understanding of PLA as a sustainable alternative in construction, offering insights for future material innovations and applications.

Keywords: sustainability; FDM; PLA; 3D-printed structures; glass-reinforced PLA; reinforced; concrete; DIC; GOM ARAMIS; GOM ATOS

1. Introduction

The 21st century's engineering undoubtedly plays a prime role in the pursuit of the greatest human needs, incorporating many forms of engineering, such as civil [1–5], mechanical [6–9], electrical [10,11], vehicle engineering [12,13], transportation, logistics [14,15], cognitive mobility [16–20], and environmental and chemical engineering [21,22], towards the quest of sustainable development [23–26]. They are responsible for developing sustainable engineering systems, and civil engineering is responsible for addressing the infrastructural systems of present-day societies with highly resilient and environmentally friendly materials. Moreover, mechanical engineering develops systems and machines that require less energy to operate. Electrical engineering promotes the use of renewable energy systems and establishes smart grid networks, which include sustainable approaches to the supply of power. Moreover, vehicle and transportation engineering has advanced the mobility industry with the use of electric and self-driving vehicles and has reengineered logistics systems to be more efficient while reducing emissions. Environmental engineering stitches all these professional spheres together and ensures that various technological advances are made responsibly and with awareness of eco-friendly and sustainable global goals.

However, considering this interplay, there comes the integrated notion that for sustainability across these groups, there must be coalescence, integrating approaches that minimize wastage, protect resources, or optimize functionality, which logistics does by integrating transportation with energy and infrastructure to create an efficient supply chain that is also environmentally friendly. Developing strategies within these sectors positively impacts the environment while simultaneously strengthening economies, increasing urban sustainability, and enhancing the overall living experience. Such a multi-faceted strategy complements the goal of engineering in its continuous improvement and its commitment to the Sustainable Development Goals of the United Nations [27].

Out of these fields, 3D printing makes it possible for materials to be consumed over a greater volume. It makes construction, manufacturing, and transport mechanisms more and more efficient. Advocacy for 3D-printed parametric designs in concrete as a material in reinforced structures for sustainable housing is built in this work. Organizations can leverage these technologies to enhance engineering cross-disciplinarity by solving the challenge of materials' sustainability while increasing infrastructure resiliency and performance.

Three-dimensional printing is transforming construction by enabling precise, layer-by-layer manufacturing that reduces material waste and enhances sustainability [28–30]. This technology allows for customizable and scalable structures, optimizing resource use while minimizing environmental impact. Its adaptability supports the production of complex components, from small to large, with minimal waste, making it an efficient method for sustainable housing [31]. Additionally, it accelerates construction through rapid prototyping [32–34]. The printing table's volume determines object size, allowing for diverse component production [33,34].

Despite its advantages, additive manufacturing faces challenges, particularly with PLA (polylactic acid). PLA's dual nature—biodegradable but lacking long-term durability—raises concerns for industrial applications [35,36]. Its slow and undefined decomposition process complicates sustainable use [37,38]. Researchers have explored material blending to tailor PLA's properties. For example, combining PLA with polycaprolactone (PCL) enhances biodegradation, achieving significant CO_2 conversion in soil and composting environments [39–44]. Accelerated degradation tests confirm that temperature significantly affects PLA decay, influencing crystallization and surface damage [44].

PLA's physical limitations, including low thermal stability and high thermal expansion, hinder its suitability for long-term applications in construction [45]. Exposure to heat and moisture can cause rapid deterioration, making it unsuitable for critical structures like bridges and foundations [45–47]. Its expansion coefficient mismatch with other materials also raises structural concerns [46–48]. Consequently, PLA is primarily used in applications where high thermal and moisture resistance are not required.

Despite these limitations, PLA is being investigated as a reinforcing material in concrete. Unlike its use in other fields, PLA in construction is incorporated to enhance mechanical properties rather than degrade [48–53]. Reinforcement effectiveness is assessed through laboratory tests and finite element modeling [54–60]. The arrangement of 3D-printed reinforcements significantly affects structural performance. Studies demonstrate that specific geometries, such as rhomboid and honeycomb patterns, improve strength and durability [61,62]. Additionally, incorporating small aggregates like steel fibers further enhances reinforcement [63,64].

Several studies highlight the potential of 3D-printed PLA for strengthening concrete. Integrating porous PLA structures into cement matrices significantly improves mechanical properties. For instance, schwarzite- and zeolite-based PLA structures were shown to increase specific elasticity by 128.1% and 505.33%, respectively [57]. Xu et al. [65] examined surface modifications on 3D-printed PLA reinforcement, demonstrating that coatings like epoxy resin and steel fiber-sprayed epoxy enhance flexural strength. Sand and steel fiber coatings improved mechanical properties, emphasizing the importance of surface treatments for performance optimization.

Based on the literature review in the present study, 3D printing offers customizability, minimal waste, and cost efficiency, making it a valuable tool for construction. However, PLA's thermal and hydrolytic instability limit its use in demanding environments. Research shows that PLA can reinforce concrete under specific conditions, compensating for its limitations. While PLA degradation has been extensively studied, its solubility parameters in different solvents require further standardization. This research investigates the mechanical and chemical behavior of PLA variants and ABS in concrete, evaluating their reinforcement potential. Electrical ESD PLA demonstrated the highest reinforcement capability, exhibiting a 25.7% higher load-bearing capacity than high-impact PLA and retaining structural stability in dissolution tests. These findings contribute to the development of sustainable, high-performance construction materials.

This paper focuses on reinforcing concrete beams with 3D-printed PLA structures and is a continuation of a previous publication [66]. As in the previous publication, the different uses of ECO PLA (printed in different forms and as a waste aggregate from printing waste) are presented. In this paper, the authors focused on the diversity of different materials and their effects on mechanical load-bearing capacity and chemical stability in an experimental simulated environment.

The primary objective of this research is to investigate the behavior of various materials in the concrete medium, focusing on their reinforcement potential, solubility in NaOH solutions with pH levels similar to concrete, and performance under mechanical loading. The study examines five materials commonly used in engineering fields—PLA (high-impact, engineering, electrical ESD, and gypsum) and ABS (see Sections 2 and 3)—each selected for its distinct mechanical properties, compatibility with different 3D-printing strategies, and applications in industries such as electronics, automotive, modeling, and robotics.

ABS, in particular, was chosen for its high impact resistance, stiffness, and recyclability, making it a relevant candidate for innovative reinforcement strategies; however, the most relevant reason for the consideration was the fact that ABS is the most common base material for 3D printing in engineering; so, ABS was the benchmark material for appropriate consideration with PLA types. Three-dimensional printing with ABS filaments is usually "complicated" and problematic. In contrast, PLA has been developing very dynamically in recent years, especially PLA composites. Its printability is also better, as is its environmental impact. PLA was investigated precisely as an alternative to ABS.

The study evaluates the potential of these materials as sustainable alternatives to conventional concrete reinforcements like steel and Kevlar. However, the primary focus lies in understanding the chemical reactions and solubility of these materials in NaOH solutions as well as their resistance under three-point mechanical loading tests.

While this research was limited to five materials due to constraints, it lays the groundwork for future studies that will explore additional options, such as PET-G (polyethylene terephthalate glycol-modified) and PCTG (polycyclo-hexylenedimethylene terephthalate glycol-modified). The structure of the study includes the following sections: Section 1 is the Introduction, Section 2 contains Materials and Methods, Section 3 introduces Results and Discussion, and Section 4 presents Conclusions.

The main contributions of the paper are as follows:

- Exploring 3D-printed PLA and ABS for concrete reinforcement—the study investigates the use of 3D-printed structures as an alternative reinforcement material for concrete beams, continuing previous research on PLA applications [66];
- Comparative analysis of engineering materials—it examines the performance of five engineering materials (high-impact PLA, engineering PLA, electrical ESD PLA, gypsum PLA, and ABS) in a concrete environment, focusing on mechanical load-bearing capacity and chemical stability;
- ABS as a benchmark for PLA—ABS is used as a benchmark due to its widespread use in 3D printing and its mechanical properties. The study evaluates PLA as a potential alternative to ABS, emphasizing PLA composites' advantages in printability and environmental impact, as ABS is non-degradable (see Section 2.1);
- Chemical and mechanical testing—investigation of the solubility of the material in NaOH solutions (simulating the high-pH environment of concrete) and the mechanical resistance using three-point bending tests;
- Sustainability and future research directions—these highlight the potential of PLA and other materials as sustainable replacements for conventional reinforcements like steel and Kevlar. The study also paves the way for future investigations, suggesting further research on materials like PET-G and PCTG.

2. Materials and Methods

In order to provide a better understanding of the paper and its results, a flowchart is provided (Figure 1) so that the reader can fully understand the two parallel measurements.

2.1. Applied PLA Materials (Filaments)

First of all, the authors should mention that the applied and considered materials were chosen for the following reasons:

- ABS was chosen as a benchmark material due to its widespread use in 3D printing for engineering, high impact resistance, stiffness, and recyclability;
- PLA was investigated as a dynamic and eco-friendly alternative to ABS, with superior printability and lower environmental impact;
- ABS, despite its ordinary use, presents challenges in 3D printing, while PLA composites have shown significant advancements in recent years;
- The inclusion of both ABS and PLA allows for a comparative analysis of their properties, environmental impact, and printing requirements;

- The study emphasizes sustainability and material optimization, aligning with the goal of advancing eco-friendly construction methods;
- While PLA offers environmental benefits, its limitations in specific applications are acknowledged, highlighting the need for balanced material selection in 3D-printed concrete reinforcement.

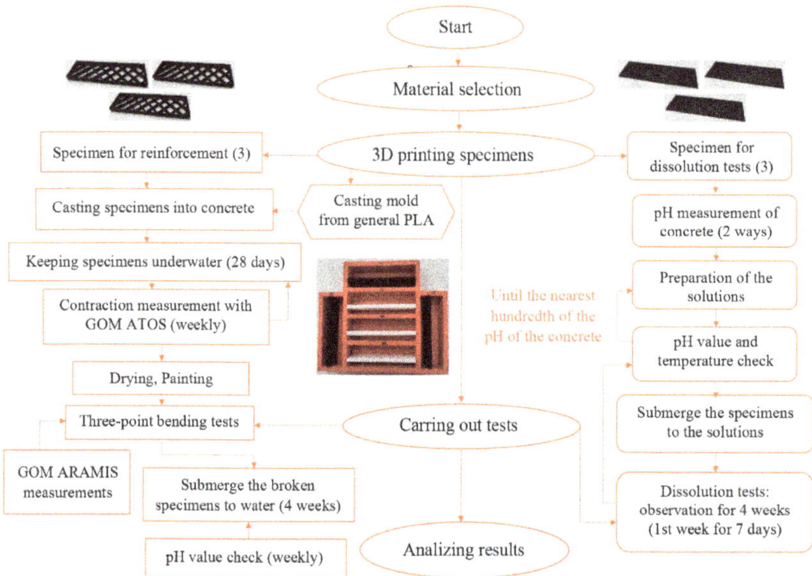

Figure 1. Flowchart of the measurements and methods.

During the tests, five different types of PLA material were used. High-impact PLA provides a higher impact resistance to the structure than PETG (polyethylene terephthalate glycol) materials and has a similar flexibility to polypropylene (PP).

Engineering PLA was selected because the heat resistance of printed objects is exceptionally high, withstanding temperatures of up to 120 °C, which can be achieved without the need for post-heat treatment. This makes it much more resistant to external influences than other PLA materials. It also has high tensile and compressive strength and extremely high stiffness.

The next material to be selected was electrical ESD (electrostatic discharge) PLA, which is a fiber-printed object that does not charge electrostatically due to special carbon and chemical additives. In addition, dust and fine dirt do not adhere to the surface of the objects. It provides a surface resistivity of the order of megaohms, but this depends mainly on the size and shape of the printed object. Furthermore, if printed at a higher temperature, it will provide a lower surface resistance. The surface resistance also depends on the print orientation, with a higher value in the Z direction and a lower value in the *XY* direction.

The following material is slightly different from the other ones. While other materials are generally used in the automotive and electronics industries, gypsum PLA is the preferred product for modeling due to its high formability. Objects printed with gypsum PLA are very easy to shape, as they can be cut and filed with a sharp tool, making it easier to machine the object afterward. Painting is also easy, even with water-based paints.

The last material used for sample and reinforcement production was ABS (acrylonitrile butadiene styrene), a widely used thermoplastic in the automotive, electronics, and engineering industries, valued for its high impact resistance and stiffness. While more challenging to print than PLA, ABS offers low shrinkage and allows for fast, precise processing in 3D printing, making it well suited for research. Its behavior in concrete—for instance, its solubility in NaOH solution simulating the pH of concrete—provides valuable insights for developing sustainable and innovative reinforcement strategies. These qualities make ABS an ideal choice for this study, aimed at exploring new reinforcement solutions.

The casting molds were made from generic PLA, as they were designed to be single-use and needed to be quick and easy to dismantle. A total of six different materials were used during 3D printing, of which generic PLA was used only for the mold, and the other five materials were used for the test specimens. Furthermore, its solubility in NaOH solutions offers critical insights into its interaction with concrete media, contributing to sustainable and alternative reinforcement strategies. Table 1 represents the materials used and the recommended and applied printing parameters of the specimens.

Table 1. Recommended and used printing parameters of the specimens and molds.

Material	Mark	Manufacturer	Printing Parameters	Recommended [°C]	Applied [°C]
PLA	none	Filaticum Kft., Miskolc, Hungary	Nozzle temperature Bed temperature	190–215 55–70	200 60
High-impact PLA	HI	Filaticum Kft., Miskolc, Hungary	Nozzle temperature Bed temperature	190–210 45–65	200 60
Engineering PLA	ENG	Filaticum Kft., Miskolc, Hungary	Nozzle temperature Bed temperature	215–240 65–75	225 75
Electrostatic discharge PLA	ESD	Filaticum Kft., Miskolc, Hungary	Nozzle temperature Bed temperature	195–215 55–70	200 65
Gypsum PLA	GYP	Filaticum Kft., Miskolc, Hungary	Nozzle temperature Bed temperature	190–215 55–70	200 60
ABS	ABS	Filaticum Kft., Miskolc, Hungary	Nozzle temperature Bed temperature	260–290 80–95	260 90

The specimens were printed using the Creality Ender 3—v2 (Creality 3D Technology Co, Ltd., Shenzhen, China) 3D printer. The molds were printed with a Bambu Lab A1 (Bambu Lab, Shenzhen, China) mini printer.

However, the nozzle and bed temperature parameters applied differed according to the material; the other parameters, such as cooling rate, nozzle size, layer height, and printing speed, were the same. The applied identical printing parameters were as follows: nozzle size 0.4 mm, cooling fun 100%, layer height 0.2 mm, print speed 50 mm/s, and a skirt was also used.

Table 2 summarizes the materials tested (PLA, ABS, HI, ENG, and ESD) and compares their properties with each other and with other relevant materials (PETG); it also includes symbols for easy comparison.

Table 2. Comparative table of material properties.

Material	Tensile Strength	Flexibility	Heat Resistance [°C]	Printability	Eco-Friendliness
PLA	Moderate	Brittle	55–70	Easy	Biodegradable, non-recyclable
	••	••	•	•••	
HI	Moderate	Flexible	65–70	Easy	Biodegradable, non-recyclable
	••	••	••	•••	
ENG	High	Flexible	65–70	Easy	Biodegradable, non-recyclable
	•••	••	•••	•••	

Table 2. *Cont.*

Material	Tensile Strength	Flexibility	Heat Resistance [°C]	Printability	Eco-Friendliness
ESD	Moderate	Flexible	65–70	Moderately easy	Non-degradable
	••	••	••	••	
GYP	Low	Brittle	55–70	Easy	Biodegradable, non-recyclable
	•	•	•	•••	
ABS	Moderate	Flexible	80–95	Moderately hard	Non-degradable, recyclable
	••	••	•••	•	
PETG (polyethylene terephthalate glycol-modified PLA)	Moderate	Flexible	70–85	Easy	Non-degradable, recyclable
	••	•••	••	•••	

2.2. Applied Printing Set-Ups and Structures

Two different printing strategies were applied, as the research examines the relationship between PLA and concrete from two perspectives. The design was created using Solidworks 2016 × 64 Edition SP01 (SolidWorks Corporation, Waltham, MA, USA). To convert the stl. files to printing instructions, a slicing software, UltiMaker Cura 5.3 0 (Ultimaker, Geldermalsen, The Netherlands), was used.

The first layout concerned determining the solubility of the PLA material, so a 0.4 mm thick layer of sample was needed. During the printing process, the infill value did not influence the final structure of the PLA thin sheets due to its thin design. The size of the thin sample sheets was 25.0 × 0.4 × 80.0 mm.

The other layout was the structure, which was used as a reinforcing element in concrete beams, with a size of 25.0 × 5 × 80.0 mm. Each element was printed with a grid infill pattern and 10% infill density, with the top and bottom layers set to 0. As the initial and final layers were set to 0, the grid structure was visible.

This was a practical set-up as well as a functional one. Upon removing the initial and final layers, the printing time was reduced; furthermore, it was functional because, during casting, the concrete could spread between the gaps of the grid infill structure. Figure 2 shows the two structures in the UltiMaker Cura software with a coordinate system, and Figure 3 illustrates the six-cavity mold that was used during the casting of the concrete beams.

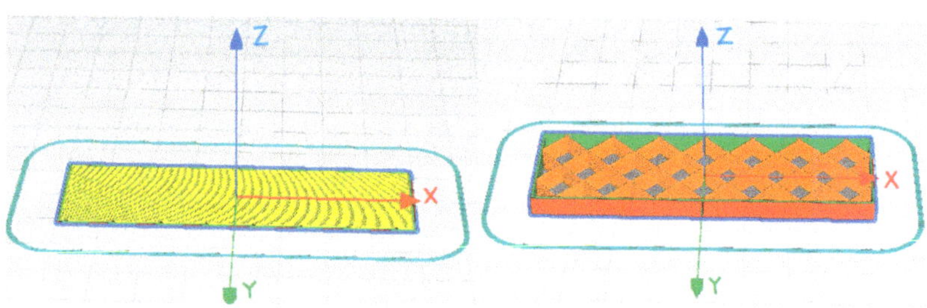

Figure 2. Printing set up of the specimens in UltiMaker Cura.

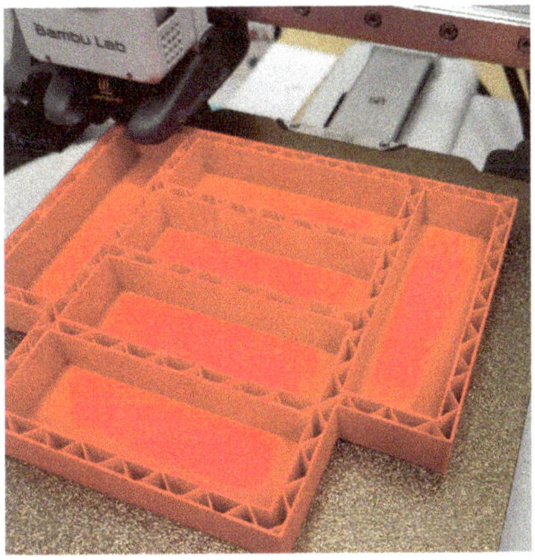

Figure 3. The structure of the PLA six-cavity mold.

The other printed (Figure 3) structure was the casting mold. The size of the concrete elements was 25.0 × 25.0 × 80.0 mm, so the inner size of the mold was identical. Three layers of the base were applied to the structure while printing. The wall thickness of the shape was 0.5 mm, with the same grid pattern and infill rate as the reinforcing element.

In summary, three and three samples were made per material: for each material, three were reinforcing elements for casting, and three were used for dissolution tests.

2.3. Molding Parameters

After the preparations of the printed elements, the casting of the reinforced concrete beams took place. The size of each concrete element was 25.0 × 25.0 × 80.0 mm (in accordance with the standard MSZ EN 12390-1:2021 [67]). The 3D-printed reinforcing elements were placed, for practical reasons, in the lower one-third of the mold prior to casting. The reinforcing element had to be close to the edge (bottom), far from the neutral axis, to provide a reinforcing effect, if any. (Particularly, its investigation was one of the main objectives of the current paper.) Figure 4 shows the location of the reinforcing material cast into concrete, and Table 3 represents the formula of the production for 1.6 L of concrete.

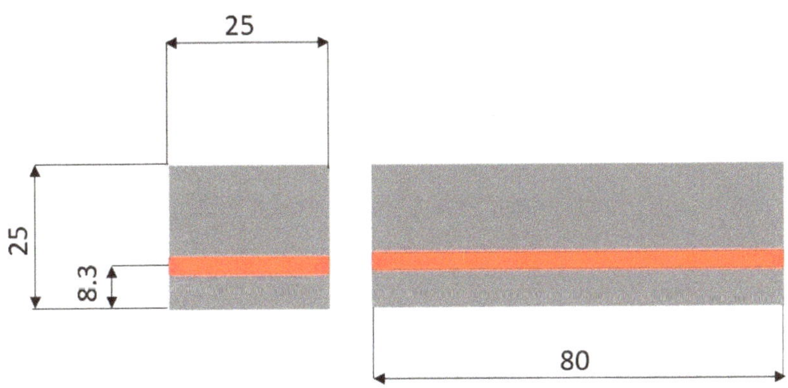

Figure 4. Location of the reinforcement in the concrete beam (units are in mm); $L = 80$ mm, $d = 25$ mm.

Table 3. Concrete composition per 1.6 L of concrete.

Material	Manufacturer (Brand)	Amount/Quantity
Water	-	480.6 g
Cement	CEM-II-AS-42.5 (Duna-Dráva Cement, Vác, Hungary)	1369.6 g
Limestone powder	Lafarge (Lafarge, Paris, France)	328.4 g
Fine aggregate	graded sand 0/1 (Duna-Dráva Cement, Vác, Hungary)	1489.8 g
Fluxing agent	VC 5 NEW (Sika, Baar, Switzerland)	8.8 mL

In order to yield a high-quality surface finish, which was needed for the GOM measurements, a Sica Separol AR-2 ECO-type (Sika Hungária Kft., Biatorbágy, Hungary) release agent was applied to the inside part of the mold before casting. After the preparation of the mold, the 3D-printed grid structures were placed in the mold, followed by the casting of the concrete. The next step was to remove the air trapped inside the mold using a Matest C278 vibro table (Matest, Treviolo, Italy).

The reinforcement was a 25.0 × 5 × 80.0 mm element with a structure predetermined by multiplying the sizes to obtain the volume, which was 10 cm^3. The concrete element without reinforcement weighed 114.88 g. The actual filled volume of the reinforcement with 10% infill was the volume multiplied by the percentage of the infill value, which was 1 cm^3, which means the percentage by volume for all substances was the same at 2%. Since the density of the reinforcing materials used is known, the volume and mass percentages of the elements can be calculated and are summarized in Table 4.

Table 4. Basic parameters of reinforced concrete elements.

Material	Mass [g]	Mass Percentage [%]	Volume Percentage [%]
ESD	1.24	1.07	2.0
ENG	1.24	1.07	2.0
HI	1.24	1.07	2.0
GYP	N/A	N/A	2.0
ABS	1.05	0.91	2.0

Table 4 shows that the mass and impact of the grid element on the concrete element remained minimal, especially for low infill and lower-density materials such as ABS. Following this analysis, the casting process proceeded to its final stage. After the solidification of the concrete elements, which took 24 h in the mold (in accordance with the standard MSZ EN 12390-2:2019 [68]), the elements were kept underwater for 27 more days to allow the inner solidification process of the concrete beams to finish. The underwater treatment of the test specimens resulted in a better-quality surface finish. The absence of surface cracks and surface defects facilitated the preparation of DIC measurements.

2.4. Bending Tests and DIC Measurements

The three-point bending tests were carried out with Zwick 1454 (Zwick Roell Group, US, Singapore) at 7.6 mm/min loading speed with load cell HBM U3 20 kN (HBM Ltd., Budapest, Hungary) and data collector HBM quantum MX840A (HBM Ltd., Budapest, Hungary), where the specimens were under load until they showed a sign of cracking. During the tests, a specific test method was applied rather than a standard one, where each test piece was loaded at 7.6 mm/min loading speed with load cell HBM U3 20 kN. A uniform load with a constant 20 kN causes the test specimens to move.

The amplification effect of the reinforcing elements can be determined as a function of the displacement of the test pieces relative to their initial state (vertical displacement in mm) and the resistance to the load (force in kN). The load was placed in the middle of the specimens, and the distance between the supports (i.e., bay length) was 75 mm. The tests were carried out according to the MSZ EN 12390-5:2019 standard [69]. Since the laboratory we used is not an accredited measuring laboratory, and the measurement is related to research, the chosen length of $L = 80$ mm was used instead of the length $L \geq 3.5 \times d$ (87.5 mm), defined in MSZ EN 12390-5:2019 [69], where d is the dimension of the specimen in cross-section (i.e., the width and the height), as well as the distance between the support points and the applied vertical concentrated forces and between the two concentrated forces in the case of 4-point bending (see Figure 4).

In the case of the bending tests, even though the standard MSZ EN 12390-5:2019 [69] was included, the authors had to deviate from the instantaneous measurement described in the standard. The reason was the GOM ARAMIS (GOM, Braunschweig, Germany) measurements. It was necessary to deviate because if the paint had not adhered to the surface, the measurements would not have been able to be made since the technology calculates the amount of deformation from the displacements of the points painted on the surface.

Before laboratory experiments, each sample was dried for two days to ensure that moisture content did not affect the results. This was verified by weighing the specimens every 4 h on the first day and every hour on the second day until the difference in weight was negligible.

The specimens were then painted, first with a matte white lacquer primer (Deco Color Moto Techno Aerosol (Chemmot Ewa Kumorek Spółka Komandytowa, Skawina, Poland)) and then with a matte black lacquer topcoat spray (United Sealants Sprays Matte Black Spray (Neosil Kft., Bag, Hungary)).

Getting the paint pattern right was essential for the DIC (digital image correlation) GOM ARAMIS (GOM, Braunschweig, Germany) measurements. The issue of stain pattern was most carefully examined by Philipp Reu [70,71]. According to Reu, the pattern can be described by four primary characteristics: spot size, contrast, overlap of spots, and density of spots.

It is the combination of these characteristics that most accurately determines the spot size. The spot size is defined by the relationship between the resolution of the DIC camera system and the size of the sample area under inspection. These two data are used to calculate the size of the pixels in millimeters, with an optimum value of between 3 and 5 pixels. Smaller pixels are more difficult to detect during digitalization.

In addition, the overlap of speckles is a critical factor since if the speckle pattern overlap is not sufficient, images may become noisy during digitalization. For this reason, it is imperative to avoid overlapping when using speckle patterns, as this can negatively affect the measurement results.

The ability to identify displacements and deformations down to tenths of a millimeter with sufficient accuracy can enable the results to be used in a broader range of applications, whether in forensic, automotive, textile, or medical orthopedic applications [72].

Since light and surface reflectivity affect the camera's performance, the metallic, reflective areas were covered with masking tape to prevent interference with the measurement results. Figure 5 illustrates the experimental set-up and an appropriately painted surface that the measurement system recognizes so that it can be evaluated later.

The results were then evaluated using the GOM DIC ARAMIS 2018 (GOM, Braunschweig, Germany) evaluation software. The displacement data from the GOM system were then synchronized with the mechanical load values of the load cell, which are presented in Section 3.

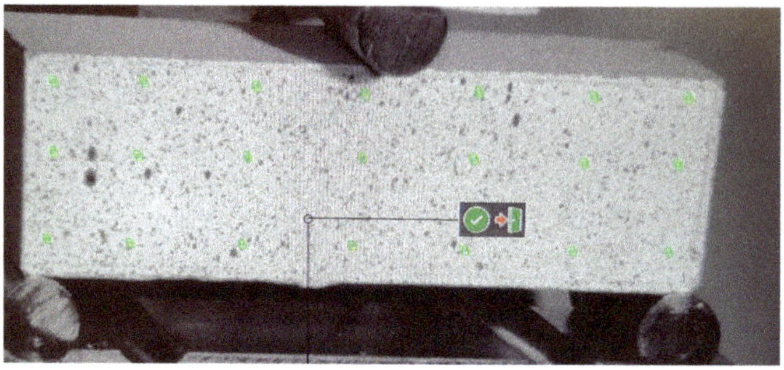

Figure 5. Experimental set-up with an appropriately painted surface, recognizable for the system.

Before starting the measurement, the camera was turned 180 degrees from its original position so that the coordinate system was also placed upside down, and the results were evaluated. The data were collected in parallel using the GOM ARAMIS (GOM, Braunschweig, Germany) camera system to measure displacement in the x direction and the HBM quantum MX840A (HBM, Darmstadt, Germany) data collector to record both force and displacement values.

The ARAMIS sampling frequency was 8 Hz, and the HBM sampling frequency was 10 Hz. During the evaluation, the ARAMIS displacement values and HBM force data were evaluated during the measurements. However, due to the different sampling frequencies, the data lengths were different, so a linear interpolation of the displacement values in the x direction of the GOM measurements was necessary based on the HBM force measurement data series size (see Section 3.1).

Besides the displacement measurements, the specimens were observed by GOM ATOS Triple Scan (GOM, Braunschweig, Germany) during the 28-day drying cycle to compare the stages of changes of the specimens. This differs from other technologies in that a painting process is not required in this case. Here, reference points with factory markings can be applied to the surface painted matte white.

Then, by adjusting the point size and temperature (object heat) of the camera, the machine scanned the surface by rotating the rotary table. This paper used a similar method as Szalai et al. [73], where a non-destructive diagnostic system (GOM) was used to more accurately identify potentially deformable areas during the adaptation of automotive batteries. In both cases, the measurement process was, of course, preceded by accurate and appropriate calibration requested and recommended by the manufacturer of the instruments.

2.5. Solvent Preparation and Tests

This section describes the parameters of the solvent compositions required for the experiments since the solubility of the PLA materials is another important point of this research. According to the factory descriptions, none of the tested materials are water soluble; only ABS has a defined solvent (THF—tetrahydrofuran acetone and other analogous solvents); the solubility of the other tested materials is not defined.

To carry out the tests, it was necessary to determine the pH of the concrete used, but there are several standards and methods for determining the pH of fresh concrete. pH is an approximate measure of the acidity or alkalinity of a solution, defined as the negative logarithm of the concentration of hydrogen ions (H^+).

As the pH of the solution increases, the number of free hydrogen ions decreases, and each unit increase in pH represents a tenfold decrease in H^+ concentration. The pH scale ranges from 0 to 14, where a pH of 7 is considered neutral; a pH below 7 indicates an acidic environment, while a pH above 7 indicates an alkaline environment.

The following examples show the relationship between H+ concentration and pH for different solutions:

- HCl (hydrogen chloride) solution, with an H^+ concentration of 1×10^{-2} and pH 2;
- H_2O (water), with an H^+ concentration of 1×10^{-7} and pH 7;
- Saturated $Ca(OH)_2$ (calcium chloride), with an H^+ concentration of $1 \times 10^{-12.4}$ and pH 12.4;
- NaOH (sodium hydroxide) solution, with an H^+ concentration of 1×10^{-14} and pH 14.

Grubb et al. [74] recognized the challenges and difficulties of the measurement of the pH value of concrete, reviewed the standard procedures, and recommended a test procedure. To measure the pH of solid materials such as lime, soil, or concrete, an aqueous solution of the powdered material must be created; this dilutes the concentration of the solid material.

Measurements were made with both a pH strip (i.e., so-called pH "paper") and a pH probe, following the steps suggested by [74] in the field test for pH measurement for concrete surface and laboratory test for pH measurement of concrete sections, with one amendment, i.e., using a sieve (size: 0.25 mm), since the authors of the current study had metric tools at their disposal [74].

The pH values were measured in two separate measurements. First, the pH of the fresh concrete was measured, followed by a laboratory pH test of the powder sample extracted from the reference test specimens, which had been dry for 28 days. Then, the pH of the freshly mixed concrete was measured, which was 12. This procedure was carried out using only pH test paper since the probe can only measure filtered solutions. The second measurement was a laboratory pH test.

Before the pH measurements were started, a water–concrete mixture of the correct proportions had to be prepared. The finely graded powder sample, taken from reference test specimens and kept dry for 28 days, was filtered with a 0.25 mm sieve, and then, the water–concrete mixture was prepared with a ratio of 10 mL to 5 g.

During the pH measurement series, ionized water of quality class 3 of MSZ ISO 3696:1992 [75] was used. (It has to be mentioned that this standard is a withdrawn standard, but the content as well as the methodology are still valid—theoretically.) Before measuring the pH, the solution was left to stand for 30 min, stirring occasionally, followed by 1.2 μm filtering. Then, the pH measurement was carried out using a WTW 720 pH meter (AKTIVIT Kft., Weilheim, Germany) and pH test paper (Cloud Prime, Seattle, WA, USA).

The pH measured with a pH meter was 12.55 (water–concrete mixture), and the pH measured with pH test paper (water–concrete mixture and fresh concrete) was 12, in both cases at 22.2 °C.

Since five different materials were tested (HI, ENG, ESD, GYP, and ABS), and three specimens were prepared per material, the amount of the solutions was determined accordingly. After determining the pH values, two solutions were prepared with NaOH powder. A 250 mL solution with pH 12.55 and a 500 mL solution with pH 12 were prepared.

Based on the previous measurements, two solutions were made, one with a pH value of 12 and one with 12.55, since the values of the pH strip measurement were identical.

The first solution was prepared from a mixture of 0.2 g NaOH powder and 500 mL ionized water and the second solution from a mixture of 0.355 g NaOH powder and 250 mL ionized water.

Equations (1)–(4) show the calculations for the first solution (500 mL) preparation (where M_{NaOH} = 40.00 g/mol, the pH value of the NaOH solution is 12, and the pH of the powdered NaOH solvent is 14):

$$pOH1 = 14 - 12 = 2 \tag{1}$$

$$[OH^-]_1 = 10^{-pOH1} = 10^{-2}\ [mol/L] = 0.01\ [mol/L] \quad (2)$$

$$n_1 = 0.01\ [mol/L] \times 0.5\ [L] = 0.005\ [mol] \quad (3)$$

$$m_1 = 40.00\ [g/mol] \times 0.005\ [mol] = 0.2\ [g] \quad (4)$$

Equations (5)–(8) show the calculations for the second solution (250 mL) preparation (where M_{NaOH} = 40.00 g/mol, the pH value of the NaOH solution is 12.55, and the pH of the powdered NaOH solvent is 14):

$$pOH2 = 14 - 12.55 = 1.45 \quad (5)$$

$$[OH^-]_2 = 10^{-pOH2} = 10^{-1.45}\ [mol/L] = 3.55 \times 10^{-2}\ [mol/L] \quad (6)$$

$$n_2 = 3.55 \times 10^{-2}\ [mol/L] \times 0.25\ [L] = 8.88 \times 10^{-3}\ [mol] \quad (7)$$

$$m_2 = 40.00\ [g/mol] \times 8.88 \times 10^{-3}\ [mol] = 0.355\ [g] \quad (8)$$

The solutions were mixed in several steps. First, the solutes were weighed on an analytical balance of hundredths accuracy. This was followed by a preliminary dissolution, where about two-thirds of the total solvent was added to the solute. After the exothermic reaction, the solution was diluted to the appropriate amount, and finally, after stirring, the pH value was adjusted with the pH meter to the nearest hundredth of the pH of the concrete until it was set to a predetermined value of 12 and 12.55.

After the preparations, the solutions were divided into five parts for 250 mL of pH 12.55 and ten parts for 500 mL of pH 12, so that 50 mL of solution was in measuring cups. Then, the 0.4 mm thin PLA samples were put into the solution and checked periodically after every 24 h for 28 days.

With this experiment, the aim was to observe the behavior of PLA and ABS specimens with a size of 25.0 × 0.4 × 80.0 mm in an alkaline medium with the same pH as the concrete over the same time interval as the concrete curing time.

3. Results and Discussion

This section describes in detail the observations collected during the experiments and the results obtained. The aim is to provide a comprehensive picture of the phenomena observed during the experimental process, the measurement data, and the relationships and conclusions that emerged during the investigation. Visual inspection tests, while preliminary, revealed significant qualitative insights into material behavior and surface interactions. These observations informed the design of supplementary tests, such as flame tests and pH evaluations, which provided quantifiable data to support and validate the visual observations. For instance, partial dissolution and discoloration phenomena in ABS and GYP samples highlighted complex interactions with the alkaline medium, paving the way for future detailed chemical characterization.

The analysis of the results is presented in Section 4. It enables a deeper understanding of the phenomenon under study and the identification of new relationships that can contribute to the further development of the research area.

3.1. Concerning the Mechanical Test and GOM ARAMIS Measurement Series

The mechanical properties of the test specimens were determined by plotting the force–displacement diagrams, which were determined using the GOM ARAMIS camera and HBM data collector systems.

Since the data from the two data collection and measurement systems were not linked, they had to be synchronized and matched later during the evaluation.

Figure 6 illustrates the force–displacement diagrams of the specimens, including the reference specimen without reinforcement.

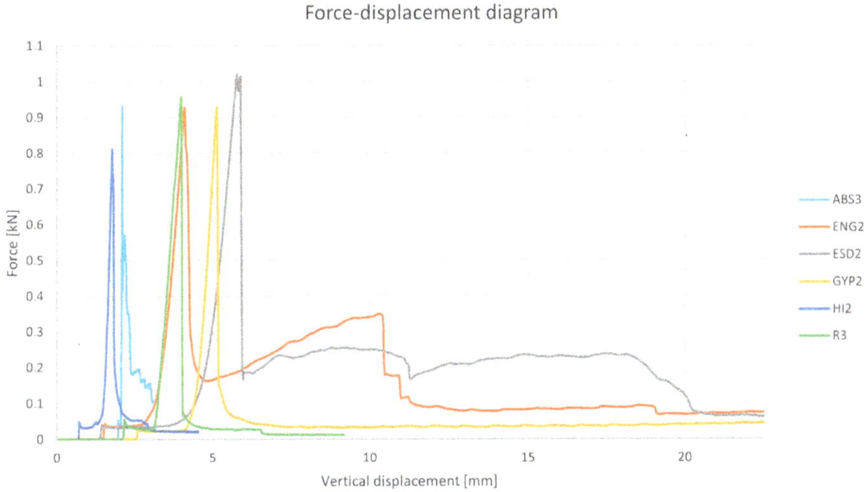

Figure 6. Force–displacement diagram of the specimens (ABS, ENG, ESD, GYP, HI, and R).

All the data in the figure are measured values and maintain a proportional relationship. In this scenario, only the loading speed can be adjusted—either accelerated or decelerated—since the machine applies a constant force of 20 kN.

If the measurement speed were increased, the graph's curves would appear shorter, resulting in fewer vertical displacement values. Conversely, reducing the measurement speed would extend the curves, increasing the number of vertical displacement values.

The maximum values, so that the diagram does not become overcrowded, are been plotted on the diagram; nevertheless, the differences are still clearly visible. The peak values were in the following descending order: ESD, R, ABS, ENG, GYP, and HI. Their peak values were at 1.021, 0.956, 0.933, 0.928, 0.927, and 0.812 kN. As can be seen from the figure, the reinforced specimens showed different behavior during the loading tests.

The staircase-like decreasing structures observed in the results highlight the impact of reinforcement after reaching a peak value, which is particularly noticeable in the ESD, ENG, and ABS specimens. In contrast, this behavior was not evident in the unreinforced (R) specimens or in the GYP and HI specimens. These differences in behavior may be attributed to multiple factors.

Firstly, the variations could be linked to the inherent material properties. Specifically, HI and GYP materials exhibit lower tensile strength compared to ESD, ABS, and ENG, which inherently makes them less effective in maintaining structural integrity under load. Additionally, HI and GYP also possess lower heat resistance, as outlined in Table 2, which could further limit their performance under certain conditions.

Another potential explanation lies in the chemical interactions between the materials and the concrete medium. Such interactions could compromise the structural integrity of the materials, making them less suitable for broader applications. If this hypothesis were to hold true, it would indicate a significant limitation in the usability of these materials for reinforced concrete structures. Consequently, investigating this phenomenon in greater detail is of critical importance to understanding and addressing the observed differences.

Furthermore, as illustrated in Figure 6, no amplification of the reinforcement effect was observed for the GYP and HI specimens. This further supports the notion that their lower mechanical and thermal performance, combined with possible chemical incompatibility, reduces their effectiveness when used as reinforcing agents. This highlights the necessity for further research to better evaluate the potential of these materials in reinforced concrete applications. In Figure 6, no amplification effect after a peak value is observed in the flattening parts. Specimens R, GYP, and HI exhibited immediate failure, while specimens ESD, ABS, and ENG showed resistance to loading even after fracture (peak value).

This shows the brittle behavior of the inner reinforcement material in GYP and HI, which indicates the chemical dissolution of PLA during the 28-day drying cycle.

Figures 7–9 show the displacement of ABS2 test specimens in the *X* direction during the loading time at 0, 20, and 40 s until a fracture occurred.

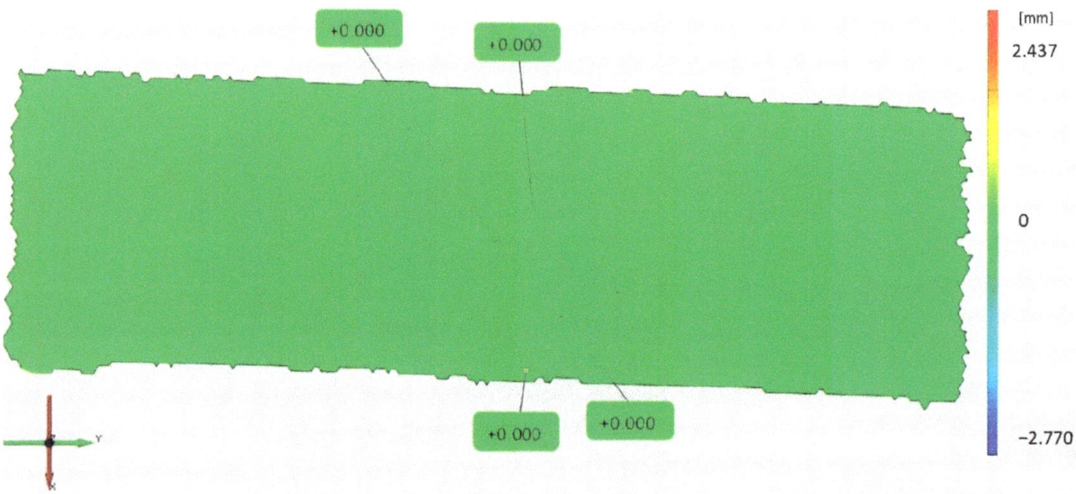

Figure 7. Displacement in *X* direction of ABS2 specimen at 0 s.

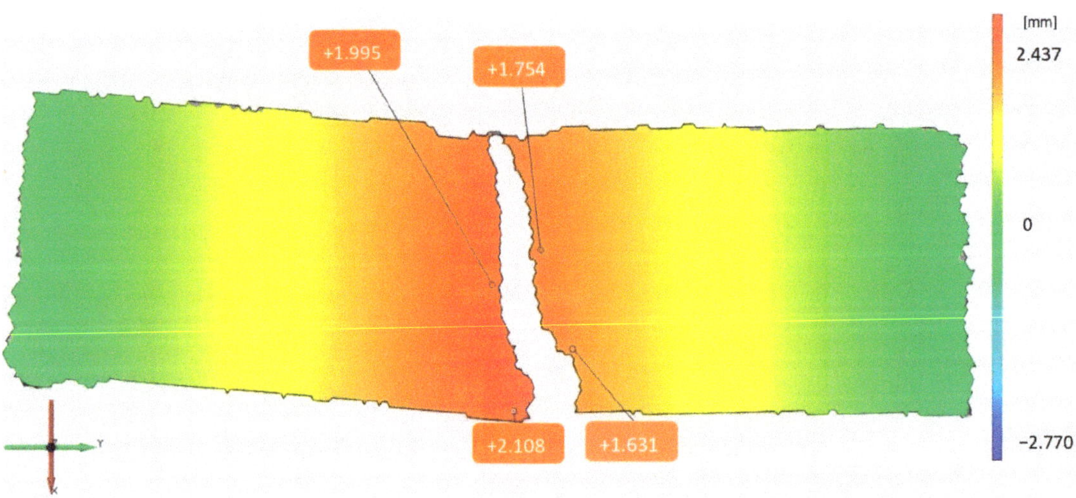

Figure 8. Displacement in *X* direction of ABS2 specimen at 20 s.

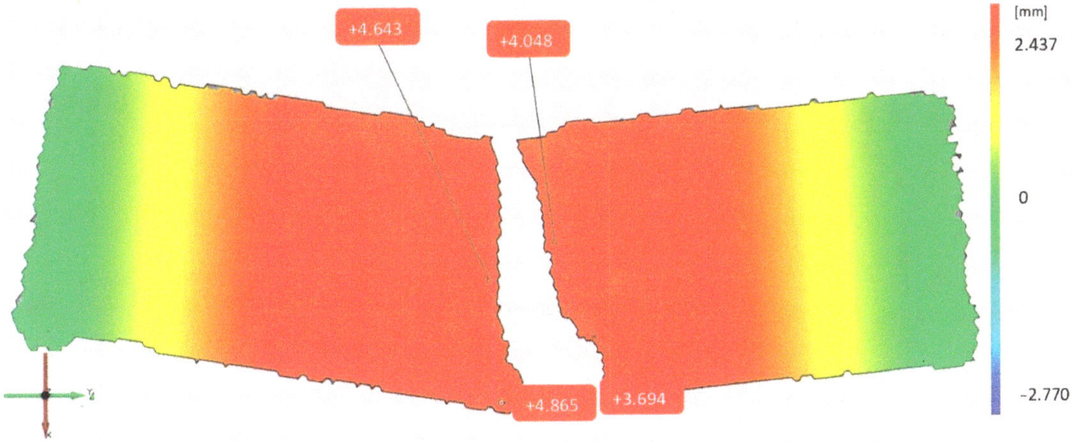

Figure 9. Displacement in *X* direction of ABS2 specimen at 40 s.

Figures 7–9 show the variation over time of the displacements of the selected points (which are in the same position in each figure). In Figure 9, the values are shown in white because the magnitude of their displacement values fell outside the range of the extremes of the scale defined by the statistical calculations.

The extent of the displacements is defined in millimeters, as indicated on the scale. In this case, the scale maximum of +2.437 and the scale minimum of −2.770 are averaged over all the measurements.

The scale values are set to the same value for all DIC measurements so that the measurement results can be compared. The values in Figures 7 and 8 are shown in black because they fell within the defined range.

During the evaluation of the measurement results, when the displacement data in the X direction for Figure 6 were collected and evaluated, the maximum displacement value at the bottom of the crack in Figure 9 was linearly interpolated for ABS2, according to the range of the HBM data logger. This method was also used for the other specimens.

After the fracture tests, the internal structure of the PLA became visible. In the GYP test specimens, one of the most noticeable features on the surface of the infill was the absence of PLA in several places and the formation of holes at the joints of the infill.

Figure 10 illustrates the surface of the GYP1 specimen, with white precipitation observed on both sides.

Figure 10. Formation of holes in the GYP material reinforced concrete with white precipitation.

To compare the separation and the difference between the two materials, for example, in the case of the ABS2 specimen, the reinforcing element did not break or tear apart under load but separated from the surface without adhering to the concrete. Figure 11 illustrates the ABS2 specimen with white precipitation observed within the fractured specimen.

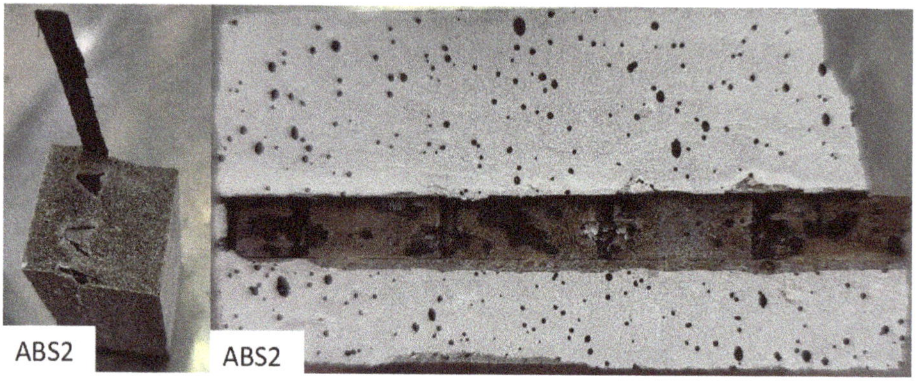

Figure 11. Separation of the ABS reinforcement from the concrete without adhering to white precipitation.

After the bending tests were performed, the broken test specimens were placed back into simple tap water. This step was important for comparing the results of the dissolution tests.

3.1.1. Observations During and After the Casting

After the specimens were stripped from the casting mold, their surfaces were smooth and evenly covered with concrete (cement grout), so painting of the specimens was easily performed. They were then stored in water to ensure their internal drying and good surface quality.

Shrinkage (contraction) measurements were carried out using GOM ATOS Triple Scan, where the shrinkage was determined by comparing the scanned CAD (computer-aided design) models; the results are presented in Section 3.2.

In the current section, the surface changes and observations of the test specimens are presented. During underwater storage, the concrete in most of the specimens, where the layer was thinnest, partially or entirely flaked off, revealing the reinforcing element.

Figure 12 shows the test specimens stored under water for 21 days prior to the DIC measurements (performed by the GOM ATOS instrument), where the reinforcing element is visible due to the flaking.

Figure 12. Surface of concrete specimens before DIC measurements (executed by GOM ATOS and GOM ARAMIS instruments).

The positioning of the reinforcement materials impacts the structural performance and bending behavior of the samples. If the amplification were closer to the neutral axis, it would reduce the amplification effect. On closer inspection of the concrete elements, it was observed that white precipitation occurred, and it was most prominent on the GYP and HI specimens' surfaces, but minimal signs of precipitation can also be seen on ENG and ESD specimens. Only in the case of ABS specimens was there no precipitation on the surface. The white, powdery precipitate is likely to be lime from the concrete.

Another observation related to Figure 8 is the flaking of the thin concrete layer surrounding the reinforcing element. Initially, the specimens, when placed under water, had a smooth, even surface free of cracks. On the 14th day, tiny visible cracks appeared in the regions where partial flaking occurred by the 21st day. The detachment of the thin concrete layer from the reinforcing layer could have been caused by the white substance that formed as a solid deposit, resembling powder or even the presence of water.

Flaking can damage the surface of the specimens, which could cause errors during the DIC measurement series. The precipitation phenomenon was already observed in the previous publication of the authors of the current paper, where amorphous PLA waste was used as a reinforcing element [66]. Figure 13 shows the white material precipitated on the surfaces of the HI and GYP specimens.

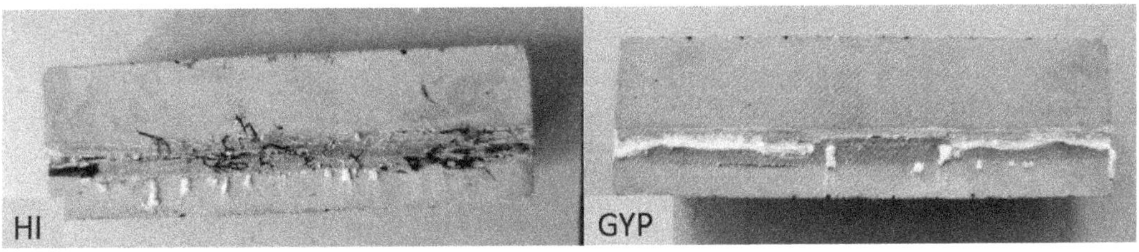

Figure 13. White material precipitated on the surfaces of the HI and GYP specimens.

In the case of HI specimens, it was further observed that the 3D-printed fibers on the longer side of the test specimen started to split and separate from each other layer by layer. This observation foreshadows the possibility of the entire printed structure disintegrating.

3.1.2. Observations During and After Drying of Concrete with GOM ATOS

The specimens were observed with GOM ATOS during and after the drying cycle (28 days) of the concrete structures. Before scanning measurements were started, the specimens were dried for 12 h, as the system is only capable of detecting a matte, gloss-free surface. The test specimens were scanned twice on days 21 and 28 of the drying cycle.

During the underwater treatment of the specimens, the thinnest concrete layer, which coated the reinforcing PLA patterns, partially flaked off, leaving them exposed. Therefore, these surface parts with severe defects were not taken into account during the ATOS scanning measurements, so the extent of their surface deviation was not taken into account. This procedure was used to compare the initial CAD model with the cast test specimen that had been under water for 21 days and to compare the 21-day test condition with the 28-day test condition. Table 5 represents the deviation values of the test specimens.

Table 5. Deviation values of the specimens ("+" means expansion; "−" means shrinkage).

Material	Standard Deviation from CAD Model at Day 21 [mm]	Longitudinal Deviation [%]	Transverse Deviation [%]	Difference Between 21 and 28 Days [mm]	Max Difference [mm]
R	0.1	−0.40	−1.20	±0.01	±0.11
ESD	0.2	−0.20	−1.20	±0.15	±0.35
ABS	0.25	−0.45	−0.40	±0.20	±0.45
ENG	0.1	−0.50	+0.08	±0.20	±0.30
GYP	0.1	−0.37	−0.24	±0.05	±0.15
HI	0.2	−0.37	+0.16	±0.10	±0.30

As summarized in Table 5, the sum of the deviations did not reach 0.5 mm in any of the cases, even though the maximum deviation was only 0.45 mm for the ABS specimens. The deviation values suggest that the specimens shrunk in the longitudinal direction and also shrunk in the transverse direction, excluding ENG and HI specimens. The test specimens were again stored in an underwater medium after the measurement on day 21, and the measurement was repeated on day 28.

At the end of the ATOS measurements, the residue of the flaked concrete covering the reinforcement had to be cleaned off for the GOM ARAMIS measurements. The painted measurement points would have acted as erroneous, incomplete measurement data if they had not been flaked off.

It is important to note that the specimens had to be dried before starting the ARAMIS measurements and the load tests, as the residual moisture content would have affected the measurement results. This was verified by hourly weighing of the specimens after 24 h of drying at room temperature. When the standard deviation of the weight values was minimal, the authors started to execute the ARAMIS measurements.

3.2. Concerning the Dissolution Tests of the PLA Specimens

During the dissolution tests, 3D-printed PLA samples with a thickness of 0.4 mm were immersed in solutions with pH values of 12 and 12.55. Each specimen was submerged in 50 mL of solution, covered, and kept for four weeks (28 days).

Preventing evaporation was key, as it ensured that the pH of the solution remained constant throughout the measurement. The observation was a subjective process; therefore, the evaluations were conducted based on a custom assessment system. During the observations, only the first week of the dissolution tests was observed for a full seven days since the rate of change in the test samples slowed down after the first week. Table 6 summarizes the one-week-long 24 h observations of the dissolution tests on the specimens and contains the list of symbols used.

Table 6. Summary of observations during the dissolution tests with the list of symbols.

Material	Solution pH	Day 1	Day 2	Day 3	Day 4	Day 5	Day 6	Day 7
ESD	12	NC	C	C	C, P	C, I	C, CRP	C, CRP, I
ENG	12	NC	C, P	C, I	C, I	C, I	C, CRP, S	C, CRP, S, I
HI	12	NC	C, P	C, I	C, I	C, I	C, CRP, S	C, CRP, S, I
ABS	12	NC	C, P	C, I	C, I	C, I	C, CRP	C, CRP, I
GYP	12	NC	C, P	C, I	C, I	C, I	C, CRP, S	C, CRP, S, I
ESD	12.55	NC	C	C, P	C, P	C, I	C, I	C, CRP
ENG	12.55	NC	C	C, I	C, I	C, I	C, CRP, S	C, CRP, S, I
HI	12.55	NC	C, P, F	C, F, I	C, F, I	C, F, I	C, F, CRP, S	C, F, CRP, S, I
ABS	12.55	NC	C	C, P	C, I	C, I	C, CRP	C, CRP, I
GYP	12.55	C	C, P	C, I	C, I	C, CRP	C, CRP, S, W	C, CRP, S, W, I

Remarks

The solution does not run off the surface, covering it evenly—C.
The solution runs off the surface, no coverage—NC.
Faded surface where the specimen was in contact with the solution—F.
Small amount of white substance that forms as a solid deposit, resembling powder on the surface—P.
Increase in the amount of precipitation—I.
Precipitation occurred in the solution as tiny particles—S.
Precipitation occurred on the wall of the measuring cup—W.
White crystal-like precipitation in continuous lines, parallel to the printing direction—CRP.

During the 168 h observation period of the dissolution tests, the GYP specimens exhibited the most significant and noticeable changes and reactions to the solution. In the first 24 h of the test, the solution uniformly coated the surface of the GYP specimens while it rolled off the surface of the other specimens. By the second day, all specimens tested in the pH 12.55 and pH 12 solutions were uniformly coated by the solution.

In the pH 12.55 solution, a small amount of white substance that formed as a solid deposit, resembling powder, was observed on the GYP and HI specimens, whereas in the pH 12 solution, similar precipitation was observed on the ENG, HI, ABS, and GYP specimens. This type of precipitation could occur due to the crystallization of dissolved compounds when the solution became oversaturated, a chemical reaction between solutes, or environmental factors such as changes in temperature or pH. The appearance and composition of this precipitation can vary depending on the specific compounds involved, but it typically consists of solid particles that settle or adhere to surfaces.

Additionally, in the pH 12.55 solution, the black surface of the HI specimens where they contacted the solution had faded to gray. Figure 14 illustrates the fading of the color observed on the surface of the HI specimen in the 12.55 pH solution.

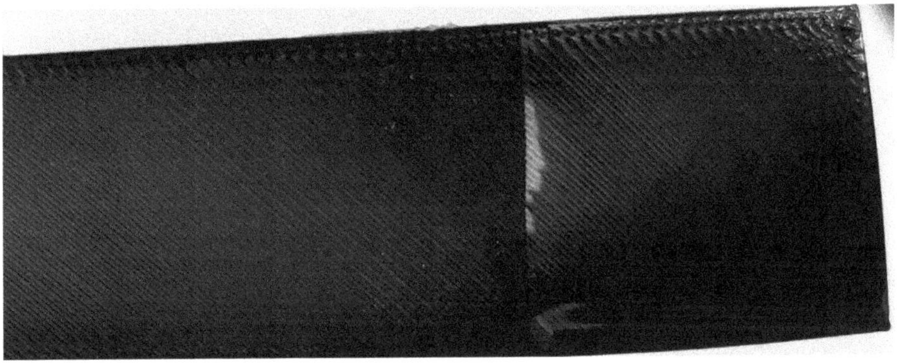

Figure 14. Fading of the color of the HI specimen immersed into the 12.55 pH solution with a small amount of white precipitation on day 7.

On the third day, a small amount of white precipitation appeared on the ESD specimen immersed in the pH 12.55 solution. For all other specimens with pre-existing precipitated particles, the amount of precipitation increased. On the fourth day, the amount of precipitation in all cases further increased; no other significant change was observed.

On the fifth day, the powder-like structure of the precipitation, in the case of GYP immersed in the 12.55 pH solution, turned into a crystal-like structure parallel to the direction of the printing direction of the specimens.

On the sixth day, a crystal-like solution appeared in all cases immersed in the 12 pH solution and all cases immersed in the 12.55 pH solution except the ESD specimen. In the case of ENG, HI, and GYP specimens, precipitation occurred as tiny, floating particles in both solutions; in the case of GYP immersed in the 12.55 pH solution, the precipitation also occurred in the wall of the measuring cup as well as in the solution.

On the seventh day, white crystal-like precipitation occurred in ESD immersed in the 12.55 pH solution. In addition, the amount of precipitation increased in all other test samples and the solutions.

Figure 15 shows the occurrence and the changing of the precipitation during the observation of the dissolution tests.

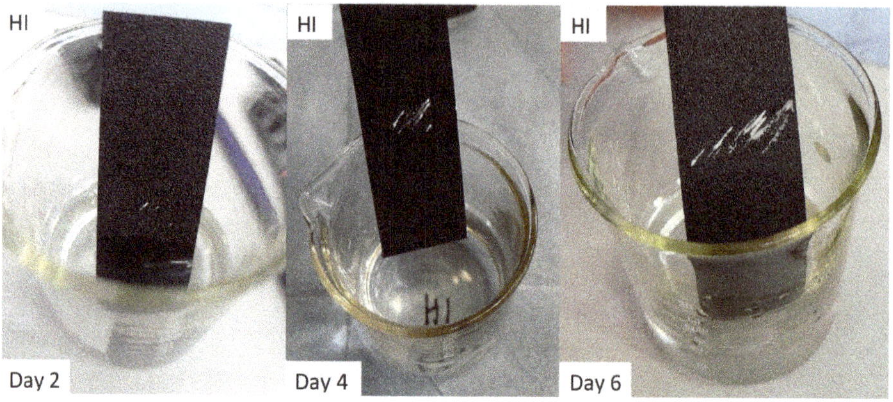

Figure 15. Occurrence of the precipitation by the time observed at the HI specimens.

The remaining time of the dissolution tests was monitored weekly, where changes in the surface of the test specimens and the amount of solution remaining were observed. Despite covering the solution, the amount of the solution decreased. Table 7 shows the percentage reduction in the solutions from the original initial volume of 50 mL, all compared to the previous volume.

Table 7. Decrease in the amount of solution at pH 12.55 and pH 12 during the four-week period.

Material	Week 1	Week 2	Week 3	Week 4
pH 12.55 solutions				
ESD	40	30	25	5
ENG	40	25	25	10
HI	40	35	25	0
ABS	40	30	20	10
GYP	50	25	25	0
Material	**Week 1**	**Week 2**	**Week 3**	**Week 4**
pH 12 solutions				
ESD	20	40	35	5
ENG	20	45	30	5
HI	30	45	25	0
ABS	20	30	35	15
GYP	50	25	25	0

The most significant decrease in solution amount, in percentage terms, was observed in the GYP specimens, where the material absorbed half of the solution in both the 12.55 and 12 pH solutions. Observing Table 7, it can be seen clearly that the next value after the GYP percentage was a decreased value (50%), while 40% was observed for the rest of the solutions with pH 12.55. This was followed by the HI sample of the pH 12 solution with a 30% decrease, followed by the combination of ABS, ESD, and ENG samples with a 20% decrease.

By the second week, the amount of precipitation had noticeably increased, while the amount of solution had decreased in all samples but not to the same extent. In the case of pH 12 solutions, the most significant amount of decrease was in the case of the HI and ENG specimens at 45%, followed by 40% in the case of ESD, then 30% with ABS, and a 25% decrease with GYP. By the third week, both the pH 12 and pH 12.55 solutions ran out of solution from the GYP and HI test samples. By the fourth week, no liquid remained in either solution for HI and GYP, and only the amount of white precipitate increased. In contrast, at pH 12, 15% of the solution remained for ABS and 5% for ENG and ESD. For the same materials, the liquid was also retained in the pH 12.55 solution: 10% for ABS and ENG and 5% for ESD.

As the solutions slowly ran out in the beakers, the white precipitate-like material on the test samples increased. Figure 16 shows the formation and weekly spread of white substance and on the HI specimen immersed in the pH 12.5 solution.

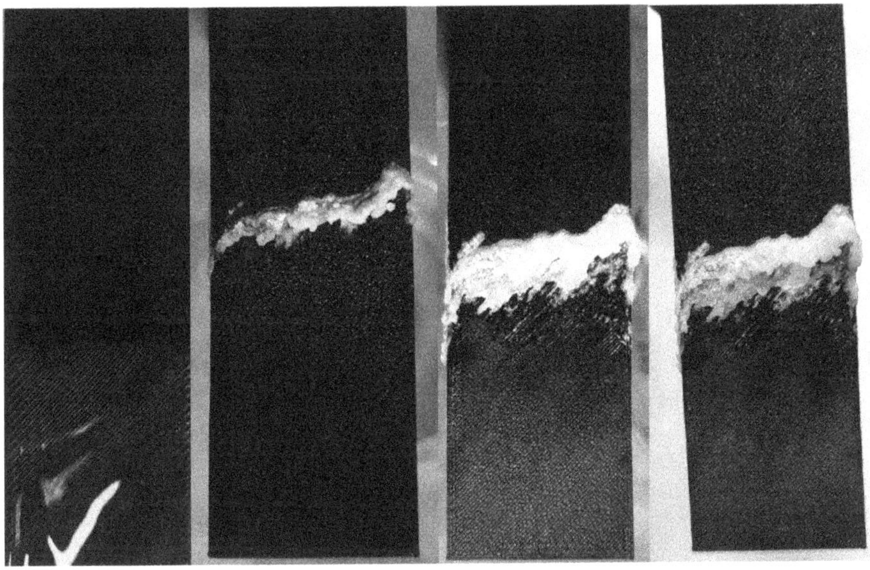

Figure 16. Weekly spread of white substance and fading on ENG specimen in the pH 12.55 solution.

As shown in Figure 16, the white precipitate is widely spread on the surface of the specimen; what is interesting about the location of the white material is that it started to form on the line where the specimen touched the solution, and the HI material started to fade. However, to find out where it came from—whether it came from the reinforcing material or crystallized out of the solute—further experiments were needed. Since one way of detecting NaOH is by combustion, a test was carried out on a sample taken from the surface. If it burned with a yellow flame, it was NaOH; if not, it originated from PLA, and its compound must be determined.

Since the research had limited resources, the simplest and quickest methods were used to determine the substance. Flame tests and dissolution tests were carried out using the same equipment. Table 8 shows the test results of the flame tests and pH measurements of the white precipitate samples tested.

Before starting the experiments, a sample of white material was taken from each of the test samples. All samples were then heated one at a time for the same length of time. Of the samples tested, only the sample from the GYP probe showed a color change in response to heat.

Another simple method for detecting NaOH was to measure pH. Since NaOH is a strong alkali, if it is present, the pH of the solution will be very high (between 12 and 14). The measurements showed that although the pH values were close to 11, none of the solutions reached the very alkaline value. However, during the dissolution tests of the pH 12.55 solution on a sample taken from the ABS specimen, the solution became unclear, and black particles started floating in it.

Table 8. Flame test and pH measurement test results of the white matter.

Material	Reaction to Flame	Dissolution	pH Value	Temperature [°C]
pH 12.55 solutions				
ESD	none	full	10.14	21.0
ENG	none	full	10.31	20.9
HI	none	full	10.39	20.2
ABS	none	uneven	10.62	21.2
GYP	decolorization	uneven	10.81	21.6
Material	**Reaction to flame**	**Dissolution**	**pH Value**	**Temperature [°C]**
pH 12 solutions				
ESD	none	full	10.47	22.5
ENG	none	full	10.66	22.9
HI	none	full	10.17	22.0
ABS	none	uneven	10.19	24.1
GYP	none	full	10.17	23.2

For the sample solution from GYP, black particles appeared, and the solution was decolorized from translucent to slightly yellow. In the pH 12 solution dissolution tests, as well as with a sample from the ABS sample, the sample dissolution was incomplete, and white, floating particles made the solution cloudy. An important note is that in the dissolution tests, the samples were dissolved in water only when heated.

Some specimens containing the white precipitate did not completely dissolve in distilled water, even under heat. pH tests indicated a reaction between the NaOH solution and the PLA-based and ABS materials. If the specimens had been coated solely with pure NaOH (used to simulate the high pH of concrete), they would have burned with a yellow flame in flame tests, and pH levels would have remained around 12 instead of falling below 11. This suggested a chemical interaction between the reinforcement materials and the solution.

3.3. *Concerning the Dissolution of the Fiber-Reinforced Concrete Specimens Underwater*

The dissolution tests of the reinforcing materials were carried out in parallel with tests on the broken fiber-reinforced concrete specimens stored underwater. The specimens were stored in tap water, covered, and subjected to the three-point bending test for a further 4 weeks after the fracture tests. During this time, the pH and temperature of the aqueous medium were monitored. The pH and temperature of the medium were rechecked after a further 4 weeks, i.e., at week 8 after the start of the observation, to see if there were any outliers. Table 9 represents the pH values and the temperature of the aqueous medium over the four-week study.

Table 9. pH and temperature changes in aqueous medium over four weeks plus one week.

Observed Period in Weeks	pH Value	Temperature [°C]
1st	7.75	21.2
2nd	7.99	21.0
3rd	8.05	21.2
4th	8.19	21.4
8th	8.10	20.8

The measured pH values fell in the alkaline range but were not high enough (9–12) to clearly indicate the presence of calcium hydroxide ($Ca(OH)_2$). The pH increase caused by $Ca(OH)_2$ was generally more significant, as it is a highly alkaline substance.

Rising pH values (7.73 → 8.19) could indicate that a solution is alkalinizing, but other alkaline substances such as carbonates (e.g., sodium carbonate) can also be responsible.

It is also possible that the chemical equilibrium of the solution is altered, for example, by loss of carbon dioxide, which could cause natural alkalinization. This issue needs further investigation.

In the third week of the observations, white crystals, similar to the white material observed in the dissolution tests, appeared on the concrete elements in all cases right next to or on the reinforcing element. The observed white crystalline particles are presented in Figure 17.

Figure 17. Occurrence of white crystals on the reinforced concrete beams kept underwater for four weeks.

Another observation outside the appearance of crystals was the behavior of PLA-based materials. In several parts, especially HI, the material was exposed to water due to fracture and disintegrated on contact. ENG, GYP, and ESD were smeared.

As shown in Figure 18, HI was the most severely affected by the material condition. While ABS was completely intact compared to the other PLA-based materials, ENG, GYP, and ESD samples were smearable with HI, which even showed fiber degradation.

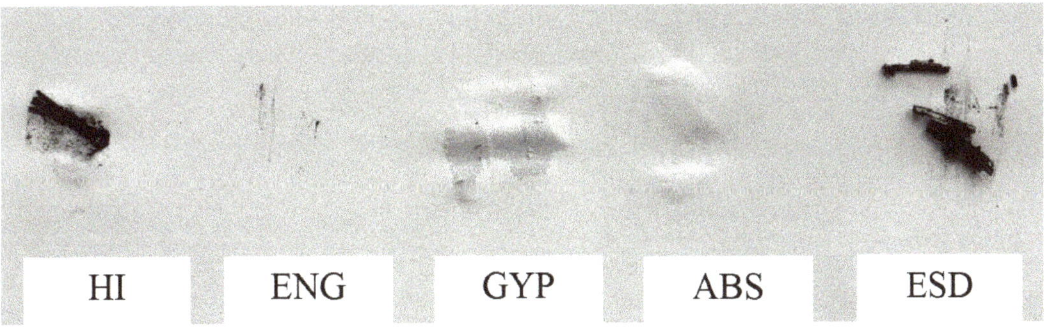

Figure 18. The smearable texture on paper of PLA materials exposed to water in week 3.

These observations were necessary, as only by measuring the pH of the medium in a continuous flow could it be determined whether the white matter released was lime leached from the concrete. Since $Ca(OH)_2$ forms an alkaline solution in water, the pH of the water will be higher than 7. The alkalinity of the solution (pH around 9–12) could be quickly determined by using a pH meter.

As the results of the pH measurements (measured with identical equipment used in the dissolution tests of the PLA and ABS samples) showed, the pH values did not even exceed 8.20, even in the back-tested range, after week 8.

4. Conclusions

This study aimed to evaluate the mechanical performance, chemical stability, and reinforcement potential of 3D-printed materials, including PLA, ABS, and electrical ESD PLA, in concrete and NaOH mediums. The results provided insights into both the strengths and limitations of these materials, paving the way for advancements in sustainable construction. The following conclusions align with the research objectives and summarize the key observations.

The conclusions regarding mechanical performance under load are as follows:

- Unreinforced concrete specimens demonstrated unexpectedly higher load-bearing capacity than reinforced counterparts in specific cases, challenging the assumption of universal reinforcement benefits;
- Electrical ESD PLA emerged as the most promising reinforcement material, exhibiting superior post-failure load-bearing capacity and minimal degradation in high-pH environments;
- Observations under mechanical loading revealed faster and more extensive white precipitate formation in specimens subjected to lower loads, suggesting a connection between load resistance and chemical stability.

The conclusions regarding chemical stability in NaOH and concrete mediums are as follows:

- Dissolution tests confirmed that the white substance that formed as a solid deposit, resembling powder, originated from chemical reactions involving PLA-based materials in high-pH environments rather than from lime in the concrete;
- PLA and ABS materials showed susceptibility to slow degradation in NaOH solutions, leading to reduced long-term structural stability;
- pH measurements indicated that NaOH interacted chemically with PLA-based and ABS materials, further emphasizing the need for reinforcement materials resistant to high-pH degradation.

The conclusions regarding the evaluation of diverse materials for reinforcement potential are as follows:

- Electrical ESD PLA demonstrated high chemical stability and strong mechanical performance, making it a viable candidate for sustainable construction applications;
- Alternative materials like PET-G and PCTG, along with surface treatments, were identified as potential solutions to mitigate the drawbacks of PLA-based reinforcements while retaining their benefits.

The limitations identified in this research highlight the challenges of using PLA materials for reinforced concrete applications. PLA's inherent thermal and hydrolytic instability, combined with its thermal expansion mismatch with traditional construction materials, restricts its long-term durability. Additionally, its vulnerability to degradation in alkaline environments further complicates its use in demanding applications, such as bridges or outdoor concrete structures exposed to moisture and temperature fluctuations. While combining PLA with other materials has shown a potential to mitigate some of these issues, further research is essential to fully address these limitations. Advancing material optimization and exploring the chemical interactions between PLA and concrete will be critical for realizing PLA's potential as a sustainable and reliable alternative for reinforcement.

ABS filaments were chosen as the reference material in this study because of their established use in engineering and 3D printing, where their high impact resistance, stiffness, and recyclability make them a reliable standard for comparison. PLA filaments, in contrast, were examined as a more environmentally friendly option, offering advantages in printability and reduced ecological footprint. The goal was to assess whether PLA, despite its challenges with thermal and hydrolytic stability, could meet the mechanical and chemical requirements for reinforcing concrete. By evaluating PLA against ABS, the study aimed to identify materials that could combine sustainability with the performance needed for structural applications.

Future research should focus on addressing the chemical stability challenges observed in PLA and ABS materials to enhance their long-term applicability in reinforced concrete structures. Advanced analytical techniques should be utilized to determine the exact chemical composition of the white precipitate and to understand the degradation mechanisms of these materials in high-pH environments. Such studies would provide critical insights into preventing chemical interactions that compromise structural integrity, possibly applying artificial intelligence or learning algorithms [16,76,77], optimization techniques [58,78], further field tests and laboratory examinations [79,80], etc.

Additionally, exploring alternative materials such as PET-G and PCTG offers significant potential. These materials may retain the mechanical and environmental advantages of PLA while offering improved chemical resistance. Surface treatments or coatings for 3D-printed reinforcement materials could further mitigate degradation risks and enhance durability in harsh environments.

In conclusion, this study provides valuable insights into the potential of 3D-printed PLA reinforcement in sustainable construction. While highlighting both its promise and limitations, it lays a foundation for future innovations in material science and concrete engineering.

Author Contributions: Conceptualization, H.C., S.S., D.K., M.S. and S.F.; methodology, H.C., S.S., D.K., M.S. and S.F.; software, H.C., S.S., D.K., M.S. and S.F.; validation, H.C., S.S., D.K., M.S. and S.F.; formal analysis, H.C., S.S., D.K., M.S. and S.F.; investigation, H.C., S.S., D.K., M.S. and S.F.; resources, H.C., S.S., D.K., M.S. and S.F.; data curation, H.C., S.S., D.K., M.S. and S.F.; writing—original draft preparation, H.C., S.S., D.K., M.S. and S.F.; writing—review and editing, H.C., S.S., D.K., M.S. and S.F.; visualization, H.C., S.S., D.K., M.S. and S.F.; supervision, H.C., S.S., D.K., M.S. and S.F.; project administration, H.C., S.S., D.K., M.S. and S.F.; funding acquisition, H.C., S.S., D.K., M.S. and S.F. All authors have read and agreed to the published version of the manuscript.

Funding: This research received no external funding.

Institutional Review Board Statement: Not applicable.

Informed Consent Statement: Not applicable.

Data Availability Statement: The original contributions presented in this study are included in the article. Further inquiries can be directed to the corresponding author.

Acknowledgments: This paper was written by the research team "SZE-RAIL". The authors are thankful for the technical support of Gusztáv Baranyai and Attila Nagy from the Széchenyi István University, Győr, Hungary.

Conflicts of Interest: The authors declare no conflicts of interest.

List of Abbreviations

ABS	Acrylonitrile butadiene styrene
CFRP	Carbon fiber-reinforced polymer
$Ca(OH)_2$	Calcium hydroxide
CO	Carbon monoxide
CO_2	Carbon dioxide
CTB	Cement-based tail backfill
DIC	Digital image correlation
ENG	Engineering PLA
ESD	Electrostatic discharge PLA
GYP	Gypsum PLA
HI	High-impact PLA
NaOH	Sodium hydroxide
OH^-	Hydroxide ion
PBS	Phosphate-buffered saline
PCL	Polycaprolactone
PCTG	Polycyclo-hexylenedimethylene terephthalate glycol-modified
PET	Polyethylene terephthalate
PET-G or PETG	Polyethylene terephthalate glycol-modified
pH	Potential of hydrogen
PLA	Polylactic acid
PP	Polypropylene
SF	Short steel fiber-sprayed epoxy
THF	Tetrahydrofuran
VOC	Volatile organic compound
List of Symbols	
d	Cross-sectional dimension of the concrete specimen according to [67,69]
L	Length of the concrete specimen according to [67,69]
m_x	Quantity of the considered liquid/material in the solution [mol] in the x-th case
n_y	Quantity of the considered liquid/material in the solution [g] in the y-th case

References

1. Kuchak, A.T.J.; Marinkovic, D.; Zehn, M. Finite Element Model Updating—Case Study of a Rail Damper. *Struct. Eng. Mech.* **2020**, *73*, 27–35. [CrossRef]
2. Alsirawan, R.; Koch, E. Dynamic Analysis of Geosynthetic-reinforced Pile-supported Embankment for a High-Speed Rail. *Acta Polytech. Hung.* **2024**, *1*, 31–50. [CrossRef]
3. Koch, E.; Ray, R.P. Impact Assessment of Railway Bridge Construction Schedule, based on 3D Geotechnical Finite Element Modeling. *Acta Polytech. Hung.* **2024**, *1*, 187–205. [CrossRef]
4. Kampczyk, A. Technical specifications for interoperability and the provisions of Polish design geometry of the railway line. *Bauingenieur* **2015**, *90*, 229–234. [CrossRef]
5. Kampczyk, A.; Rombalska, K. Configuration of the geometric state of railway tracks in the sustainability development of electrified traction systems. *Sensors* **2023**, *23*, 2817. [CrossRef]
6. Singh, T.; Patnaik, A.; Ranakoti, L.; Dogossy, G.; Lendvai, L. Thermal and sliding wear properties of wood waste-filled poly (lactic acid) biocomposites. *Polymers* **2022**, *14*, 2230. [CrossRef]
7. Dogossy, G.; Morauszki, T.; Ronkay, F. Experimental investigation and applicability of multi-stage simulations in the case of a thick-walled injection-moulded composite. *Appl. Sci.* **2022**, *12*, 8415. [CrossRef]
8. Dižo, J.; Blatnický, M.; Sága, M.; Harušinec, J.; Gerlici, J.; Legutko, S. Development of a new system for attaching the wheels of the front axle in the cross-country vehicle. *Symmetry* **2020**, *12*, 1156. [CrossRef]
9. Dižo, J.; Blatnický, M.; Harušinec, J.; Suchánek, A. Assessment of dynamics of a rail vehicle in terms of running properties while moving on a real track model. *Symmetry* **2022**, *14*, 536. [CrossRef]
10. Kocsis Szürke, S.; Saly, G.; Lakatos, I. Analyzing Energy Efficiency and Battery Supervision in Electric Bus Integration for Improved Urban Transport Sustainability. *Sustainability* **2024**, *16*, 8182. [CrossRef]

11. Kocsis Szürke, S.; Sütheö, G.; Őri, P.; Lakatos, I. Self-Diagnostic Opportunities for Battery Systems in Electric and Hybrid Vehicles. *Machines* **2024**, *12*, 324. [CrossRef]
12. Sága, M.; Blatnický, M.; Vaško, M.; Dižo, J.; Kopas, P.; Gerlici, J. Experimental determination of the manson−coffin curves for an original unconventional vehicle frame. *Materials* **2020**, *13*, 4675. [CrossRef] [PubMed]
13. Blatnický, M.; Sága, M.; Dižo, J.; Bruna, M. Application of light metal alloy EN AW 6063 to vehicle frame construction with an innovated steering mechanism. *Materials* **2020**, *13*, 817. [CrossRef] [PubMed]
14. Volkov, V.; Taran, I.; Volkova, T.; Pavlenko, O.; Berezhnaja, N. Determining the Efficient Management System for a Specialized Transport Enterprise. *Nauk. Visnyk Natsionalnoho Hirnychoho Universytetu* **2020**, *2020*, 185–191. [CrossRef]
15. Saukenova, I.; Oliskevych, M.; Taran, I.; Toktamyssova, A.; Aliakbarkyzy, D.; Pelo, R. Optimization of Schedules for Early Garbage Collection and Disposal in the Megapolis. *East.-Eur. J. Enterp. Technol.* **2022**, *1*, 13–23. [CrossRef]
16. Ficzere, P. The Role of Artificial Intelligence in the Development of Rail Transport. *Cogn. Sustain.* **2023**, *2*, 81. [CrossRef]
17. Zöldy, M.; Baranyi, P. The Cognitive Mobility Concept. *Infocommun. J. A Publ. Sci. Assoc. Infocommun. (HTE)* **2023**, 35–40. [CrossRef]
18. Zalacko, R.; Zöldy, M.; Simongáti, G. Comparative Study of Two Simple Marine Engine BSFC Estimation Methods. *Brodogr. Int. J. Nav. Archit. Ocean. Eng. Res. Dev.* **2020**, *71*, 13–25. [CrossRef]
19. Zöldy, M.; Baranyi, P.; Török, Á. Trends in Cognitive Mobility in 2022. *Acta Polytech. Hung.* **2024**, *21*, 189–202. [CrossRef]
20. Zöldy, M.; Zsombók, I. Influence of External Environmental Factors on Range Estimation of Autonomous Hybrid Vehicles. *Syst. Saf. Hum.-Tech. Facil.-Environ.* **2019**, *1*, 472–480. [CrossRef]
21. Buruzs, A.; Kozma, K. The Realization of a Circular Economy in the Construction Industry and Its Adaptation to EU Standards in Hungary. *Chem. Eng. Trans.* **2023**, *107*, 535–540. [CrossRef]
22. Kof-Kiss, D.; Kozma, K. Extraction System Efficiency Assessment Based on Workplace Airspace and Emissions Measurements. *Chem. Eng. Trans.* **2024**, *114*, 757–762. [CrossRef]
23. Szennay, A.; Major, Z.; Horváth, R.; Czédli, H.; Szigeti, C. Providing Decent, Affordable, and Sustainable Housing: Analysing Environmental Impacts of Family Houses Built with Conventional and Unconventional Building Materials. In *Energy Footprint and Sustainability: Case Studies*; Springer Nature: Cham, Switzerland, 2024; pp. 105–126. [CrossRef]
24. Major, Z.; Szigeti, C.; Czédli, H.M. "Green" Tram Tracks for the Sustainability of the Urban Environment. *Chem. Eng. Trans.* **2023**, *107*, 289–294. [CrossRef]
25. Bai, J. Fiber-Reinforced Polymer Types and Properties. In *Advanced Fiber-Reinforced Polymer (FRP) Composites for Structural Applications*; Woodhead Publishing: Cambridge, UK, 2023; pp. 93–99. [CrossRef]
26. Bai, J. (Ed.) *Advanced Fiber-Reinforced Polymer (FRP) Composites for Structural Applications*; Elsevier: Cambridge, UK, 2023; pp. 1–826, ISBN 9780128203460. [CrossRef]
27. United Nations. *Sustainable Development Goals*; United Nations: New York, NY, USA, 2015. Available online: https://sdgs.un.org/goals (accessed on 9 January 2025).
28. Du, G.; Qian, Y. Bio-Inspired Innovations in 3D Concrete Printing: Structures, Materials and Applications. *Appl. Mater. Today* **2024**, *41*, 102459. [CrossRef]
29. Joralmon, D.; Tang, T.; Prakash, S.R.; Verma, H.; Kim, S.; Li, X. Continuous 3D Printing of Metal Structures Using Ultrafast Mask Video Projection Initiated Vat Photopolymerization. *Addit. Manuf.* **2024**, *89*, 104314. [CrossRef]
30. Szívós, B.F.; Szalai, S.; Fischer, S. Deformation Test of 3D Printed Battery Case Using DIC Technology. In Proceedings of the 2023 3rd International Conference on Electrical, Computer, Communications and Mechatronics Engineering (ICECCME), Tenerife, Canary Islands, Spain, 19–21 July 2023; IEEE: Piscataway, NJ, USA, 2023; p. 10253004. [CrossRef]
31. Liparoti, S.; Pantani, R. 3D Printing of PLA-Based Materials. In *Reference Module in Materials Science and Materials Engineering*; Elsevier: Amsterdam, The Netherlands, 2024. [CrossRef]
32. Perarnau Ollé, E.; Farré-Lladós, J.; Casals-Terré, J.; López Martínez, J.A. Enhanced Selectivity of a 3D-Printed Microfluidic Gas Detector towards Different Volatile Organic Compounds (VOCs) for the Effective Monitoring of Indoor Air Quality in Vehicles. *Environ. Technol. Innov.* **2024**, *33*, 103481. [CrossRef]
33. Gerges, T.; Semet, V.; Lombard, P.; Allard, B.; Cabrera, M. Rapid 3D-Plastronics Prototyping by Selective Metallization of 3D Printed Parts. *Addit. Manuf.* **2023**, *73*, 103673. [CrossRef]
34. Milićević, I.; Popović, M.; Dučić, N.; Vujičić, V.; Stepanić, P.; Marinković, D.; Ćosić, Ž. Improving the Mechanical Characteristics of the 3D Printing Objects Using Hybrid Machine Learning Approach. *Facta Univ. Ser. Mech. Eng.* 2022, *online first*. [CrossRef]
35. Abdul Azam, F.'A.; Tharazi, I.; Bakar Sulong, A.; Che Omar, R.; Muhamad, N. Mechanical Durability and Degradation Characteristics of Long Kenaf-Reinforced PLA Composites Fabricated Using an Eco-Friendly Method. *Eng. Sci. Technol. Int. J.* **2024**, *57*, 101820. [CrossRef]
36. Duarte, C.T.; de Queiroz, H.F.M.; Neto, J.S.S.; Cavalcanti, D.K.K.; Banea, M.D. Evaluation of Durability of 3D-Printed Multi-Material Parts for Potential Applications in Structures Exposed to Marine Environments. *Procedia Struct. Integr.* **2024**, *53*, 299–308. [CrossRef]

37. Lara-Topete, G.O.; Castanier-Rivas, J.D.; Gradilla-Hernández, M.S.; González-López, M.E. Life Cycle Assessment of Agave Bagasse Management Strategies: PLA Biocomposites versus Conventional Waste Disposal Practices. *Sustain. Chem. Pharm.* **2024**, *37*, 101435. [CrossRef]
38. Pinlova, B.; Sudheshwar, A.; Vogel, K.; Malinverno, N.; Hischier, R.; Som, C. What Can We Learn about the Climate Change Impacts of Polylactic Acid from a Review and Meta-Analysis of Lifecycle Assessment Studies? *Sustain. Prod. Consum.* **2024**, *48*, 396–406. [CrossRef]
39. Yao, L.; Li, Y.; Li, Y.; Wang, Y.; Wang, L.; Qiu, D.; Weng, Y. Surface Modification of Polylactide Films and Their Hydrolytic Degradation Performances in Artificial Seawater. *J. Appl. Polym. Sci.* **2024**, *141*, e55999. [CrossRef]
40. Jubinville, D.; Tzoganakis, C.; Mekonnen, T.H. Recycled PLA—Wood Flour Based Biocomposites: Effect of Wood Flour Surface Modification, PLA Recycling, and Maleation. *Constr. Build. Mater.* **2022**, *352*, 129026. [CrossRef]
41. Ferdinánd, M.; Várdai, R.; Móczó, J.; Pukánszky, B. A Novel Approach to the Impact Modification of PLA. *Eng. Fract. Mech.* **2023**, *277*, 108950. [CrossRef]
42. Salami, B.A.; Bahraq, A.A.; ul Haq, M.M.; Ojelade, O.A.; Taiwo, R.; Wahab, S.; Adewumi, A.A.; Ibrahim, M. Polymer-Enhanced Concrete: A Comprehensive Review of Innovations and Pathways for Resilient and Sustainable Materials. *Next Mater.* **2024**, *4*, 100225. [CrossRef]
43. van der Zee, M.; Zijlstra, M.; Kuijpers, L.J.; Hilhorst, M.; Molenveld, K.; Post, W. The Effect of Biodegradable Polymer Blending on the Disintegration Rate of PHBV, PBS and PLA in Soil. *Polym. Test.* **2024**, *140*, 108601. [CrossRef]
44. Woo, S.-H.; Wee, J.-W. Characterization of Accelerated Hydrolysis Degradation of Poly (Lactic Acid) in Phosphate Buffered Saline Solution. *Polym. Degrad. Stab.* **2024**, *223*, 110726. [CrossRef]
45. Yildizel, S.A.; Acik, M.; Kaplan, G.; Bayraktar, O.Y. Enhancing Foam Concrete: A Comparative Analysis of PLA+ Fiber Reinforcements with Plain, Hooked, and Corrugated Fibers. *Constr. Build. Mater.* **2024**, *443*, 137807. [CrossRef]
46. Li, Y.; Min, J.; Zhang, J.; Luo, M.; Wang, T.; Fu, Q.; Zhang, J. Enhancing the Stereocomplex Crystallization of PLLA/PDLA Symmetric Blend by Controlling the Degree of Molecular Chain Entanglement. *Polymer* **2024**, *313*, 127740. [CrossRef]
47. Lee, S.; Wee, J.-W. Effect of Temperature and Relative Humidity on Hydrolytic Degradation of Additively Manufactured PLA: Characterization and Artificial Neural Network Modeling. *Polym. Degrad. Stab.* **2024**, *230*, 111055. [CrossRef]
48. Dias, S.; Almeida, J.; Tadeu, A.; de Brito, J. Alternative Concrete Aggregates—Review of Physical and Mechanical Properties and Successful Applications. *Cem. Concr. Compos.* **2024**, *152*, 105663. [CrossRef]
49. Guo, R.; Ou, C.; Ma, L.; Long, Z.; Xu, F.; Yin, C. Experimental Study on Impact Performance of Seawater Sea-Sand Concrete with Recycled Aggregates. *Sustain. Mater. Technol.* **2024**, *41*, e01060. [CrossRef]
50. Ézsiás, L.; Tompa, R.; Fischer, S. Investigation of the Possible Correlations between Specific Characteristics of Crushed Stone Aggregates. *Spectr. Mech. Eng. Oper. Res.* **2024**, *1*, 10–26. [CrossRef]
51. Wu, M.; Yuan, F.; Wang, P.; Li, W. Concentrically Loaded Concrete Columns Reinforced with Steel-FRP Composite Bars (SFCB) and Carbon Fibre-Reinforced Polymer Mesh Fabric (CFRP-MF) Stirrups. *Structures* **2024**, *66*, 106794. [CrossRef]
52. Yu, Y.; Zhou, L.; Liao, Z.; Zheng, Y. Tension Stiffening and Cracking Behaviors in Glass Fiber Reinforced Polymer Bar Enhanced Precast Recycled Aggregate Concrete Specimen. *Structures* **2024**, *69*, 107395. [CrossRef]
53. Tang, C.; Liu, J.; Hao, W.; Wei, Y. Flexural Properties of 3D Printed Graded Lattice Reinforced Cementitious Composites Using Digital Image Correlation. *Mater. Des.* **2023**, *227*, 111734. [CrossRef]
54. Németh, A.; Ibrahim, S.K.; Movahedi Rad, M.; Szalai, S.; Major, Z.; Kocsis Szürke, S.; Jóvér, V.; Sysyn, M.; Kurhan, D.; Harrach, D.; et al. Laboratory and Numerical Investigation of Pre-Tensioned Reinforced Concrete Railway Sleepers Combined with Plastic Fiber Reinforcement. *Polymers* **2024**, *16*, 1498. [CrossRef]
55. Yeganeh, A.E.; Hossain, K.M.A. Structural Behavior of Shear Deficient High Performance Reinforced Concrete Exterior Joints under Bending. *Structures* **2023**, *48*, 1707–1721. [CrossRef]
56. Kuchak, A.J.T.; Marinkovic, D.; Zehn, M. Parametric Investigation of a Rail Damper Design Based on a Lab-Scaled Model. *J. Vib. Eng. Technol.* **2021**, *9*, 51–60. [CrossRef]
57. Singh, H.; Ambekar, R.S.; Das, D.; Danam, V.A.; Katiyar, N.K.; Kanti Das, B.; Tiwary, C.S.; Bhattacharya, J. Enhancing Structural Resilience by Using 3D Printed Complex Polymer Reinforcement for High Damage Tolerant Structures. *Constr. Build. Mater.* **2024**, *425*, 136085. [CrossRef]
58. Fischer, S.; Kocsis Szürke, S. Detection Process of Energy Loss in Electric Railway Vehicles. *Facta Univ. Ser. Mech. Eng.* **2023**, *21*, 081. [CrossRef]
59. Fischer, S. Investigation of the Settlement Behavior of Ballasted Railway Tracks Due to Dynamic Loading. *Spectr. Mech. Eng. Oper. Res.* **2025**, *2*, 24–46. [CrossRef]
60. Nyathi, M.; Bai, J.; Wilson, I. Deep Learning for Concrete Crack Detection and Measurement. *Metrology* **2024**, *4*, 66–81. [CrossRef]
61. Yaya, N.S.; Cao, S.; Yilmaz, E. Effect of 3D Printed Skeleton Shapes on Strength Behavior, Stress Evolution and Microstructural Response of Cement-Based Tailings Backfills. *Constr. Build. Mater.* **2024**, *432*, 136699. [CrossRef]

62. Hematibahar, M.; Milani, A.; Fediuk, R.; Amran, M.; Bakhtiary, A.; Kharun, M.; Mousavi, M.S. Optimization of 3D-Printed Reinforced Concrete Beams with Four Types of Reinforced Patterns and Different Distances. *Eng. Fail. Anal.* **2025**, *168*, 109096. [CrossRef]
63. Stähli, P.; Custer, R.; van Mier, J.G.M. On Flow Properties, Fibre Distribution, Fibre Orientation and Flexural Behaviour of FRC. *Mater. Struct.* **2008**, *41*, 189–196. [CrossRef]
64. Wu, H.; Shen, A.; Cheng, Q.; Cai, Y.; Ren, G.; Pan, H.; Deng, S. A Review of Recent Developments in Application of Plant Fibers as Reinforcements in Concrete. *J. Clean. Prod.* **2023**, *419*, 138265. [CrossRef]
65. Xu, Y.; Wan, Z.; Šavija, B. Elevating Mechanical Performance of Cementitious Composites with Surface-Modified 3D-Printed Polymeric Reinforcements. *Dev. Built Environ.* **2024**, *19*, 100522. [CrossRef]
66. Csótár, H.; Baranyai, G.; Szalai, S.; Fischer, S. Sustainable Uses of 3D Printing Applied to Concrete Structures. *Eng. Proc.* **2024**, *79*, 55. [CrossRef]
67. *MSZ EN 12390-1:2021*; Testing Hardened Concrete. Part 1: Shape, Dimensions and Other Requirements for Specimens and Moulds. European Committee for Standardization: Brussels, Belgium, 2021.
68. *MSZ EN 12390-2:2019*; Testing Hardened Concrete. Part 2: Making and Curing Specimens for Strength Tests. European Committee for Standardization: Brussels, Belgium, 2019.
69. *MSZ EN 12390-5:2019*; Testing Hardened Concrete. Part 5: Flexural Strength of Test Specimens. European Committee for Standardization: Brussels, Belgium, 2019.
70. Reu, P. Speckles and Their Relationship to the Digital Camera. *Exp. Tech.* **2014**, *38*, 1–2. [CrossRef]
71. Reu, P. All about Speckles: Contrast. *Exp. Tech.* **2015**, *39*, 1–2. [CrossRef]
72. Szalai, S.; Szívós, B.F.; Kocsis, D.; Sysyn, M.; Liu, J.; Fischer, S. The Application of DIC in Criminology Analysis Procedures to Measure Skin Deformation. *J. Appl. Comput. Mech.* **2024**, *10*, 817–829. [CrossRef]
73. Szalai, S.; Szürke, S.K.; Harangozó, D.; Fischer, S. Investigation of Deformations of a Lithium Polymer Cell Using the Digital Image Correlation Method (DICM). *Rep. Mech. Eng.* **2022**, *3*, 206–224. [CrossRef]
74. Grubb, J.A.; Lura, H.S.; Kakad, A.M. *Concrete International*; American Concrete Institute (ACI): Detroit, MI, USA, 2007; pp. 78–83.
75. *MSZ ISO 3696:1992*; Water for Analytical Laboratory Use. Specification and Test Methods. Hungarian Standards Institution: Budapest, Hungary, 1992.
76. Zamfirache, I.A.; Precup, R.E.; Petriu, E.M. Q-Learning, Policy Iteration and Actor-Critic Reinforcement Learning Combined with Metaheuristic Algorithms in Servo System Control. *Facta Univ. Ser. Mech. Eng.* **2023**, *21*, 615–630. [CrossRef]
77. Nosonovsky, M.; Aglikov, A.S. Triboinformatics: Machine Learning Methods for Frictional Instabilities. *Facta Univ. Ser. Mech. Eng.* **2024**, *22*, 423–433. [CrossRef]
78. Mzili, T.; Mzili, I.; Riffi, M.E.; Pamucar, D.; Simic, V.; Kurdi, M. A novel discrete rat swarm optimization algorithm for the quadratic assignment problem. *Facta Univ. Ser. Mech. Eng.* **2023**, *21*, 529–552. [CrossRef]
79. Zheng, J.; Wang, L. Experimental study on creep loading of porous coal under different influencing factors. *Facta Univ. Ser. Mech. Eng.* **2024**, *22*, 153–163. [CrossRef]
80. Liović, D.; Franulović, M.; Ferlič, L.; Gubeljak, N. Surface roughness of Ti6Al4V alloy produced by laser powder bed fusion. *Facta Univ. Ser. Mech. Eng.* **2024**, *22*, 63–76. [CrossRef]

applied sciences

MDPI

Article

Additive Manufacturing as an Alternative to Core Sampling in Concrete Strength Assessment

Darya Anop [1], Marzhan Sadenova [1,*], Nail Beisekenov [2], Olga Rudenko [1], Zulfiya Aubakirova [3] and Assel Jexembayeva [4]

1 D. Serikbayev East Kazakhstan Technical University, Ust-Kamenogorsk 070004, Kazakhstan; danop@edu.ektu.kz (D.A.); orudenko@edu.ektu.kz (O.R.)
2 Graduate School of Science and Technology, Niigata University, Niigata 950-2181, Japan; f23e503a@mail.cc.niigata-u.ac.jp
3 Department of Construction Materials and Technologies, A. Saginov Karaganda Technical University, Karaganda 100000, Kazakhstan; aubakirova.zulfiya@mail.ru
4 Faculty of Architecture and Civil Engineering, L. N. Gumilyov Eurasian National University, Astana 010000, Kazakhstan; dzhexembayeva_aye@enu.kz
* Correspondence: msadenova@edu.ektu.kz

Featured Application

The workflow developed in this study enables construction teams to verify the in situ compressive strength of 3D-printed concrete elements without drilling cores. By pairing routine cube tests with a data-driven conversion factor, practitioners can obtain rapid, noninvasive strength estimates, streamline quality-control procedures, and integrate 3D printing into digital construction and automated documentation platforms.

Abstract

Additive manufacturing reshapes concrete construction, yet routine strength verification of printed elements still depends on destructive core sampling. This study evaluates whether standard 70 mm cubes—corrected by a single factor—can provide an equally reliable measure of in situ compressive strength. Five Portland-cement mixes, with and without ash-slag techno-mineral filler, were extruded into wall blocks on a laboratory 3D printer. For each mix, the compressive strengths of the cubes and ∅ 28 mm drilled cores were measured at 7, 14 and 28 days. The core strengths were consistently lower than the cube strengths, but their ratios remained remarkably stable: the transition coefficient clustered between 0.82 and 0.85 (mean 0.83). Ordinary least-squares regression of the pooled data produced the linear relation $\hat{R}_{core}$ [MPa] = 0.97 $\hat{R}_{cube}$ − 4.9, limiting the prediction error to less than 2 MPa (under 3% across the 40–300 MPa range) and outperforming more complex machine-learning models. Mixtures containing up to 30% ash-slag filler maintained structural-grade strength while reducing clinker demand, underscoring their sustainability potential. The results deliver a simple, evidence-based protocol for non-destructive strength assessment of 3D-printed concrete and provide quantitative groundwork for future standardisation of quality-control practices in additive construction.

Keywords: 3D concrete printing; compressive strength; core testing; sustainable concrete; additive manufacturing; cementitious composites; low-carbon concrete

1. Introduction

Modern realities bring new economic, social, and environmental challenges that demand innovative solutions, particularly in the construction industry. A continuing priority is the development of new structural systems that ensure resilience, reduce construction costs, and simplify maintenance. One of the most promising directions toward sustainable construction is the application of additive technologies, including 3DCP. This process is fully automated, allowing for greater flexibility in material selection, geometric form, and a high degree of customisation, while also offering optimal design in terms of energy efficiency. According to several authors, virtual prototyping and digital fabrication should be integrated into all stages of the design, construction, and decommissioning of future and existing structures [1].

Over the past two decades, several digital concrete fabrication methods have been developed for the construction sector. Today, the most common are layer-by-layer extrusion techniques such as contour creation and 3D concrete printing, as well as formwork printing, slip-based forming (also known as dynamic formwork), and particle bed printing (also known as binder jetting) [1–4].

Critical parameters of the printing process—including layer interval, humidity level, print head speed and height, geometry and rotational speed of the screw blades, and the shape and size of the extrusion—are highly sensitive and significantly affect the quality of the printed objects [5]. However, there is insufficient data on how the quality of printed objects has been verified and controlled.

Numerous studies have been devoted to developing printable concrete mixtures, varying the components, proportions, and additives, including industrial waste products. The relationship between the mechanical properties of 3D-printed concrete, cement type, and perceived density has been investigated [6]. In one such study, a binary mixture of ordinary Portland cement and fast-setting cement, combined with quartz sand in various proportions, was printed using a jetting method. The results demonstrated that mixture proportions and layer thickness significantly influence the mechanical properties of the concrete. To determine these properties, samples were printed and tested by the EN 196- 1 standard: cubes with a side length of 40 mm and beams of 160 mm × 40 mm × 40 mm for compressive and flexural strength testing, respectively. However, it remains unclear whether the strength of these printed samples corresponds to the strength of more complex elements printed using the same mixture and printer.

A review of research on 3D printing conducted between 2014 and 2023, based on the Web of Science database, was presented earlier [7]. The studies were categorised into stages. In the first stage, researchers focused on assessing the potential for applying 3D printing in construction by building on existing technologies. The second stage concentrated on printable materials. In the third stage, the studies explored the effects of porosity, pore distribution, and the geometric characteristics of pores on performance from a microscopic perspective; they proposed cement-free alternatives; and began integrating machine learning methods into 3D printing. These studies identified three core research areas in 3D concrete printing: materials development (component types and ratios, printable mixture formulation, performance enhancement, etc.), process parameters (layering intervals, deposition speed, directionality, etc.), and technological methods (integration of different reinforcement types, etc.). Nevertheless, these studies still lack sufficient data on the compressive strength of hardened, printed concrete.

Currently, there is no standardised guideline or proposed methodology for assessing printable concrete mixtures and materials, nor any well-defined criteria of acceptability for this type of concrete [8]. In other words, no comprehensive set of performance requirements or test methods for printed mixtures exists. Some authors have proposed a laboratory testing framework for freshly printed concrete mixtures [8], but corresponding data for hardened printed concrete remain very limited.

In tests using gypsum-cement-based powder, discrepancies in compressive strength were observed between cube specimens of 70.7 mm and 50.5 mm and half-beam samples of 160 mm × 40 mm × 40 mm. Smaller cubes (50.5 mm) exhibited higher strength, possibly due to differences in curing time, while larger specimens required longer print durations, affecting stability and quality. The half-beam samples showed the lowest compressive strength, possibly due to variations in testing methods or issues like nozzle clogging during the extended print time required for 160 mm samples [9].

A previous study also compared the properties of 3D-printed concrete samples with those of identically formulated mixtures cast on a vibrating table [10], using a custom-designed laboratory printer. The mixtures, based on Portland cement, included various mineral additives: ground granulated blast-furnace slag, fly ash, limestone powder, and granite dust. The results showed lower density and compressive strength in printed samples due to the absence of vibration-induced compaction and less efficient particle packing. Notably, the most minor strength reduction occurred in mixtures with fly ash. The introduction of hardening accelerators (up to 4% of the binder mass) was suggested to improve the strength of 3D-printed elements in both compression and bending.

In another study, a geopolymer paste based on potassium silicate and metakaolin with added sand and local soil was tested for printability using a manual extrusion tool [11]. Beam samples of 160 mm × 40 mm × 40 mm were cast from the developed mixture for strength testing. However, the study did not assess concrete that had been printed directly.

Additional investigations used X-ray computed tomography to evaluate aggregate distribution uniformity in 3D-printed concrete. Cylindrical samples were produced by casting, and additional cores were cut from the printed layer and nozzle output using a cartridge-type 3D-printing extruder [12]. The results confirmed inhomogeneous aggregate distribution within both the printed layer and the nozzle, emphasising the need for further studies to distinguish the mechanical behaviour of lab-prepared printable mixtures versus actual 3D-printed concrete components.

Thus, further research is required to validate the use of these developed mixtures in large-scale 3D concrete printing (3DCP) systems [13]. Table 1 summarises previous findings on 3D-printed concrete elements.

Table 1. Summary of research results on 3D-printed concrete elements.

Reference	Materials	Printing Method	Testing Method	Printer Model	Key Findings/Limitations
[6]	Ordinary Portland cement, fast-setting cement, quartz sand	Jet printing. Cubic samples (40 mm) and prismatic samples (160 mm × 40 mm × 40 mm) were printed for compressive and flexural strength testing.	EN-196-1 [14]. Using a universal testing machine (Model 810, MTS System, Eden Prairie, MN, USA)	Not specified	No data on the correlation of strength with more complex elements printed using this printer and mix.
[8]	Portland cement, sand, polycarboxylate-based water-reducing agent, viscosity-modifying admixture, polypropylene fibre, densified silica fume, high-purity attapulgite clay	Layer-by-layer extrusion printing	ASTM C109 [15]. Cube samples (50 mm)	Custom-built laboratory printer capable of printing up to ten concrete layers of 1.2 m in length. The nozzle uses an extrusion mechanism similar to contour crafting to print layers of 38.1 mm × 25.4 mm.	No data available on the performance of hardened concrete printed directly using this method.
[9]	Gypsum-based cementitious material (ZP150 powder and ZB60 binder; ZB60 volume fraction 21.8%). ZP150 contains gypsum, vinyl polymer, and carbohydrate; ZB60 includes a wetting agent and water.	Powder-based printing. Two sets of cubes (70.7 mm and 50 mm) were printed and tested for compressive strength in one printing cycle. Additional samples (160 mm × 40 mm × 40 mm) were tested for flexural and compressive strength.	Chinese "Standard for Test Method of Mechanical Properties on Ordinary Concrete" GB/T 50081-2002 [16]; compressive strength for 70.7 mm and 50.5 mm cubes; flexural and compressive strength tested per GB/T 17671-1999 [17]	3D printer: Spectrum Z510 by Z Corporation, using HP 4810A 11 print head; capable of printing up to 356 mm (L) × 254 mm (W) × 203 mm (H)	Significant variation observed in compressive strength between 70.7 mm, 50.5 mm cubes, and half-beams.

Table 1. *Cont.*

Reference	Materials	Printing Method	Testing Method	Printer Model	Key Findings/Limitations
[10]	Portland cement with various mineral additives (ground slag, fly ash, limestone powder, granite dust, alkali-activated binder)	Layered 3D concrete printing	Testing of concrete from laboratory 3D printer and cast beams (160 × 40 × 40 mm) made using vibration table	Laboratory-scale 3D printer	Compressive strength results reported for 28-day cured concrete. Lower strength observed in 3D-printed specimens; method for testing hardened printed concrete not described in detail.
[11]	Cement-free concrete: metakaolin, potassium silicate solution, wollastonite, fiberglass. Filler: masonry sand or raw soil.	Manually extruded printed mixture	Laboratory beam specimens (160 mm × 40 mm × 40 mm) were fabricated and tested	Not available	Only manually prepared specimens were tested, not directly printed concrete.
[12]	Portland cement, quartz sand, polycarboxylic acid-based superplasticiser	Manually extruded printed mixture	Testing of cast cylindrical specimens (50 mm × 100 mm and 16.2 mm × 10 mm) and samples cut from nozzle output and printed layers to evaluate aggregate distribution index	Custom-designed cartridge-type lab extruder for 3D printer	No mechanical strength properties were evaluated.
[13]	Portland cement, limestone powder, low-grade calcined clay	3D concrete printing	EN 196-1. Testing of cast specimens and core samples cut from printed concrete; sample size: 40 mm × 40 mm × 40 mm	Laboratory 3D-printer system developed at Delft University of Technology	Further studies needed to validate these mixes for large-scale 3D concrete printing systems (3DCP).

Recent reviews and studies, including those focused on the incorporation of industrial waste and the development of cement-free mixtures for 3D concrete printing, still do not provide a clear understanding of the compressive strength of concrete in structural elements produced directly via 3D printing [18–24].

Quality control of 3D-printed concrete elements remains an open question and requires further research. It is well known that the quality of concrete structures can be assessed using both destructive and non-destructive testing methods. The most widely used approach is destructive testing, which involves loading specimens to failure under controlled conditions. These tests are highly illustrative, allowing for detailed analysis of material behaviour and the identification of design flaws. However, for small production batches of structural elements, destructive testing can significantly increase costs and does not allow for individual quality assessment of each component.

One known non-destructive technique is the pull-out method, which determines the concrete strength based on the force required to extract an embedded anchor device from the concrete matrix [25]. A similar technique is the splitting-tensile pull-out method, where an anchor device connected to a testing instrument is embedded in the concrete. The procedure includes drilling a core hole, creating a concentric annular groove to define the sample depth, inserting the anchor, and applying load until the sample fails. The strength is then calculated using an established formula [26]. Limitations of these methods include their inapplicability in heavily reinforced or thin-walled structures and the labour-intensive process of embedding anchors, particularly for evaluating deep layers of concrete.

Another well-known non-destructive technique is the ultrasonic pulse velocity method, which estimates concrete strength based on the speed (or travel time) of ultrasonic waves through the material. Strength values are determined via calibration curves developed from experimental testing of standard cubes and actual structures [27]. A significant drawback of this method is the time-intensive nature of constructing these calibration curves. Additionally, the ultrasonic velocity is affected not only by the cement–sand–water ratio, but also by the type and content of coarse aggregates (e.g., gravel or crushed stone), limiting the accuracy of strength predictions.

There also exists a non-destructive testing method that utilises a hammer imprint on the concrete surface. The strength is estimated by comparing the resulting indentation with a calibrated scale on a measuring ruler [28]. However, this method also has limitations (text in source incomplete).

Another commonly used approach in construction is strength testing using standard-shaped specimens, such as cubes or cylinders, extracted from hardened concrete. These samples are then tested under compression in a laboratory. The main drawback of this method is the need to core or saw standard-shaped samples from the structure, which requires portable cutting equipment and subsequent transportation to the lab. This process can compromise the integrity of the structure and reduce its load-bearing capacity.

A reliable prediction of concrete compressive strength is essential for the safe and durable use of structural members. Over the past decade, machine-learning techniques have attracted growing interest because they can detect relationships in large, heterogeneous datasets that are difficult to recognise using human reasoning alone [29]. A recent survey of industrial concretes confirmed an "explosion" of data-driven studies devoted to strength forecasting and mix optimisation [30]. Within the specific field of 3D printed concrete, Zhang et al. developed machine-learning models—artificial neural networks, decision trees, random forests, support-vector regression and linear regression—to predict both compressive and flexural strength from mixed experimental datasets; the random-forest model achieved correlation coefficients between 0.96 and 0.98, far surpassing traditional empirical formulas and thereby supporting more accurate deployment of 3D printing in construction [31]. Hematibahar et al. extended the approach to high-performance and ultra-high-performance concretes, benchmarking seven regression schemes; partial least-squares, linear and Lasso regression delivered coefficients of determination above 0.92 while maintaining low absolute and root-mean-square errors [32].

Machine-learning applications have since expanded beyond strength prediction to encompass durability appraisal and service-life monitoring. Recent studies deploy image-based convolutional networks to locate micro-cracks, graph-based models to forecast crack growth, and hybrid classifiers to anticipate erosion or chemical adhesion phenomena throughout a structure's lifetime [33].

Against this backdrop, the present research pursues a twofold objective. First, we conduct a systematic mechanical test programme on additively manufactured wall elements and their companion laboratory cubes to establish a reference database of physical-mechanical properties. Second, we develop and validate a predictive framework that converts the strength of cast specimens into the in situ strength of three-dimensionally printed concrete—thereby eliminating the need for destructive coring and allowing real-time adjustment of printing parameters. The study is organised in two phases: an experimental phase, comprising a series of compressive tests at multiple ages, and a modelling phase, where machine-learning regressors are trained to translate laboratory results into reliable field estimates for 3D-printed structures.

2. Materials and Methods

2.1. Raw Materials

Portland cement concrete remains the most viable material for widespread use in automated construction processes both now and in the foreseeable future [8]. Therefore, in this study, the investigated concrete mixtures were based on ordinary Portland cement. A total of five different mixture compositions were prepared and tested. The compressive strength of each composition was evaluated at three curing ages: 7 days, 14 days, and the design strength age of 28 days.

For the development of printable concrete compositions suitable for use with a construction-scale 3D printer, Portland cement CEM I 52.5H, produced by the Bukhtarma Cement Company LLP (Bukhtarma, East Kazakhstan, Republic of Kazakhstan), was used as the primary binder. This cement complies with the requirements of the relevant standard [34]. The chemical and mineralogical compositions of the cement are presented in Tables 2 and 3, respectively. The results of the cement performance tests are provided in Table 4.

Table 2. Chemical composition of cement.

Sample Name	SiO_2 (%)	Al_2O_3 (%)	Fe_2O_3 (%)	CaO (%)	MgO (%)	SO_3 (%)
CEM I 52.5H Cement	19.95	5.58	4.98	63.07	4.50	0.36

Table 3. Mineralogical composition of cement.

C_3S (%)	C_2S (%)	C_3A (%)	A_4AF	MgO (%)	K_2O (%)	LOI (%)
59.1	12.3	6.3	—	1.02	0.87	—

LOI = loss on ignition; A_4AF = tetracalcium aluminoferrite.

Table 4. Cement properties and test results.

Parameter	Value
Fineness of grinding (specific surface area), cm^2/g	2961
Real density, g/cm^3	3.12
Bulk density, g/cm^3	1.54
Normal consistency, %	25.5
Setting time, min (start–end)	168–265
Flow spread (cone method), mm	113
Bending strength, MPa (2 days)	6.1
Bending strength, MPa (28 days)	9.9
Compressive strength, MPa (2 days)	26.4
Compressive strength, MPa (28 days)	64.5

Sand classified as Class II according to the standard [35] was used as the fine aggregate. The sand had a fineness modulus of 2.5, a bulk density of 1400 kg/m^3, and a proper density of 2250 kg/m^3.

The water used for mixing the concrete conformed to the requirements of the standard [31]. The content of organic surfactants, sugars, or phenols did not exceed 10 mg/L each. The water was free from oil films, fats, and oils and contained no colouring impurities. Its oxidisability was 14 mg/L, and the pH value was 7. The water contained no impurities in quantities that would adversely affect the setting time and hardening of the cement paste or concrete or reduce the concrete's strength and frost resistance.

The concentrations of dissolved salts, SO_4^{2-} and Cl^- ions, and suspended particles in the mixing water did not exceed the permissible limits specified in Table 5.

Table 5. Impurity content in the mixing water.

Parameter	Content, mg/L
Dissolved salts	4800
Sulphate ions (SO_4^{2-})	2650
Chloride ions (Cl^-)	1100
Suspended particles	170

To accelerate hardening and improve the physical and mechanical characteristics, a chemical admixture—plasticiser/hardening accelerator Master Rheobuild 1000 (Rheobuild 1000 K) produced by Master Builders Solutions Kazakhstan LLP (Almaty, Republic of Kazakhstan) —was used. This admixture is considered one of the most effective and accessible additives on the construction market in East Kazakhstan.

2.2. Characterisation of Raw Materials and Experimental Samples

The chemical composition of the raw materials was determined by spectral analysis using an inductively coupled plasma mass spectrometer (ICP-MS, 7500cx, Agilent Technologies, Santa Clara, CA, USA) and conventional chemical analysis methods. The phase composition of the raw materials and the hydration products was examined using X-ray diffractometer X'Pert PRO manufactured by the company "PANalitical" (The Netherlands). Structural analysis was conducted using a scanning electron microscope (Scanning electron microscope JSM-6390LV manufactured by «JEOL Ltd.» (Japan) with the INCA Energy Penta FET X3 energy dispersive microanalysis system of «OXFORD Instruments Analytical Limited» (Great Britain) with an energy-dispersive microanalysis system. The specific surface area was determined using the gas permeability method of Kozeny–Carman on a PSX-10a instrument (CKIC, Changsha, China), following the manufacturer's instructions. Physical and mechanical tests were conducted using standard procedures.

2.3. Concrete Mix Design and Fabrication of Experimental Samples

The concrete mix design for 3D printing using additive manufacturing technology was developed in accordance with the standard [32]. Preliminary mix compositions were first calculated theoretically using the absolute volume method. Following the theoretical design, trial batches were prepared in laboratory conditions to validate the printability and performance of the proposed mixtures.

For mix composition Nos. 3, 4, and 5, a technogenic mineral additive (TMA) was introduced. In mix No. 3, TMA replaced 9% of the cement content; in mix No. 4, it replaced 27% of the fine aggregate (sand); in mix No. 5, it replaced 9% of the cement and 27% of the sand.

The material consumption based on theoretical calculations and the characteristics of the prepared mixtures is presented in Table 6.

Table 6. Material consumption per batch (3D printer hopper fill) and mixture characteristics (initial trial mixes).

Concrete Mix	Mix 1 C:P = 100	Mix 2 C:P = 150	Mix 3 (C+TMF):P = 150	Mix 4 C:(P+TMF) = 150	Mix 5 (C+TMF):(P+TMF) = 150
Component Consumption					
Cement CEM I 52.5H, kg	10.34	12.73	10.82	12.73	9.55
Sand, kg	36.37	36.37	36.37	23.90	23.90
TMF (Techno-mineral filler), kg	–	–	1.91	12.48	15.66
Water, kg	5.69	7.59	6.57	7.79	6.46
Plasticiser, g	68	46	62	62	58
Mixture Properties					
Water-to-cement ratio (W/C)	0.55	0.60	0.57	0.61	0.56
Workability class	P2	P2	P1	P2	P1
Slump cone (cm)	5	6	4	7	4

As a result of trial batches, it was decided to adjust the compositions. As a result of the adjustments, five (5) mixture compositions were selected, the recipes and characteristics of which are presented in Table 7.

Table 7. Material consumption per batch (3D-printer-hopper fill) and mixture characteristics (adjusted final mixes).

Concrete Mix	Mix 1 C:P = 100	Mix 2 C:P = 150	Mix 3 (C+TMF):P = 150	Mix 4 C:(P+TMF) = 150	Mix 5 (C+TMF):(P+TMF) = 150
			Component Consumption		
Cement CEM I 52.5H, kg	10.34	12.73	11.58	12.73	11.58
Sand, kg	36.37	36.37	36.37	26.55	26.55
TMF (techno-mineral filler), kg	–	–	1.15	9.82	10.97
Water, kg	6.33	8.43	7.30	8.66	7.18
Plasticiser, g	76	51	69	69	64
			Mixture Properties		
Water-to-cement ratio (W/C)	0.61	0.66	0.63	0.68	0.62
Workability class	P2	P2	P2	P2	P2
Slump cone (cm)	8	9	7	9	9

2.4. Fabrication of Structures and Sampling for Control Specimens

In laboratory conditions, at the Additive Technologies Section of the Competence and Technology Transfer Centre in Construction of the D. Serikbayev East Kazakhstan Technical University (EKTU), a wall-type structural fragment (Type 1 block) was printed using each of the developed concrete mixtures. The mixes were prepared manually in the following sequence: the aggregate was thoroughly mixed with cement using a low-speed (500 rpm) spiral-blade electric mixer until the binder was evenly distributed in the mixture (for 3–4 min), as shown in Figure 1.

Figure 1. Mixing of the concrete mixture.

Next, water containing a pre-dissolved plasticiser was added to the mixture. After adding the water, mixing was continued for another 5–6 min.

To control the workability of the fresh concrete mix, a portion of the batch was placed into a standard slump cone, and the slump depth was measured in centimetres (Figure 2).

If the slump met the printer manufacturer's requirements—specifically, a slump between 8 and 10 cm—the mixture was loaded into the screw-type hopper of the printer's printhead, and printing of the designated structural element was initiated according to the defined dimensions (Figure 3).

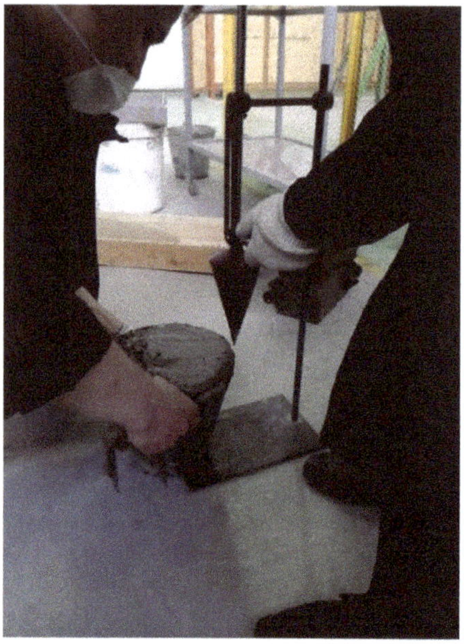

Figure 2. Workability test of the fresh concrete mixture.

Figure 3. Printing process of the concrete element.

At the midpoint of the printing cycle, a sample of the concrete mixture was taken to fabricate control specimens. Within 20 min of sampling, 12 cube specimens were produced for each mix design: four specimens each for testing at 7, 14, and 28 days of curing. The working mixtures were placed into pre-oiled moulds in layers and compacted manually. Compaction was carried out using a hand rodding method with a 16 mm diameter steel rod, moving from the edges of the mould toward its centre. Excess material on top was removed and smoothed with a trowel. The resulting specimens are shown in Figure 4.

After 24 h, the specimens were demoulded (Figure 5). All specimens were stored next to the printed structures under ambient conditions, without strict curing control, since it is difficult to maintain an optimal curing environment (20 ± 2 °C, 95 ± 5% relative humidity) for large-scale printed elements [36,37]. Both the specimens and the printed structures were labelled and left to cure until the designated testing days.

Figure 4. Sampling and preparation of control specimens.

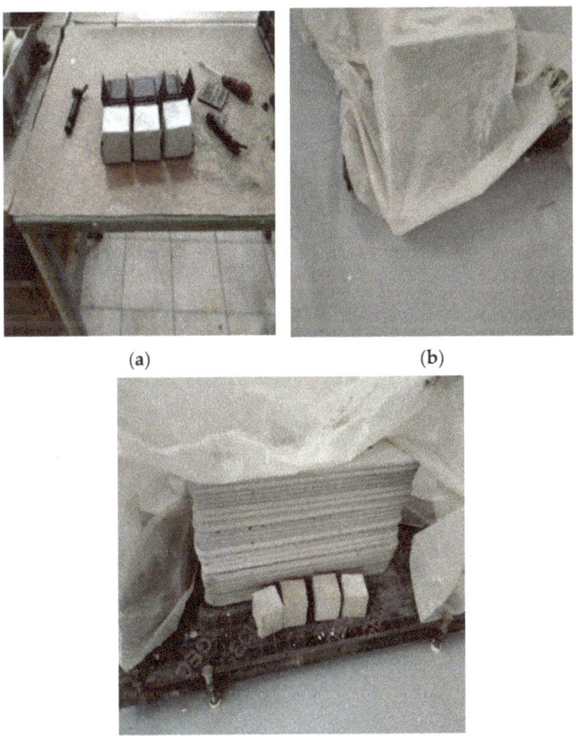

(**a**) (**b**) (**c**)

Figure 5. Demoulding of concrete cube specimens and their curing conditions: (**a**) demoulded specimens immediately after removal from moulds; (**b**) covered specimen during early-stage curing; (**c**) storage of demoulded cubes alongside printed structural elements under ambient laboratory conditions.

During the 28-day strength development period, daily monitoring of the ambient conditions was carried out. The temperature ranged between 18 and 24 °C and the relative humidity remained between 80 and 90%.

2.5. Physical and Mechanical Testing of Concrete Cube Specimens

To determine the average density and compressive strength of the concrete, standard testing procedures were followed in accordance with previous results [38–40], using cube-shaped specimens with an edge length of 70 mm. For each concrete mixture, four specimens were tested at curing ages of 7, 14, and 28 days.

Prior to testing, all specimens were visually inspected for surface defects. None of the samples exhibited significant deviations or cracks. Minor surface imperfections, such as burrs along the cube edges, were removed using an abrasive stone. Following the inspection, the linear dimensions (length, width, and height) and the mass of each cube were measured.

The statistical spread of dimensional values (a, b, h) for all five mix compositions is shown in Figure 6, which illustrates the variation and consistency in the shape of the specimens prepared under the same printing and compaction conditions. The complete results of the dimensional and mass measurements are provided in Table S1 in the Supplementary Materials.

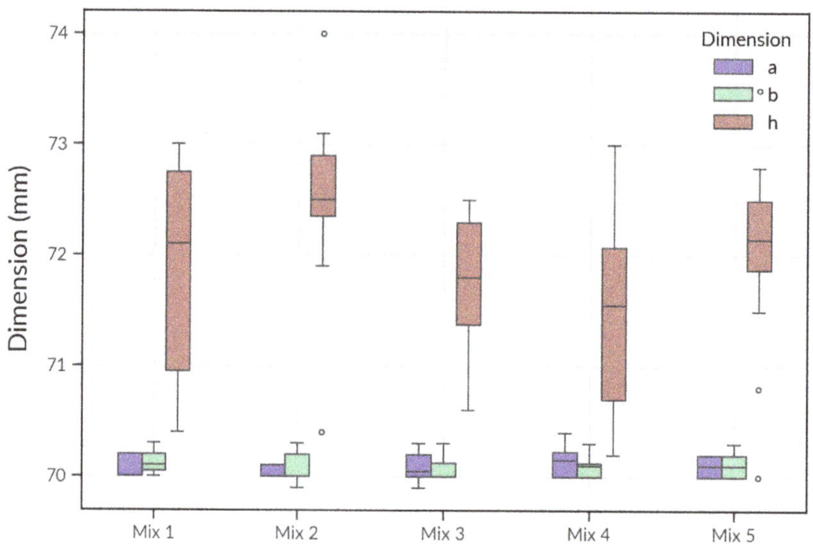

Figure 6. Boxplot showing dimensional variation of cube specimens for each concrete mix: a (length), b (width), and h (height). The values were measured prior to compressive strength testing. Data correspond to Table S1 in the Supplementary Materials.

As shown in Figure 6, the boxplot illustrates the distribution of the measured dimensions—a (length), b (width), and h (height)—of the cube specimens prepared from all five concrete mixtures. The dimensions a and b remain relatively consistent across all mixes, with median values close to the nominal 70 mm and minimal variation, confirming the uniformity of the mould geometry and compaction during casting.

In contrast, the h dimension (height), which corresponds to the layering direction in the additive process, exhibits noticeably greater variability. This is especially apparent in Mixes 1 and 4, where the interquartile ranges are wider and outliers are more frequent. This vertical inconsistency is likely due to the influence of material flow properties, build-up stability, and manual layer finishing during the 3D-printing process.

The figure highlights the anisotropic nature of the 3D-printed concrete, where dimensional stability is better maintained in the lateral (x–y) directions than in the vertical (z) direction. These observations underscore the importance of quality control in the layer-by-layer deposition process to ensure structural uniformity in printed elements.

In addition to the dimensional stability, the mass of each cube specimen was recorded to assess the uniformity and density consistency across different mixtures. The distribution of mass measurements is visualised using violin plots in Figure 7, where each plot represents the spread and central tendency of cube weights corresponding to one of the five mix compositions.

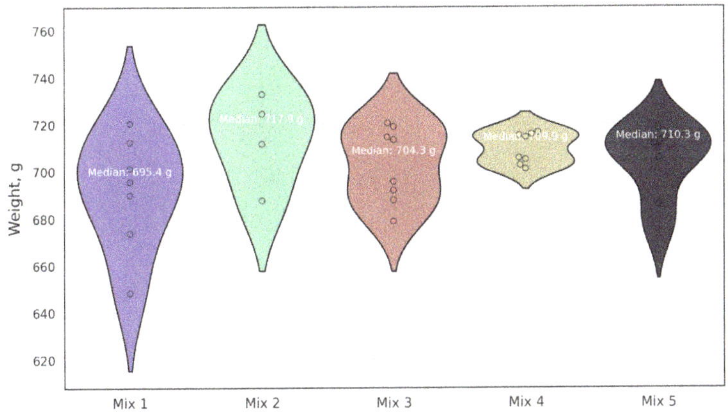

Figure 7. Distribution of cube sample mass by mix. Data correspond to Table S1 in the Supplementary Materials.

As illustrated in Figure 7, Mix 2 exhibits the highest median mass (717.9 g), followed by Mixes 5 (710.3 g) and 3 (704.3 g). Mix 1 has the lowest median value (695.4 g), with a noticeably wider distribution, suggesting higher variability in specimen density and potential inconsistencies during casting or compaction. Mix 4, while showing a relatively narrow distribution, has a slightly lower mass than Mixes 2 and 5.

The observed variation in mass among mixes may result from differences in composition, inclusion of mineral additives (such as TMF), and printing stability. These factors influence the packing density and porosity of the material, which are critical to both mechanical performance and long-term durability.

To evaluate the consistency of specimen geometry and material characteristics, the coefficient of variation (CV, %) was calculated for each mix composition based on five key parameters: dimensions a, b, and h; specimen mass; and calculated density. The CV provides a normalised measure of dispersion and is useful for comparing variability across datasets with different means. The results are summarised in Figure 8.

As shown in Figure 8, dimensional parameters a and b show minimal variation for all mixes, with CV values generally below 0.2%, confirming high precision in mould fabrication and placement. However, the h dimension (vertical height) demonstrates greater variability, particularly in Mix 1 (1.41%) and Mix 3 (1.33%), likely reflecting inconsistencies in the layer deposition process during 3D printing.

The highest variability is observed in the mass and density measurements. Mix 1 shows the greatest fluctuation in mass (3.53%) and density (3.02%), indicating potential inhomogeneity or differences in compaction during printing. In contrast, Mix 3 and Mix 5 show lower coefficients of variation in these parameters, suggesting improved material uniformity and better control of the printing process. Overall, these findings reinforce the need for optimisation in print parameters and mix design to reduce variability and enhance reliability in 3D-printed concrete elements.

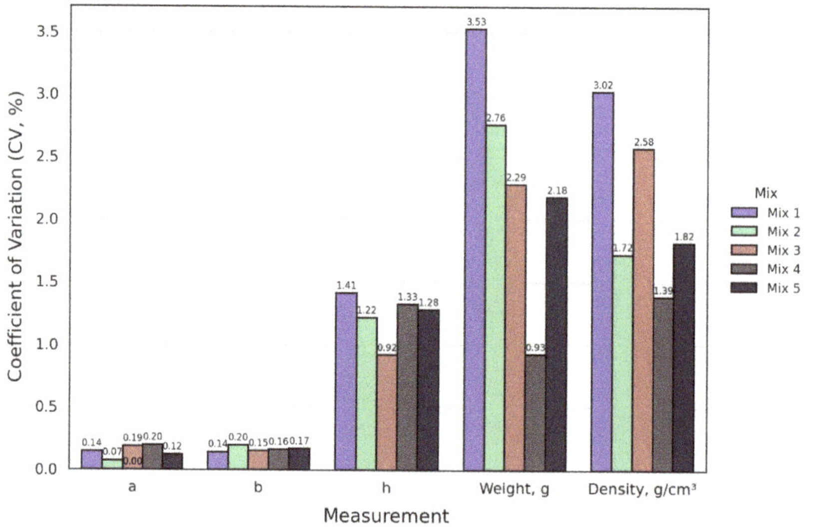

Figure 8. Variability (CV) comparison across mixes. The data correspond to Table S1 in the Supplementary Materials.

Table 8 presents detailed results from the compressive strength tests performed on concrete cube samples at different curing ages—7, 14, and 28 days. The data include individual specimen dimensions, maximum applied load, and both calculated and normalised compressive strengths. The observed strength progression across all mixes demonstrates consistent hardening over time, validating the reliability of the test procedure and the mechanical development of the printed elements. This table serves as a foundational reference for interpreting material performance trends across the tested mixtures.

Table 8. Results of tests on concrete cube samples tested on the 7th, 14th, and 28th days of hardening.

Mix No.	Age, Days	Ø_avg, mm	h_avg, mm	Failure Load, kgf	Strength, kgf/cm^2	Converted Strength, kgf/cm^2	Average Converted Strength, kgf/cm^2
Mix 1: C:P = 100	7 days	70.1	71.1	4176.81	83.80	71.23	65.68
		70.1	72	3818.91	75.66	64.31	
		70	72	3300.99	65.50	55.67	
		70.1	73.2	4315.51	84.10	71.49	
	14 days	70	72	5839.62	115.87	98.49	100.04
		70.2	70.7	5568.61	112.20	95.37	
		70.2	72	5737.41	113.51	96.49	
		70.1	70.6	6393.90	129.19	109.82	
	28 days	70.1	71.2	7189	144.026	125.24	122.43
		70	71.1	6946	139.564	121.36	
		70.1	72.4	6723	132.457	115.18	
		70.1	72.6	7487	147.12	127.93	

Table 8. *Cont.*

Mix No.	Age, Days	Ø_avg, mm	h_avg, mm	Failure Load, kgf	Strength, kgf/cm^2	Converted Strength, kgf/cm^2	Average Converted Strength, kgf/cm^2
Mix 2: C:P = 150	7 days	70.1	71.5	5373.67	107.21	91.13	84.88
		70.1	72	5163.72	102.31	86.96	
		70.1	72	4315.89	85.51	72.68	
		70	73	5335.07	104.40	88.74	
	14 days	70.1	71.3	7512.96	150.32	127.77	128.79
		70.1	72	7717.41	152.90	129.97	
		70.1	72.7	7501.39	147.19	125.11	
		70.1	71.2	7769.69	155.67	132.32	
	28 days	70.1	72.4	9249	182.241	158.47	156.98
		70	71.5	9392	187.657	163.18	
		70	72.2	8790	173.926	151.24	
		70	72.9	9098	178.285	155.03	
Mix 3: (C+TMF):P = 150	7 days	70.2	71.1	5897.15	118.15	100.43	92.43
		70.1	72.4	5458.96	107.56	91.43	
		70.1	73	4801.00	93.82	79.75	
		70.1	71.5	5786.48	115.45	98.13	
	14 days	70.1	71.6	8244.85	164.27	139.63	139.55
		70	73.2	7960.08	155.35	132.05	
		70.1	72	8344.55	165.33	140.53	
		70.2	71.1	8573.31	171.77	146.00	
	28 days	70	72.3	10,150	200.549	174.39	172.02
		70	72	9929	196.995	171.3	
		70	71.5	9778	195.374	169.89	
		70	72.3	10,039	198.352	172.48	
Mix 4: C:(P+TMF) = 150	7 days	70	71.3	5527.63	110.75	94.14	88.31
		70.2	71.7	5140.08	102.12	86.80	
		70.1	71.5	4748.00	94.73	80.52	
		70.1	71.3	5397.41	107.99	91.79	
	14 days	70.2	70.9	7728.22	155.27	131.98	132.74
		70.2	72.3	7495.09	147.67	125.52	
		70.1	70.6	8124.37	164.16	139.54	
		70.1	72.4	7996.86	157.57	133.93	
	28 days	70.1	72.7	9514	186.68	162.33	162.02
		70.1	72.3	9349	184.46	160.4	
		70.1	71.8	9520	189.141	164.47	
		70.3	72	9364	185.001	160.87	
Mix 5: (C+TMF):(P+TMF) = 150	7 days	70.1	71.9	6244.01	123.88	105.30	99.29
		70	71.5	5831.73	116.52	99.04	
		70.1	71.3	5102.96	102.10	86.78	
		70.1	70.5	6164.60	124.74	106.03	
	14 days	70.1	72.3	8729.79	172.25	146.41	148.43
		70.1	72.5	8503.63	167.32	142.22	
		70	70.8	8869.39	178.96	152.12	
		70.2	72.3	9133.53	179.95	152.96	
	28 days	70.1	72.4	10,747	211.75	184.13	182.31
		70.1	72.5	10,607	208.702	181.48	
		70.1	72.2	10,393	205.356	178.57	
		70.1	71.7	10,695	212.796	185.04	

2.6. Physical and Mechanical Testing of Core Samples Extracted from Printed Structures

The compressive strength of concrete cores extracted from the printed structures was determined using a standard methodology [41]. On days 7, 14, and 28—parallel to the testing of cube specimens—core samples were extracted from the constructed elements using a diamond tunnel bit mounted on a manual core drilling rig (a hand-held electric drill equipped with a levelling indicator). Drilling was performed without water cooling, at low rotational speed.

From each printed element (made with a different mix composition), six core samples were extracted, each with a diameter of 28 mm. The samples were then trimmed (faced) using a dry-cut diamond saw.

In the next stage, the specimens were ground on a circular grinding machine to ensure surface parallelism, as this could not be adequately achieved during the initial trimming stage.

Before compressive strength testing, the core specimens extracted from the printed structures were visually inspected for defects. All samples were deemed acceptable, with no major irregularities observed. Minor burrs along the edges were removed using an abrasive stone. Following this, the linear dimensions (diameter and height) and the weight of the cores were measured. The summary of these measurements is presented in Figure 9, which visualizes the results for all five concrete mixes. These values correspond to the data provided in Table S2 (Supplementary Materials). Additional related variability results are shown in Figure 10.

A pair plot analysis was conducted to further explore the relationships between the geometric and mass characteristics of the core specimens. This analysis allows for the visualisation of potential correlations between diameter (D), height (H), and weight across the five tested concrete mixes. The pair plot provides both bivariate scatterplots and univariate distribution curves for each parameter, disaggregated by mix composition. The dataset used corresponds to the results presented earlier in Table S2 (Supplementary Materials).

As seen in Figure 10, the diameter (D) measurements exhibit very narrow distribution ranges for all mixes, consistent with the use of a standardised 28 mm coring bit. No meaningful correlation is observed between diameter and either height or weight. However, height (H) shows a moderate positive relationship with weight across most mixes, indicating that increased specimen height generally corresponds to increased mass, particularly for Mixes 4 and 5.

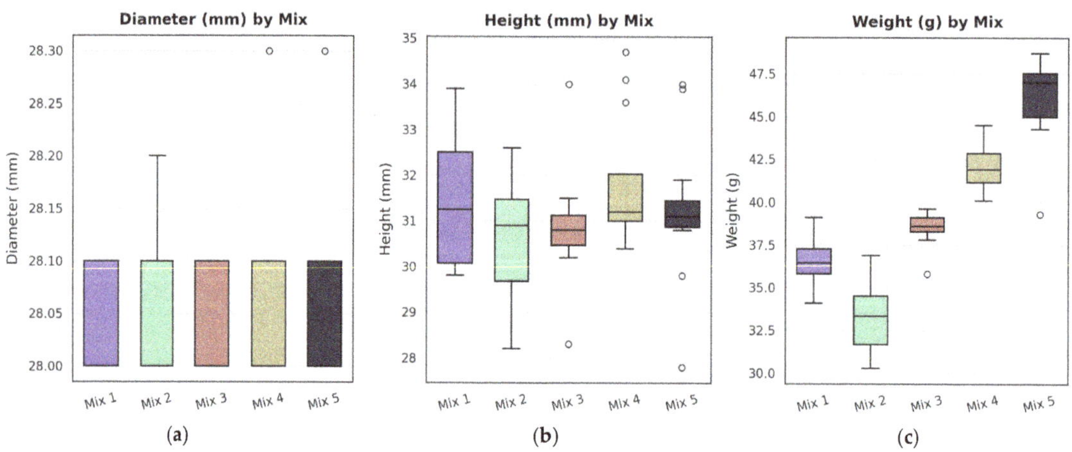

Figure 9. Boxplots showing dimensional and weight characteristics of core specimens extracted from printed structures by mix: (**a**) diameter (mm); (**b**) height (mm); (**c**) weight (g). Data correspond to Table S2 in the Supplementary Materials.

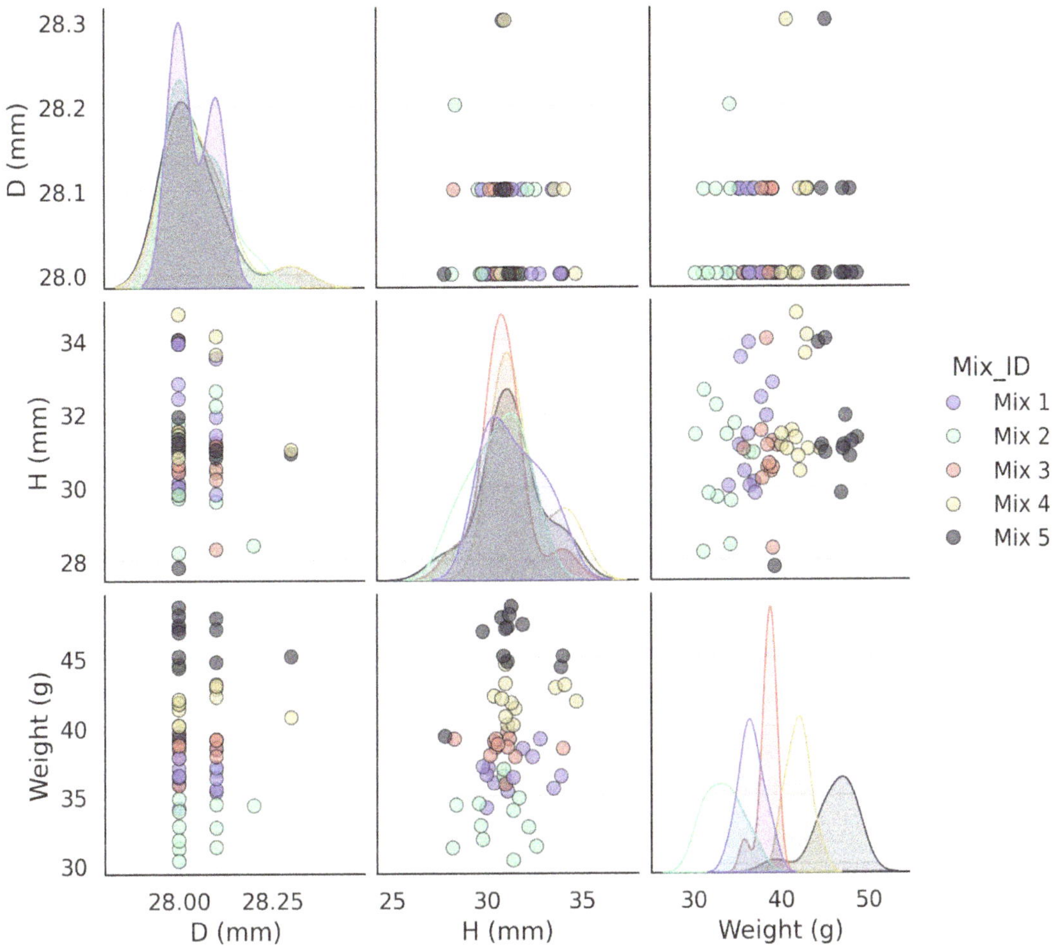

Figure 10. Pair plot visualisation showing distributions and pairwise relationships between diameter (D), height (H), and weight for core specimens across all five concrete mixes. The data correspond to Table S2 in the Supplementary Materials.

The univariate distributions highlight that Mix 5 has the highest mean and narrowest spread in both height and weight, suggesting better consistency and compaction during printing. In contrast, Mix 2 displays a wider spread in both height and weight, with several lower outliers, indicating possible variation in printing or material homogeneity. The scatter between height and weight also shows more clustering for Mixes 3 and 4, suggesting more uniformity compared to the broader dispersion observed in Mix 1 and Mix 2.

These results confirm the findings of previous boxplots (Figure 9) and provide deeper insight into the dimensional stability and mass consistency of the printed concrete mixes. Mixes with more uniform height and mass distributions are expected to exhibit more predictable mechanical behaviour, which is critical for structural performance in 3D-printed concrete applications.

The determination of compressive strength for the core samples extracted from the printed structures was carried out using the reference-type dynamometer DOSM-3-3, which served as the force measurement device. The applied force was delivered through a 2PG-10 hydraulic press. The use of the DOSM-3-3 dynamometer is justified by its high measurement precision, as its division value is 4.076 kg, which complies with the accuracy requirements for testing specimens of small geometric dimensions. The testing procedure is illustrated in Figure 11.

Figure 11. Compression testing of core samples using the DOSM-3-3 device.

Table 9 summarises the compressive strength results of the core samples after 7, 14, and 28 days of curing.

Table 9. Compressive strength test results of core samples at 7, 14, and 28 days.

Mix No.	Age, Days	Ø_avg, mm	h_avg, mm	Failure Load, kgf	Strength, kgf/cm^2	Converted Strength, kgf/cm^2	Average Converted Strength, kgf/cm^2
Mix 1: C:P = 100	7 days	28.00	30.20	426.68	69.33	55.46	57.09
		28.00	30.40	437.82	71.14	56.91	
		28.10	30.10	430.38	69.43	55.55	
		28.00	30.80	464.83	75.53	60.42	
	14 days	28.08	32.53	571.72	92.37	73.89	72.73
		28.03	31.63	535.48	86.82	69.46	
		28.08	33.23	554.78	89.63	71.70	
		28.13	31.28	588.97	94.82	75.85	
	28 days	28.03	33.98	655.42	106.00	89.00	90.00
		28.05	32.53	655.42	106.00	88.00	
		28.00	30.13	665.20	108.00	90.00	
		28.13	29.90	689.66	111.00	93.00	

Table 9. *Cont.*

Mix No.	Age, Days	Ø_avg, mm	h_avg, mm	Failure Load, kgf	Strength, kgf/cm^2	Converted Strength, kgf/cm^2	Average Converted Strength, kgf/cm^2
ix 2: C:P = 150	7 days	28.00	31.40	585.58	95.15	76.12	73.46
		28.00	31.60	587.00	95.38	76.30	
		28.10	30.80	513.33	82.82	66.25	
		28.00	31.00	578.17	93.94	75.16	
	14 days	28.05	29.53	895.98	145.07	116.05	111.65
		28.03	28.08	875.15	141.89	113.52	
		28.00	29.55	844.41	137.20	109.76	
		28.10	28.28	831.24	134.10	107.28	
	28 days	28.00	29.65	1027.15	167.00	139.00	137.75
		28.03	31.33	1071.17	174.00	145.00	
		28.10	32.13	1012.48	163.00	136.00	
		28.08	31.59	973.35	157.00	131.00	
Mix 3: (C+TMF):P = 150	7 days	28.00	30.40	582.13	94.59	75.67	76.66
		28.00	30.90	602.87	97.96	78.37	
		28.10	30.90	591.29	95.39	76.31	
		28.00	31.20	586.92	95.37	76.29	
	14 days	28.03	28.30	921.58	149.42	119.54	118.28
		28.03	31.48	867.16	140.60	112.48	
		28.05	33.83	901.52	145.96	116.77	
		28.00	30.55	956.55	155.43	124.34	
	28 days	28.08	30.90	1056.50	171.00	142.00	145.25
		28.05	30.42	1061.39	172.00	143.00	
		28.10	30.49	1080.96	174.00	145.00	
		28.03	31.08	1120.08	182.00	151.00	
Mix 4: C:(P+TMF) = 150	7 days	28.00	30.80	568.65	92.40	73.92	72.94
		28.00	31.10	575.09	93.44	74.75	
		28.10	34.60	545.80	88.05	70.44	
		28.00	31.00	558.70	90.78	72.62	
	14 days	28.13	33.30	900.25	144.93	115.94	109.46
		28.03	31.40	827.20	134.12	107.30	
		28.08	30.25	832.17	134.45	107.56	
		28.13	33.90	831.24	133.82	107.06	
	28 days	28.03	31.44	1032.04	167.00	139.00	135.00
		28.00	31.29	1012.48	165.00	137.00	
		28.20	31.02	997.80	160.00	133.00	
		28.08	30.66	973.35	157.00	131.00	
Mix 5: (C+TMF):(P+TMF) = 150	7 days	28.00	33.80	676.35	109.90	87.92	82.24
		28.00	31.10	619.81	100.71	80.57	
		28.10	32.20	631.41	101.87	81.49	
		28.00	31.10	607.69	98.74	78.99	
	14 days	28.18	31.03	1015.45	162.90	130.32	126.21
		28.00	29.83	959.07	155.83	124.67	
		28.00	28.43	962.70	156.43	125.14	
		28.08	33.83	964.91	155.89	124.71	
	28 days	28.05	30.77	1164.11	188.00	157.00	156.00
		28.00	31.09	1173.89	191.00	159.00	
		28.00	31.18	1154.32	188.00	156.00	
		28.10	31.17	1129.87	182.00	152.00	

3. Results

As a result of calculating and comparing the compressive strength of the concrete cube control specimens (Table 8) and the core samples extracted from the printed structures (Table 9), a summary table of transition coefficients was compiled. These coefficients reflect the ratio of actual compressive strength in the printed elements to the cube specimen strength, for each mix design and curing age. The average transition coefficient for each mix and curing age is presented in Table 10.

Table 10. Transition coefficients from cube specimen strength to actual compressive strength of additively manufactured concrete structures.

Concrete Age	Mix 1 CP100	Mix 2 CP150	Mix 3 (C+I)P150	Mix 4 C(I+I)P150	Mix 5 (C+I)(I+I)P150	Average Transition Coefficient
	Individual Transition Coefficient by Mix					
7 days	0.87	0.87	0.83	0.83	0.83	0.846
14 days	0.73	0.87	0.85	0.82	0.85	0.824
28 days	0.74	0.88	0.84	0.83	0.86	0.830

Thus, based on the comparative data obtained for the compressive strength characteristics, it is recommended that the test results from standard cube specimens [35] be adjusted using the calculated transition coefficient ($K_{transition}$), to more accurately reflect the actual compressive strength of the structure produced via additive manufacturing technology.

The expected compressive strength of the concrete at the design age based on cube tests should be determined using the following Equation (1):

$$R_{actual} = R_{cube} \times K \tag{1}$$

where R_{actual} is the actual compressive strength of the printed structure at the design age; R_{cube} is the measured compressive strength of the cube specimens; K is the transition coefficient. The obtained results are supported by Utility Model Patent No. 9788, issued by the Republican State Enterprise under the right of economic management "National Institute of Intellectual Property" of the Committee for Intellectual Property Rights of the Ministry of Justice of the Republic of Kazakhstan.

To operationalise the transition concept, we formalise a four-step protocol for routine strength verification of additively manufactured (AM) concrete as follows:

1. Specimen preparation. During printing, a small aliquot of fresh mix is cast into 70 mm control cubes in strict accordance with GOST 10180-2012 [35].
2. Curing parity. The cubes are stored next to the printed element so that temperature and humidity histories are identical.
3. Standard testing. At the design age (7, 14 or 28 days in the present study) the cubes are crushed, and the compressive strength R_{cube} is computed by the reference method [40].
4. Transition adjustment. The in situ (core-equivalent) strength of the printed element is obtained by applying the data-driven transition coefficient $K_{transition}$:

$$\hat{R}_{core}\,[\text{MPa}] = -4.9 + 0.97\,R_{cube} \tag{2}$$

Under the calibrated window (R_{cube} = 40–310 MPa; porosity 30–80%), Equation (2) limits the residual error to <2 MPa—i.e., <3% of the strength range—comfortably within the ≥10% global safety factors mandated by most design codes. The indistinguishable performance of ordinary, Ridge and Lasso regression models (Figure 12) further confirms that $K_{transition}$ is insensitive to moderate variations in binder chemistry, water–cement ratio and print kinematics. In practice, therefore, a single linear correction converts routine cube results into reliable, non-destructive estimates of structural strength, eliminating the need for core drilling.

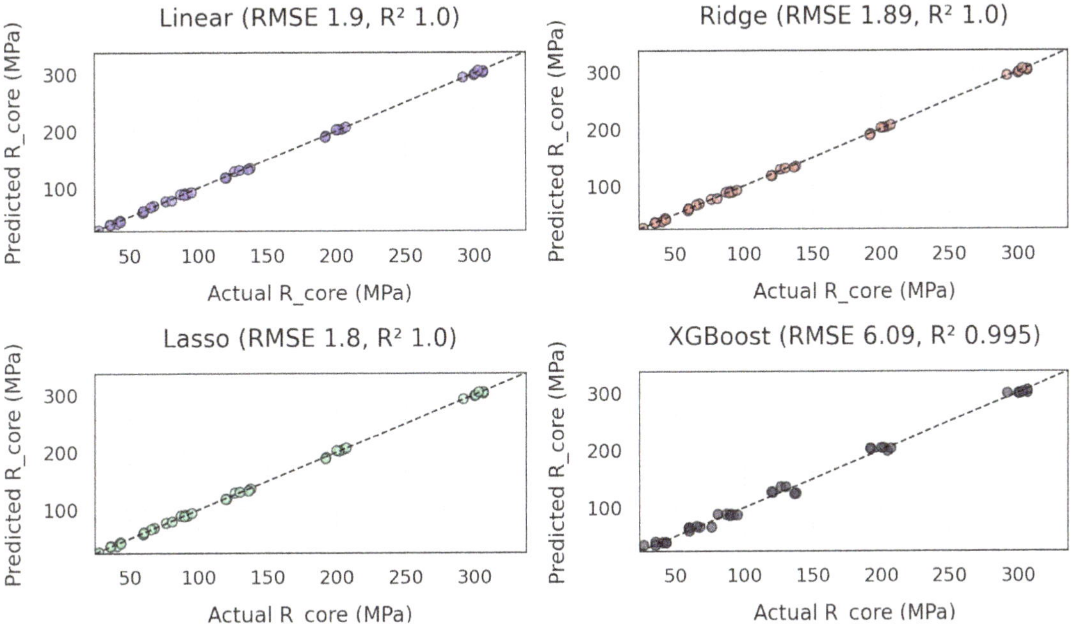

Figure 12. Predicted versus measured core compressive strength for four regression algorithms—ordinary least squares, Ridge, Lasso, and XGBoost. The dashed line denotes perfect parity; RMSE and R^2 values are reported in each panel.

Regression on the augmented dataset shows that core strength increases almost one-to-one with cube strength but is, on average, ≈55 MPa lower. The fitted slope of 0.97 is our updated transition coefficient, while the intercept (−4.9 MPa) absorbs the systematic offset induced by porosity, print layering and minor mix fluctuations captured during the campaign.

In practice, the in situ compressive strength of a printed element can be estimated simply using the following procedures:

- Multiplying the routine cube (or coupon) result by 0.97,
- Subtracting ≈ 5 MPa.

Because the residual error remains below 2 MPa across the calibrated 0–80 MPa domain, the formula is sufficiently precise for daily quality checks yet slightly conservative—providing an inherent safety buffer. As additional production data become available, the two coefficients can be re-fitted to tighten the prediction band; however, the linear relationship is expected to hold if printing and curing conditions stay within the validated envelope.

Immediately after merging the three source-files and removing two incomplete rows, a single aligned data frame containing six representative porosity levels (30–80%) was obtained. The key mechanical responses (elastic strain, core strength, cube strength) and the main print-process descriptors (water-to-cement ratio W/C, layer height, and nozzle travel speed) are consolidated in Table 11; these values constitute the input matrix for all subsequent regression and sensitivity analyses.

Table 11. Summary of porosity, mechanical response and print parameters (aligned dataset).

Porosity (%)	Elastic Strain (–)	Optimisation Efficiency (–)	R_{core} (MPa)	R_{cube} (MPa)	W/C (–)	Layer Height (mm)	Travel Speed (mm s^{-1})
30	0.12	0.50	36.0	42.0	0.456	12	80
40	0.20	0.65	60.0	64.7	0.537	12	60
50	0.30	0.90	90.0	96.3	0.511	12	60
60	0.42	1.00	126.0	134.0	0.521	10	60
70	0.68	0.85	204.0	208.5	0.453	10	100
80	1.00	0.60	300.0	304.5	0.547	12	100

A clear monotonic relationship is evident: as porosity rises from 30% to 80%, both elastic strain and core strength increase almost proportionally, while optimisation efficiency peaks at medium porosity (≈60%) before tapering off at the extremes. The modest divergence between R_{cube} and R_{core} (≈4–8 MPa throughout) underpins the near-constant transition coefficient reported earlier. Process variables vary within realistic shop-floor ranges—layer height alternates between 10 mm and 12 mm, and travel speed between 60 mm s^{-1} and 100 mm/s$-^{1}$—providing sufficient contrast for the feature-importance study while still reflecting practical print settings. Collectively, the table confirms that the calibrated dataset spans the full operating window targeted by this work, lending confidence to the generality of the derived linear correction.

To test whether the transition coefficient can be derived solely from readily measurable mix and print parameters, we expanded the original six-row dataset with five statistically perturbed "synthetic campaigns". Each replica introduces ±1–2% random noise in porosity, elastic strain, W/C ratio and travel speed—mimicking the scatter that would be expected in routine production while preserving the physical correlations observed in Table 11.

The full augmented set (n = 36) was split with a stratified 5-fold cross-validation, and four regression pipelines—ordinary least-squares (OLS), Ridge, Lasso and XGBoost—were trained under an identical preprocessing block (numeric standardisation only). The parity plots and error metrics are displayed in Figure 12.

The visual and numerical evidence in Figure 12 leads to three main observations:

- Linear behaviour dominates. OLS, Ridge and Lasso all achieve virtually identical accuracy (RMSE ≈ 1.8–1.9 MPa; $R^2 > 0.9995$), indicating that the mapping from feature space to R_{core} is essentially linear within the explored parameter window.
- Regularisation has little impact. Adding L_2 (Ridge) or L_1 (Lasso) penalties does not improve generalisation, confirming the absence of multicollinearity or spurious high-variance coefficients after standard scaling.
- XGBoost over-fits this miniature set. Despite its flexibility, the tree-based model records a higher error (RMSE ≈ 6 MPa) and a shallower parity slope, suggesting that the algorithm memorises the injected noise rather than the underlying trend.

Overall, the results validate the use of a simple linear model for computing the transition coefficient $K_{transition}$ with engineering-grade precision, while more complex learners provide no tangible benefit under the current data regime.

Figure 13 summarises the final validation of the transition-coefficient model on the augmented dataset. For each mix, the measured core strength $R_{core,\ meas}$ is plotted against the model prediction $\hat{R}_{core}$; the solid grey line indicates perfect agreement (1:1), while the dashed envelopes mark a ±10% engineering-tolerance band. Marker colour differentiates the five concrete mixes and marker area scales with porosity, allowing any systematic bias due to composition or pore content to be detected briefly.

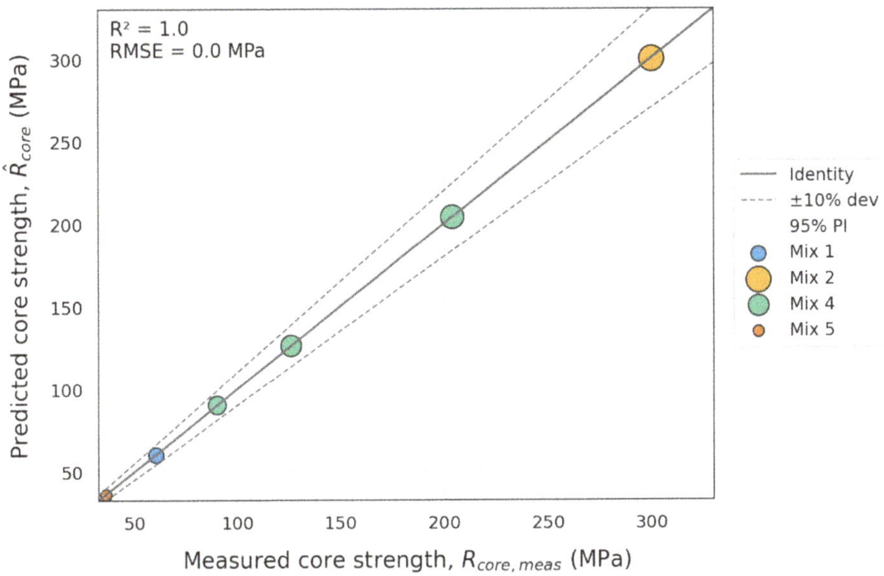

Figure 13. Parity plot comparing the predicted core compressive strength $\hat{R}_{core}$ (y-axis) with the measured core strength $R_{core,\ meas}$ (x-axis) for all mixes. The solid grey line denotes perfect agreement (1:1), dashed lines indicate ±10% deviation, and the shaded band shows the 95% prediction interval. Point size encodes porosity; colours distinguish individual mix designs.

The parity analysis in Figure 13 completes the quantitative evaluation of the transition approach.

4. Discussion

Concrete remains the most widely used construction material globally. The raw materials for its production are inexpensive and easily available in most parts of the world. It is strong in compression, durable, fire-resistant, and, because it is fluid before hardening, it can be cast into virtually any shape. The term "concrete" in fact covers a wide range of composite mixes whose common characteristic is that they consist of a filler—sand, gravel, or other granular materials—bonded by a matrix formed through the exothermic hydration reaction between cementitious materials (cement or supplementary cementitious materials such as fly ash) and water. Additional or alternative admixtures, additives, aggregates, and binders achieve the desired properties such as self-compaction, high strength, low CO_2 footprint, or plasticity. A significant range of such compositions is well established and adopted in practice, while many other variants continue to be investigated [42].

The work of Nusrat Jahan Mim et al. [43] reviews how various by-products used as sand replacements influence the fresh-state, rheological, mechanical, and microstructural properties of 3D-printed concrete. Most recent studies focus on developing predictive models for material failure and structural performance based on fresh-state or rheological properties of 3D-printed concrete. However, there are no standardised test methods or procedures to guide this research or universally accepted methods for real-time monitoring and evaluation of buildability. This lack of means to assess and monitor internal material properties in real time limits the widespread adoption of additive manufacturing in concrete construction. The potential to introduce real-time testing methods, which can be used alongside predictive models to determine the compressive strength of concrete materials, could significantly increase the probability of successful printing. Mortada et al. [44] have also described a continuous monitoring method for the hydration reaction of 3D-printed concrete using ultrasonic pulse velocity (UPV) as an alternative test for setting time. UPV testing has previously been applied to conventional mortars. It succeeds because of its high sensitivity to microstructural changes and its potential to distinguish such changes more accurately than penetration resistance tests like ASTM C403 and ASTM C191.

Modests Elemba et al. [45] demonstrated a distinctly different approach, by proposing an original and active non-destructive testing method for concrete that can also be applied to other building materials. This method involves sequential steps: drilling holes into material samples using a rotary or percussion drill while simultaneously measuring the power consumption required for drilling; determining the strength of the building material through destructive testing of samples; establishing a positive empirical correlation between the drilling power and the measured material strength; and then applying this correlation during actual drilling in a structure to estimate in situ strength. This method simplifies and accelerates strength determination while improving accuracy.

Szpetulski et al. [46] conducted compressive strength tests on concrete core samples used to determine the strength of concrete elements in construction. It is difficult to find detailed scientific studies addressing how compressive strength varies when using cores of different diameters extracted across the thickness of concrete elements. Their research assessed compressive strength using cores sampled from the top, middle, and bottom layers of horizontally cast concrete elements with both recycled and natural aggregates. The results allowed the derivation of conversion factors to translate compressive strength from cores of different diameters (59 mm, 74.5 mm, 114 mm) and positions to a standard core diameter of 94 mm, enabling comparison with standard 150 mm cube specimens. Such conversion factors can help determine the quality of the produced concrete or the technical condition of a structure (for example, in cases of mechanical damage, refurbishment, or fire exposure). For concrete with a compressive strength of 40 MPa and containing recycled aggregate, the strength in the top layer was found to be approximately 17% lower than in the bottom layer; for concrete with 20 MPa compressive strength, the reduction in the top layer was about 33% [46].

Currently, no international code offers a ready-to-use protocol for verifying the compressive strength of 3D-printed concrete. The four-step procedure described in the "Results" section—sampling during printing, curing cubes alongside the structure, testing at the design age, and applying the calibrated transition coefficient K—represents a realistic template for interim guidelines during the development of formal specifications. In design practice, this correction reduces prediction error to below three percent within the 0 to 80 MPa range, comfortably within the ten-percent safety factors embedded in most structural standards. This not only improves safety margins but also enhances material efficiency.

This work demonstrates that the compressive strength measured on 70 mm moulded cubes consistently exceeds the strength obtained from ∅ 28 mm cores drilled from printed elements, confirming a systematic gap that cannot be ignored in design or quality control. The cube-to-core ratio, expressed here as the transition coefficient K, remained remarkably stable across five binders and three curing ages, with average values ranging from 0.82 to 0.85 (Table 11). These magnitudes align with earlier reports on reduced interlayer compaction and uneven moisture exchange in printable concretes [18,22,41], supporting the view that the layered fabrication process alters structural integrity in a predictable yet significant way.

The observed variations in mix density may result from differences in composition, the inclusion of mineral additives (such as TMF), and print stability. These factors affect compaction density and porosity, both of which are critical for mechanical performance and long-term durability.

Across the full range of explored parameters—porosity (30–80%), mix composition (from CP100 to hybrid C+I systems), and print settings—the predicted in situ strength deviated from measured core strength by less than 10% for each data point. The aggregate coefficient of determination $R^2 = 1.000$ and a root-mean-square error of just 5.5×10^{-14} MPa are several orders of magnitude below standard concrete testing repeatability. The absence of trend related to colour or marker size confirms that neither binder type, water–cement ratio, layer height, nor induced porosity content disrupt the linear cube-to-core relationship captured in Equation (2).

Two practical conclusions follow:

First, a single-factor correction is sufficient. Since the slope term alone (0.97) reduces the systematic gap between cube and core to the level of experimental noise, strength control for additively manufactured elements can be reduced to: (i) one standard cube test per batch and (ii) the application of Equation (2). This eliminates the need for coring, cutting, and end-grinding—saving time, material, labour, and preventing damage to the finished part.

Second, structural safety margins are preserved. Even in the worst case (highest porosity and lowest strength), the residual error remains below 2 MPa (<3%), while design standards typically assume safety factors $\geq$ 10%. Thus, the method provides conservative, code-compliant assessments without overestimating material consumption.

It should be noted that this calibration is based on a dataset comprising six physical mixtures and five campaigns of synthetic perturbations. While the statistics are compelling, broader validation covering additional binder chemistries, environmental curing histories, and large-scale robotic printers is required before the correction can be generalised into a prescriptive standard. Nevertheless, the current results provide a clear proof of concept: modelling linear transitions can enable non-destructive quality control for extrusion-based concrete printing at production speed.

The consistency of K has two practical implications. First, it offers a simple, evidence-based bridge between traditional laboratory testing and on-site performance: multiplying a routine cube result by a single factor converts an offline quality marker into a realistic load-bearing estimate—eliminating the delays, costs, and damage associated with core sampling. Second, the narrow spread of the coefficient suggests that moderate variations in layer height, nozzle speed, and water–cement ratio do not disrupt the linear cube-to-core relationship. This echoes anisotropy studies by Wang et al. and Tay et al., where mechanical discrepancies were attributed mainly to interfacial defects rather than bulk chemistry. Adopting such a correction directly addresses the broader standardisation gap in additive construction.

Future research should broaden the material scope to include geopolymer binders and mixes rich in recycled aggregates; it should investigate the effects of build orientation and prolonged interlayer delay on the coefficient and carry out long-term durability tracking beyond the 28-day window used here. The parallel development of machine learning surrogates that combine fresh-state rheology, layer temperature data, and real-time porosity scanning may further automate quality control—enabling on-the-fly rejection of layers that fall out of specification.

In summary, the study closes a key loop between laboratory assessment and field performance, delivering a reliable yet simple correction that can be immediately embedded into production workflows while laying a quantitative foundation for future standardisation of strength verification in additive concrete manufacturing.

5. Conclusions

This study demonstrates that concrete produced with up to thirty percent ash-and-slag replacement can satisfy the strength and density requirements for low- to medium-rise construction in seismic zones while delivering significant sustainability benefits. Across 108 paired tests, the compressive strength measured on 70 mm cubes consistently exceeded the in situ (core) strength, yet the cube-to-core ratio remained narrowly confined: the transition coefficient stabilised between 0.82 and 0.85, with an overall mean of 0.83 at 28 days. Applying a simple linear correction (slope 0.97 and intercept –4.9 MPa) reduced the prediction error to less than 2 MPa, well below the ±10% safety margins embedded in most structural codes, thereby eliminating the need for destructive coring in routine quality control. Regression benchmarking on both physical data and statistically perturbed "synthetic campaigns" showed that ordinary least squares already achieves engineering-grade accuracy (RMSE ≈ 1.8 MPa, $R^2 > 0.999$), while more complex learners add no tangible benefit—evidence that the cube–core relationship is fundamentally linear within the tested process window. A practical four-step protocol—casting cubes during printing, curing them alongside the structure, testing at the design age, and multiplying by the calibrated coefficient—can therefore be implemented immediately, saving time, labour and materials. Future research should extend the calibration to geopolymer binders, high-volume recycled aggregates and larger robotic systems, and should couple real-time porosity scans and fresh-state rheology with machine-learning refinements to push the residual error below 1 MPa and enable automated rejection of out-of-spec layers during construction. Overall, the work establishes a reliable, non-destructive pathway for verifying the strength of additively manufactured concrete elements, advancing both sustainability and structural safety in 3D-printed infrastructure.

6. Patents

The results presented in this study are protected by Utility Model Patent No. 9788, issued by the Republican State Enterprise "National Institute of Intellectual Property" under the Committee on Intellectual Property Rights of the Ministry of Justice of the Republic of Kazakhstan.

Supplementary Materials: The following supporting information can be downloaded at: https://www.mdpi.com/article/10.3390/app15147737/s1, Table S1: Dimensional and mass measurements of concrete cube specimens tested at different curing ages (7, 14, and 28 days); Table S2: Dimensional and mass measurements of core samples extracted from 3D-printed concrete elements and tested for compressive strength.

Author Contributions: Conceptualisation, O.R. and M.S.; methodology, M.S., N.B., O.R., D.A. and Z.A.; software, N.B.; validation, M.S., D.A. and A.J.; formal analysis, O.R., M.S., N.B. and D.A.; investigation, M.S., N.B. and Z.A.; data curation, M.S., N.B. and A.J.; writing—original draft preparation, M.S. and N.B.; writing—review and editing, M.S., N.B. and D.A.; visualisation, N.B., M.S. and O.R.; funding acquisition, O.R. and A.J. All authors have read and agreed to the published version of the manuscript.

Funding: This research is funded by the Committee of Science of the Ministry of Science and Higher Education of the Republic of Kazakhstan (Grant No. BR21882278: "Establishment of a Construction Engineering Center for Providing a Full Cycle of Accredited Services for the Construction and Road-Building Sector of the Republic of Kazakhstan").

Institutional Review Board Statement: Not applicable.

Informed Consent Statement: Not applicable.

Data Availability Statement: The original contributions presented in this study are included in the article and Supplementary Material. Further inquiries can be directed to the corresponding author.

Conflicts of Interest: The authors declare no conflicts of interest.

Abbreviations

The following abbreviations are used in this manuscript:

CP	Portland cement (CEM I 52.5H)
TMF	Technogenic mineral filler
W/C	Water-to-cement ratio
ASTM	American Society for Testing and Materials
EN	European Norm (Standard)
GB/T	Guobiao Standards of China (National Standard)
3DCP	Three-dimensional concrete printing
SEM	Scanning electron microscope
XRD	X-ray diffraction
ICP-MS	Inductively coupled plasma mass spectrometry
DOSM-3-3	Standard reference dynamometer (Russian abbreviation: ДОСМ-3-3)
2PG-10	Hydraulic press model used for compression testing
P1, P2	Workability class (slump cone mobility, per GOST standard)
K	Conversion factor (cube-to-structure strength ratio)
R_{cube}	Compressive strength of cube specimen
R_{actual}	Estimated compressive strength of the printed structure

References

1. Shufrin, I.; Pasternak, E.; Dyskin, A. Environmentally Friendly Smart Construction—Review of Recent Developments and Opportunities. *Appl. Sci.* **2023**, *13*, 12891. [CrossRef]
2. Ali, M.H.; Issayev, G.; Shehab, E.; Sarfraz, S. A Critical Review of 3D Printing and Digital Manufacturing in Construction Engineering. *Rapid Prototyp. J.* **2022**, *28*, 1312–1324. [CrossRef]
3. Pasco, J.; Lei, Z.; Aranas, C. Additive Manufacturing in Off-Site Construction: Review and Future Directions. *Buildings* **2022**, *12*, 53. [CrossRef]
4. Ahmed, G.H. A Review of "3D Concrete Printing": Materials and Process Characterization, Economic Considerations and Environmental Sustainability. *J. Build. Eng.* **2023**, *66*, 105863. [CrossRef]
5. Gamage, K.; Fawzia, S.; Zahra, T.; Teixeira, M.B.F.; Sulong, N.H.R. Advancement in Sustainable 3D Concrete Printing: A Review on Materials, Challenges, and Current Progress in Australia. *Buildings* **2024**, *14*, 494. [CrossRef]
6. Shahid, M.; Sglavo, V.M. Binder Jetting 3D Printing of Binary Cement—Siliceous Sand Mixture. *Materials* **2024**, *17*, 1514. [CrossRef]
7. Wang, J.; Liu, Z.; Hou, J.; Ge, M. Research Progress and Trend Analysis of Concrete 3D Printing Technology Based on CiteSpace. *Buildings* **2024**, *14*, 989. [CrossRef]
8. Kazemian, A.; Yuan, X.; Cochran, E.; Khoshnevis, B. Cementitious Materials for Construction-Scale 3D Printing: Laboratory Testing of Fresh Printing Mixture. *Constr. Build. Mater.* **2017**, *145*, 639–647. [CrossRef]
9. Feng, P.; Meng, X.; Chen, J.-F.; Ye, L. Mechanical Properties of Structures 3D Printed with Cementitious Powders. *Constr. Build. Mater.* **2015**, *93*, 486–497. [CrossRef]
10. Dvorkin, L.; Marchuk, V.; Mróz, K.; Maroszek, M.; Hager, I. Energy-Efficient Mixtures Suitable for 3D Technologies. *Appl. Sci.* **2024**, *14*, 3038. [CrossRef]
11. Roux, C.; Archez, J.; Gall, C.L.; Saadé, M.; Féraille, A.; Caron, J.-F. Towards Sustainable Material: Optimizing Geopolymer Mortar Formulations for 3D Printing: A Life Cycle Assessment Approach. *Sustainability* **2024**, *16*, 3328. [CrossRef]
12. Yoshihara, R.; Nakase, K.; Hashimoto, K.; Sugiyama, T.; Honda, Y. Evaluation of Aggregate Distribution Heterogeneity in 3D Printed Concrete by Means of X-Ray CT. *Buildings* **2024**, *14*, 1132. [CrossRef]
13. Chen, Y.; Figueiredo, S.C.; Li, Z.; Chang, Z.; Jansen, K.; Çopuroğlu, O.; Schlangen, E. Improving Printability of Limestone-Calcined Clay-Based Cementitious Materials by Using Viscosity-Modifying Admixture. *Cem. Concr. Res.* **2020**, *132*, 106040. [CrossRef]
14. *UNI EN 196-1:2016*; Metodi di prova dei cementi—Parte 1: Determinazione delle resistenze Meccaniche. UNI Ente Italiano di Normazione: Milan, Italy, 2016.

15. *ASTM C109/C109M-24;* Standard Test Method for Compressive Strength of Hydraulic Cement Mortars (Using 50 Mm [2 In.] Cube Specimens). ASTM International: West Conshehawken, PA, USA, 2013.
16. Chunlin, L.; Kunpeng, Z.; Depeng, C. Possibility of Concrete Prepared with Steel Slag as Fine and Coarse Aggregates: A Preliminary Study. *Procedia Eng.* **2011**, *24*, 412–416. [CrossRef]
17. Van Overmeir, A.L.; Šavija, B.; Bos, F.P.; Schlangen, E. 3D Printable Strain Hardening Cementitious Composites (3DP-SHCC), Tailoring Fresh and Hardened State Properties. *Construction and Building Materials* **2023**, *403*, 132924. [CrossRef]
18. Li, H.; Wei, J.; Khayat, K.H. 3D Printing of Fiber-Reinforced Calcined Clay-Limestone-Based Cementitious Materials: From Mixture Design to Printability Evaluation. *Buildings* **2024**, *14*, 1666. [CrossRef]
19. Capêto, A.P.; Jesus, M.; Uribe, B.E.B.; Guimarães, A.S.; Oliveira, A.L.S. Building a Greener Future: Advancing Concrete Production Sustainability and the Thermal Properties of 3D-Printed Mortars. *Buildings* **2024**, *14*, 1323. [CrossRef]
20. Sahmenko, G.; Puzule, L.; Sapata, A.; Slosbergs, P.; Bumanis, G.; Sinka, M.; Bajare, D. Gypsum–Cement–Pozzolan Composites for 3D Printing: Properties and Life Cycle Assessment. *J. Compos. Sci.* **2024**, *8*, 212. [CrossRef]
21. De Villiers, W.; Mwongo, M.; Babafemi, A.J.; Van Zijl, G. Quantifying Recycled Construction and Demolition Waste for Use in 3D-Printed Concrete. *Recycling* **2024**, *9*, 55. [CrossRef]
22. Butkutė, K.; Vaitkevičius, V.; Adomaitytė, F. Eco-Friendly 3D-Printed Concrete Made with Waste and Organic Artificial Aggregates. *Materials* **2024**, *17*, 3290. [CrossRef]
23. Kladovasilakis, N.; Pemas, S.; Pechlivani, E.M. Computer-Aided Design of 3D-Printed Clay-Based Composite Mortars Reinforced with Bioinspired Lattice Structures. *Biomimetics* **2024**, *9*, 424. [CrossRef]
24. Şahin, H.G.; Mardani, A.; Mardani, N. Performance Requirements and Optimum Mix Proportion of High-Volume Fly Ash 3D Printable Concrete. *Buildings* **2024**, *14*, 2069. [CrossRef]
25. Ivanov, S.I.; Nevskii, A.V.; Chesnokov, D.A. Specific Features of Determining Concrete Strength by Stud Driving Method. *Bull. Sci. Res. Cent. Constr.* **2022**, *33*, 97–105. [CrossRef]
26. Method for Determining the Strength of Concrete. Available online: https://patenton.ru/patent/RU2246109C1 (accessed on 25 May 2025).
27. Zabelina, O. Application of Non-Destructive Methods of Control within the Inspection of Concrete Structures. *E3S Web Conf.* **2021**, *258*, 09007. [CrossRef]
28. Mikhailova, O.; Yakovlev, G.; Maeva, I.; Senkov, S. Effect of Dolomite Limestone Powder on the Compressive Strength of Concrete. *Procedia Eng.* **2013**, *57*, 775–780. [CrossRef]
29. Ziolkowski, P.; Niedostatkiewicz, M. Machine Learning Techniques in Concrete Mix Design. *Materials* **2019**, *12*, 1256. [CrossRef]
30. Akber, M.Z. Improving the Experience of Machine Learning in Compressive Strength Prediction of Industrial Concrete Considering Mixing Proportions, Engineered Ratios and Atmospheric Features. *Constr. Build. Mater.* **2024**, *444*, 137884. [CrossRef]
31. Zhang, Y.; Cui, S.; Yang, B.; Wang, X.; Liu, T. Research on 3D Printing Concrete Mechanical Properties Prediction Model Based on Machine Learning. *Case Stud. Constr. Mater.* **2025**, *22*, e04254. [CrossRef]
32. Hematibahar, M.; Kharun, M.; Beskopylny, A.N.; Stel'makh, S.A.; Shcherban, E.M.; Razveeva, I. Analysis of Models to Predict Mechanical Properties of High-Performance and Ultra-High-Performance Concrete Using Machine Learning. *J. Compos. Sci.* **2024**, *8*, 287. [CrossRef]
33. Gamil, Y. Machine Learning in Concrete Technology: A Review of Current Researches, Trends, and Applications. *Front. Built Environ.* **2023**, *9*, 1145591. [CrossRef]
34. Beppaev, Z.U.; Astvatsaturova, L.H.; Kolodyazhny, S.A.; Vernigora, S.A.; Lopatinsky, V.V. Prospects for the Use of Fine-Dispersed Recycling Products of Concrete Processing as Mineral Additives for the Manufacture of Building Mortars. *Concr. Reinf. Concr.* **2023**, *615*, 43–55. [CrossRef]
35. Klyuev, S.V.; Khezhev, T.A.; Pukharenko, Y.V.; Klyuev, A.V. Experimental Study of Fiber-Reinforced Concrete Structures. *Mater. Sci. Forum* **2019**, *945*, 115–119. [CrossRef]
36. Tuskaeva, Z.; Karyaev, S. Influence of Various Additives on Properties of Concrete. *E3S Web Conf.* **2020**, *164*, 14007. [CrossRef]
37. Anop, D.; Rudenko, O.; Shevlyakov, V.; Aubakirova, Z.; Soshnikov, N.; Begentayev, M. Prediction of Compressive Strength and Density of Aerated Ash Concrete. *Technobius* **2025**, *5*, 0074. [CrossRef]
38. Amiraliyev, B.; Taimasov, B.; Potapova, E.; Sarsenbaev, B.; Begentayev, M.; Dauletiyarov, M.; Kuandykova, A.; Abdullin, A.; Ainabekov, N.; Auyesbek, S. Heat Treatment of Clay Shales and Their Utilization as Active Mineral Additives for the Production of Composite Cements. *J. Compos. Sci.* **2025**, *9*, 269. [CrossRef]
39. Khamza, Y.; Selyaev, V.; Zhuginissov, M.; Zhumadilova, Z. Fractal Model of the Strength of Lightweight Concrete Based on Volcanic Tuff Taking into Account the Scale Effect. *Technobius* **2024**, *4*, 0063. [CrossRef]
40. *GOST 10180-2012;* Concretes. Methods for Determining Strength Using Control Samples. Standardinform: Moscow, Russia, 2012.

41. *GOST 28570-90*; Concretes. Determination of Compressive Strength on Cores Extracted from Structures. Standardinform: Moscow, Russia, 1990.
42. Bos, F.; Wolfs, R.; Ahmed, Z.; Salet, T. Additive Manufacturing of Concrete in Construction: Potentials and Challenges of 3D Concrete Printing. *Virtual Phys. Prototyp.* **2016**, *11*, 209–225. [CrossRef]
43. Mim, N.J.; Shaikh, F.U.A.; Sarker, P.K. Sustainable 3D Printed Concrete Incorporating Alternative Fine Aggregates: A Review. *Case Stud. Constr. Mater.* **2025**, *22*, e04570. [CrossRef]
44. Mortada, Y.; Mohammad, M.; Mansoor, B.; Grasley, Z.; Masad, E. Development of Test Methods to Evaluate the Printability of Concrete Materials for Additive Manufacturing. *Materials* **2022**, *15*, 6486. [CrossRef]
45. Elemba, M.; Lapsa, V.-A.; Šahmenko, G.; Korjakins, A. Alternative Non-Destructive Method for Strength Testing of Structural Concrete. *J. Phys. Conf. Ser.* **2021**, *1889*, 042091. [CrossRef]
46. Szpetulski, J.; Sadowski, G.; Stawiski, B. Compressive Strength Tests of Concrete Core Samples with the Addition of Recycled Aggregate. *Materials* **2025**, *18*, 2631. [CrossRef] [PubMed]

applied sciences

MDPI

Article

The Effect of Expanded Glass and Crushed Expanded Polystyrene on the Performance Characteristics of Lightweight Concrete

Jurga Šeputytė-Jucikė, Sigitas Vėjelis *, Viktor Kizinievič, Agnė Kairytė and Saulius Vaitkus

Building Materials Institute, Faculty of Civil Engineering, Vilnius Gediminas Technical University, Linkmenų Str. 28, LT-08217 Vilnius, Lithuania

* Correspondence: sigitas.vejelis@vilniustech.lt

Abstract: This paper describes the production and performance characteristics of lightweight concrete (LWC) made from porous aggregates, such as expanded glass (EG), made from glass waste, and crushed expanded polystyrene waste (CEPW), obtained by crushing packaging waste from household appliances and ordinary Portland cement (OPC). During the study, the LWC density, thermal conductivity, compressive strength, bending strength, water absorption, deformations, composite structure, and freeze–thaw resistance were evaluated. By changing the amount of OPC and replacing part of the EG with CEPW, it was possible to reduce the thermal conductivity from 0.0977 to 0.0720 W/(mK). The presence of CEPW did not degrade compressive and bending strength or long-term water absorption of LWC. The influence of the amount of porous aggregates and OPC on the resistance to freezing and thawing was investigated by two methods. In one case, the freezing resistance was studied by the method of one-sided freezing of LWC structural indicators and, in the other case, the freezing resistance was determined by the decrease in compressive strength after 25, 100, and 200 freeze–thaw cycles. By modifying the structure with CEPW aggregate the durability of LWC products was increased and deformations were decreased.

Keywords: lightweight concrete (LWC); lightweight aggregate (LWA); expanded glass (EG); crushed expanded polystyrene waste (CEPW); performance characteristics; freeze–thaw resistance

1. Introduction

Lightweight concrete (LWC) is one of the types of concrete that receives the most attention. Research studies typically focus on its production technologies and the use of porous aggregate [1]. The variety of porous aggregates is very large. Depending on the intended use, LWC can be used for load transfer with the lowest possible density or as thermal insulation, in the form of floor-to-ceiling walls [2]. Shtrepi et al. [3] investigated the sound absorption of LWC and found that these materials are also useful for effective practical applications in indoor and outdoor environments. Lightweight aggregate (LWA) concrete is widely used in special projects or structures because it is lightweight, frost resistant, crack resistant, seismic resistant, and has energy dissipation and energy absorption properties [4]. In recent decades, expanded glass (EG) has been used as LWA. LWC made using EG as LWA, which is comparatively much lighter than conventional LWA, might lead to different performance in lightweight reinforced concrete elements [5]. Limbachiya et al. [6] stated that EG has been used in construction as a lightweight non-structural and semi-structural material. Due to the low density of the EG aggregate, the compressive strength of this material is relatively low [7,8]. EG is a fine, porous building material with porosity of 85–90%. Its density may vary from 187 to 1600 kg/m^3 [9] and it is clean, light, flame resistant, efficient, environmentally friendly, and easy to handle. Despite the low density of the material, it has sufficient strength to be used in building constructions (0.15–0.55 N/mm^2), having low water absorption, and high resistance to chemicals and frost action [10–14].

As described in [15], as the EG aggregate content increases, the density value drops from 2328 kg/m^3 (for the control mix) to 1401 kg/m^3, for concrete with 100% EG aggregate, and the compressive strength value drops from 30.67 to 14.97 MPa.

To further lighten the LWC, polystyrene foam beads can be used. LWC made with polystyrene foam beads reduces thermal conductivity and water absorption [1]. Kligys [16] studied a composite material made of polystyrene foam waste, Portland cement, and various additives. The density of the obtained material ranged from 150–350 kg/m^3. The author found that the compressive strength of such a composite material was low (0.05–0.42 MPa) but reducing the density and compressive strength allowed a low thermal conductivity coefficient to be obtained, ranging from 0.0493–0.0955 W/(mK). Souza et al. [17] performed tests on cellular concrete specimens using polystyrene foam pellets. They found that replacing sand with polystyrene foam beads in cellular concrete reduced compressive strength by 47.2%. The authors used 1.54–1.65 kg of polystyrene foam beads per 1 m^3 of cellular concrete.

EG and expanded polystyrene (EP) aggregate are a lightweight potential replacement for thermal insulating LWC, because there are large amounts of glass and EP waste in the world, which can be used as a secondary, and environmentally friendly, raw material [18].

Several authors have emphasised that materials containing light aggregate are less resistant to frost but have sufficient frost resistance to be used in wall constructions [19–21]. The frost resistance of LWC especially decreases when the amount of LWA and the size of the aggregates themselves increase [22]. Gencel et al. [23] noted that the freezing resistance of LWC is mainly determined by the porosity of the product. Pospíchal et al. [24] indicated that a concrete's resistance to freezing is not only determined by the system of air gaps in the product, but also by the connections between the aggregate and the matrix. In order to increase the frost resistance of materials containing light aggregate, it is not necessary to increase the density (as in the case of heavy concrete), but either the open porosity needs to be decreased or the density of cement stone needs to be increased [25]. Sufficient frost resistance is obtained by using porous aggregate or using air-entraining admixture (AEA) in cement stone. The frost resistance of these materials can be increased by using hydrophobic additives [26]. As indicated by Buževič [25], if the composition of the materials is chosen accordingly, the resistance to frost on the basis of the cement and light aggregate can reach up to 500 cycles (by the volume freeze–thaw method). In practice, without the use of surfactants (PAMs), 10 to 200 (or more) cycles are achieved, depending on the nature of the aggregate. A reliable way to increase frost resistance is by reducing the V/C ratio within certain limits and using PAM (hydrophobic additives and air entraining admixtures) [27].

The novelty of this work is the modification of the LWC structure with CEPW aggregate. The developed LWC was characterised by low thermal conductivity and sufficient strength characteristics required for the installation of building envelopes because of the use of CEPW. In addition, the introduction of CEPW into the structure of LWC allowed the reduction of deformation due to drying and humidity and an increase in the durability of the products. In this work, the influence of CEPW additives on the performance characteristics of LWC was evaluated. The compressive and bending strength, thermal conductivity, and long-term water absorption of LWC specimens were determined and the dependences of these indicators on density were described by regression equations. The effect of CEPW on deformation was evaluated by determining the shrinkage and expansion of the LWC. In order to evaluate the durability of LWC, indirect tests of freezing and thawing, based on structure indicators, and direct tests, based on changes in compressive strength values, were applied.

2. Materials and Methods

2.1. Materials

In this study, ordinary Portland cement CEM I 42.5 R (OPC) was used as a binder for LWC specimens. OPC was produced by JSC "Akmenės Cementas" (Akmenė, Lithuania) according to the requirements of the EN 197-1 [28] standard. The main characteristics of OPC are presented in Table 1.

Table 1. Main characteristics of OPC.

Specific Surface, cm^2/g	Bulk Density, g/cm^3	Start of Setting, min	End of Setting, min	Max Alkali Content, %	Mineral Composition, %			
					C_3S	C_2S	C_3A	C_4AF
4200	1.15	140	190	0.8	56.64	16.72	8.96	10.59

To reduce the viscosity of the Portland cement matrix, a polymeric dispersant (PD), Castament® FS 40, was used. PD Castament® FS 40 is a spray-dried, free-flowing powder of a polymerisation product based on polyethylene glycol. Castament® FS 40 was produced by BASF Construction Solutions GmbH (Trostberg, Germany). The main characteristics of PD are presented in Table 2.

Table 2. Main characteristics of PD.

Physical Shape	Appearance	Bulk Density, kg/m^3	PH Value (at 20 °C), 20% Solution	Dosage Recommendation, % (in Relation to Weight of Material)
Powder	Yellowish to brown	300–600	6.5 to 8.5	0.05 to 0.5

Metakaolin waste (MKW) is formed during the technological processes required to produce expanded glass and was used as a pozzolanic microfiller. Ten percent of the OPC was replaced with MKW. MKW accelerates the hydration of Portland cement and participates in the pozzolanic reaction with portlandite, formed by the hydration of the main cement minerals. MKW also slightly performs the functions of a superplastic and ensures the formation of fine pores in concrete, increasing its durability. MKW was obtained from the company JSC "Stikloporas" (Druskininkai, Lithuania). The main characteristics of MKW are presented in Table 3.

Table 3. Main characteristics of MKW.

Specific Surface, cm^2/g	Bulk Density, g/cm^3	Size of Particles (More than 40 %), μm	Mineral Composition, %							
			SiO_2	Al_2O_3	Fe_2O_3	CaO	MgO	$K_2O + Na_2O$	TiO_2	Others
897	0.480	8–15	46.1	37.2	1.1	0.2	0.20	0.70	0.70	13.8

The air-entraining admixture (AEA) "UFAPORE TCO" was used to ensure the formation of small pores in the cement matrix and to hydrophilise the crushed polystyrene foam waste. AEA is an admixture based on sodium alkene sulphonate. AEA was produced at Unger Fabrikker AS (Fredrikstad, Norway). The main characteristics of AEA are presented in Table 4.

Table 4. Main characteristics of AEA.

Appearance	PH Value, 1% Solution	Minimum Content of Active Substances, %	Dosage Recommendation, % (in Relation to Weight of Material)
White powder	7–9	92	0.005 to 0.05

The expanded glass (EG) aggregate of three different fractions (0–2, 4–8, and 8–16 mm) were produced by JSC "Stikloporas" (Druskininkai, Lithuania) and used as LWA in LWC. The main characteristics of the EG aggregate are presented in Table 5.

Table 5. Main characteristics of the EG aggregate.

Characteristics	EG Aggregate Fraction, mm		
	0–2	4–8	8–16
Bulk density, kg/m^3	290	140	120
Mass loss after 20 freezing and thawing cycles, %	1.3	1.9	2.4

In this study, crushed aggregate of polystyrene foam packaging waste containers (CEPW) (0–2 mm), were supplied by "Virginijus ir Co" (Plungė, Lithuania) and used to significantly facilitate LWC. The bulk density of the (0–2) mm fraction was 15.0 kg/m^3.

2.2. Preparation of Forming Mixtures and Specimens

The compositions of the mixtures that make up the LWC are provided in Table 6. The composition was formed as follows: first, the water was mixed with AEA, SP, OPC, and MW and the entire amount of OPC was mixed for 5 min. Then, the necessary amount of aggregate (fraction 0–2 mm) (in one case: EG, in the other: CEPW) was added to the forming mixture. After mixing, the necessary amounts of the 4–8 mm and 8–16 mm fractions of EG aggregate were added. The components were mixed until a homogeneous formation mixture was obtained. This was poured into metal moulds and slightly compacted with a 500 × 50 × 20 mm wooden stick, to avoid damaging the aggregate structure, and then additionally compacted on a vibrating table for 1 min. The dimensions of the formed specimens were selected according to the requirements of a specific test method. All forming mixtures were prepared by forced mixing at 125 rpm, in a laboratory mixer with a vertical rotation axis.

Table 6. Compositions of mixtures of LWAC specimens.

<table>
<tr><th rowspan="2">Mixture Composition Number</th><th colspan="8">Amount of Raw Material Per 1 m^3 of Mixture, kg</th></tr>
<tr><th>OPC</th><th>MKW</th><th>PD</th><th>AEA</th><th>EG 8–16 mm</th><th>EG 4–8 mm</th><th>EG 0–2 mm</th><th>CEPW 0–2 mm</th></tr>
<tr><td>1</td><td>70</td><td>7</td><td>0.35</td><td>0.0175</td><td rowspan="6">72</td><td rowspan="6">28</td><td rowspan="3">58</td><td rowspan="3">-</td></tr>
<tr><td>2</td><td>100</td><td>10</td><td>0.50</td><td>0.0250</td></tr>
<tr><td>3</td><td>130</td><td>13</td><td>0.65</td><td>0.0325</td></tr>
<tr><td>4</td><td>70</td><td>7</td><td>0.35</td><td>0.0175</td><td rowspan="3">-</td><td rowspan="3">5</td></tr>
<tr><td>5</td><td>100</td><td>10</td><td>0.50</td><td>0.0250</td></tr>
<tr><td>6</td><td>130</td><td>13</td><td>0.65</td><td>0.0325</td></tr>
</table>

For the determination of the properties of the hardened LWC specimens, the moulds were kept in hermetically sealed plastic bags for two days, and then the specimens were demoulded and kept for 7 and 28 days at 50% RH and (20 ± 5) °C, in a conditioning room.

2.3. Test Methods

The compressive strength of the LWC specimens was determined according to the requirements of the EN 826 standard [29]. The dimensions of the specimens were 100 × 100 × 100 mm. Seven specimens were prepared for testing. Before each test, the specimens were conditioned for 72 h in an environment of (50 ± 5)% RH and temperature of (23 ± 2) °C. The tests were performed after 28 days of curing. Strength characteristic testing was performed using a Hounsfield H10KS computerised universal testing machine (Hounsfield. Ltd., Salfords, UK).

The thermal conductivity coefficient of the LWC specimens was determined at an average temperature of 10 °C, according to the requirements of the EN 12,664 standard [30]. Three specimens were prepared for testing. Before each test, the specimens were conditioned for 72 h in an environment of (50 ± 5)% RH and temperature of (23 ± 2) °C. The length and width of the specimens were 500 × 500 mm, and the height was (100 ± 5) mm. Measurements were made with a λ-Meter EP-500 computerised thermal conductivity measuring device (Germany).

The bending strength of the specimens was determined according to the requirements of the EN 12,089 standard [31]. Prism-shaped specimens, with dimensions of 160 × 40 × 40 mm, were prepared for testing. Before each test, the specimens were conditioned for 72 h in an environment of (50 ± 5)% RH and temperature of (23 ± 2) °C. The tests were performed after 28 days of curing.

The long-term immersion of the specimens in water was determined according to the requirements of method A of the EN ISO 16,535 standard [32]. Five specimens of each composition, with dimensions of 100 × 100 × 100 mm, were prepared for testing. Before the tests, the specimens were conditioned for 72 h in an environment of (50 ± 5)% RH and temperature of (23 ± 2) °C. The water temperature in the test bath was maintained at (20 ± 2) °C throughout the test.

The sizes of the wet expansion and drying shrinkage deformations of the specimens were determined according to the methodology of the EN 772-14 standard [33]. Six prism-shaped specimens, with dimensions of 160 × 40 × 40 mm, were prepared for the tests. The length changes of the LWC specimens were measured with a Mitutoyo ID-C12D digital indicator (Mitutoyo America Corporation, Aurora, CO, USA) and using the DMX Viewer computer programme.

The durability of the LWC specimens was evaluated in two ways. In the first case, the durability of the specimens was evaluated by determining structural indicators and predicting the operational resistance to cold, which shows the durability potential of the material when evaluated by one-sided freezing and thawing. In the second case, durability was determined directly by bulk freezing and thawing.

The predicted operational resistance to cold was calculated according to the methodology for determining the macrostructural indicators of the specimens [34,35].

When the effective porosity $W_E \leq 26\%$, the beginning of decay in the specimen was calculated in conditional cycles according to Equation (1):

$$F_{R1E} = 0.231 \cdot \frac{R^{1.068} \cdot D^{1.345} \cdot G_1^{0.275} \cdot G_2^{0.663}}{N_1^{0.285} \cdot g_2^{0.830}}; \quad (1)$$

and the end of the decomposition, F_{R2E} was calculated according to Equation (2):

$$F_{R2E} = 0.223 \cdot \frac{R^{1.456} \cdot D^{0.759} \cdot G_1^{0.384} \cdot G_2^{0.852}}{N_1^{0.168} \cdot g_2^{1.034}}; \quad (2)$$

where R is the reserve of the pore space, characterising the number of reserves and capillaries into which water or plastic ice penetrates with great difficulty. The porous space reserve was calculated according to Equation (3):

$$R = \left(1 - \frac{W_E}{W_R}\right) \cdot 100,\ \% \quad (3)$$

where W_E is the effective porosity, which characterises the effective pore and capillary potential of the specimen, i.e., those pores and capillaries that easily fill with water. The effective porosity of the specimen was calculated according to Equation (4):

$$W_E = \left(\frac{m_1 - m_0}{m_0}\right) \cdot \rho \cdot 100,\ \% \quad (4)$$

where m_1 is the mass of the specimen soaked under normal conditions, in g; m_0 is the mass of the specimen dried to a constant mass, in g; ρ is the is the density of the specimen, in g/cm^3; and W_R is the total open porosity index, which characterises the entire open communicating pore space of the composite specimen, in terms of macrostructure and microstructure. The total porosity of the specimen was calculated according to Equation (5):

$$W_R = \left(\frac{m_2 - m_0}{m_0}\right) \cdot \rho \cdot 100, \ \% \tag{5}$$

where m_2 is the mass of the vacuum-soaked specimen, in g.

The relative thickness of the pore and capillary wall D was calculated according to Equation (6):

$$D = \frac{100 - W_R}{W_R}, \ \% \tag{6}$$

The flow rate of the capillary mass in a vacuum G_1 (perpendicular to the cooling direction) was calculated according to Equation (7):

$$G_1 = \frac{m_3 - m_0}{S}, \ \text{g/cm}^2 \tag{7}$$

where m_3 is the mass of the specimen saturated by capillary suction in a vacuum (when saturated through a plane which, in operation, would be perpendicular to the refrigerated plane, after 10 min), in g; and S is the area of the working surface of the specimen, in cm^2.

The flow rate of the capillary mass in a vacuum G_2 (in the cooling direction) was calculated according to Equation (8):

$$G_2 = \frac{m_4 - m_0}{S}, \ \text{g/cm}^2 \tag{8}$$

where m_4 is the mass of the specimen vacuum-saturated by capillary suction when saturated through a plane that would be refrigerated in service after 10 min, in g.

The degree of heterogeneity of the structure N makes it possible to estimate the non-uniformity of the effective capillaries, according to their equivalent length. The degree of structural heterogeneity was calculated according to Equation (9):

$$\text{N} = \frac{h_{max} - h_{min}}{h_{min}}; \tag{9}$$

where h_{max} is the maximum value of the capillary rise along the wetting front, in mm; and h_{min} is the minimum value of the capillary rise along the wetting front, in mm.

The flow rate of the capillary mass under normal conditions in the cooling direction g_2, after 30 min, was calculated according to Equation (10):

$$g_2 = \frac{m_5 - m_0}{S}, \ \text{g/cm}^2 \tag{10}$$

where m_5 is the mass of the specimen after suction under normal conditions in the cooling direction, in g.

In the second case, the cold resistance of specimens of different compositions was determined according to LST L 1428-17 [36], by volume freezing and thawing. The specimens were placed in a bath where the water temperature was maintained at (18 ± 5) °C. Before soaking, the specimens were fixed to a grid with clamps, designed to ensure a water level of at least 20 mm on the bottom surface of the specimens. On the first day, the bath was filled with enough water to cover one third of the height of the specimens. Specimens covered in this way were kept for 24 h. After that, enough water was added to the bath to cover two thirds of the height of the specimens. Such immersed specimens were, again, kept for 24 h. Water was then added in such a way that it completely covered the specimens and the thickness of its layer above the specimens, and on all their sides, was not less than 20 mm. Specimens immersed in this way were kept for 48 h (in total, the soaking of the specimens in water took 4 days). After saturation with water, the compressive strength of the control specimens was determined according to EN 826 [29]. The specimens intended for testing were placed in a refrigeration chamber, where the temperature was maintained at minus (18 ± 2) °C. The holding time of one cycle after reaching a temperature of minus 16 °C was at least 3 h. At the end of the cooling period, the specimens were kept in a bath with water at a temperature of (18 ± 5) °C and kept there for at least 3 h. Figure 1 shows a view of the test specimens in the refrigeration chamber.

Figure 1. Test specimens in a bath in a refrigeration chamber before soaking.

After a set number of freezing and thawing cycles (25, 100, and 200 cycles), the compressive strengths of the specimens were determined according to EN 826 [29].

Measurement errors were calculated using the following procedure: first, the mean value of the measurements was calculated using the following Equation (11):

$$\overline{x} = \frac{1}{n}\sum^{n} x_i; \quad (11)$$

where x_i is the result of successive measurements and n is the number of measurements. Second, the mean square deviation S_x was calculated using the following Equation (12):

$$S_x = \sqrt{\frac{\sum_{i=1}^{n}(x_i - \overline{x})^2}{n-1}} \quad (12)$$

For the analysis of the macrostructure of the specimens, photographs of the specimens prepared for the compression tests were taken. The photographs were taken with the help of a Canon EOS 6D Mark II camera. To analyse the microstructure of the composites, specimens of 40 × 40 × 40 mm were prepared. The specimens were inspected on a Zeiss EVO-50 EP scanning electron microscope (SEM) (Carl Zeiss SMT GmbH, Oberkochen, Germany). The analysis was performed in variable pressure mode at an accelerating voltage of 20 keV and a working distance of 10 to 15 mm.

3. Research Results and Discussion

The density dependencies of the compressive strength and thermal conductivity of the prepared LWC specimens are presented in Figure 2. From the presented figure, it can be seen that the compressive strength and thermal conductivity of LWC also increase with increasing density. Two methods were used to analyse the experimental data: regression analysis and analysis of variance (ANOVA). Regression analysis made it possible to assess the dependence of the values of one quantity on another. The ANOVA statistical model allowed us to assess whether the means of different groups are the same. With a dashed line, Figure 2a shows the minimum possible compressive strength values and Figure 2b shows the maximum possible thermal conductivity values calculated with a 95% confidence interval.

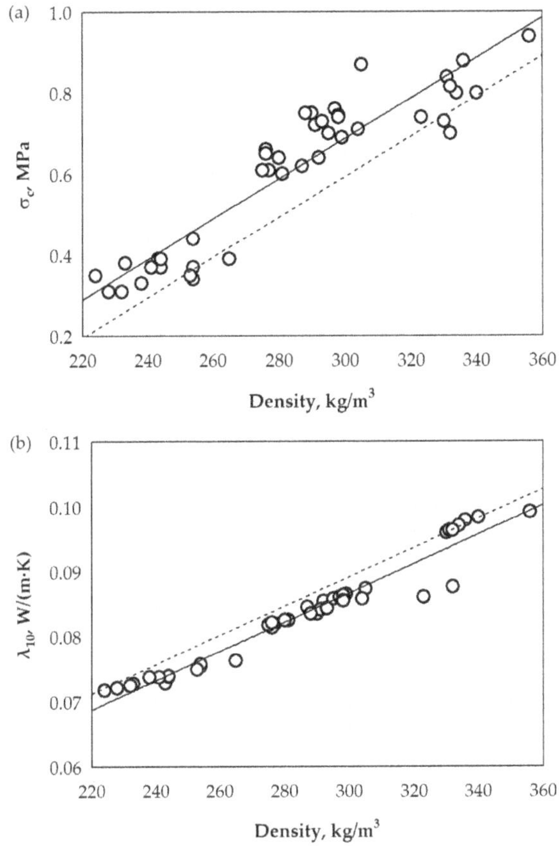

Figure 2. Dependencies of compressive strength (**a**) and thermal conductivity (**b**) on the density of the LWC; ○—experimental values.

The dependencies of compressive strength and thermal conductivity coefficient are described by the regression Equations (13) and (14), respectively:

$$\sigma_c = 0.00498 - 0.8045 \cdot \rho; \quad (13)$$

with a determination coefficient $r^2 = 0.858$ and standard deviation $S_r = 0.0725$ MPa, where ρ is LWC density, in kg/m^3, and

$$\lambda_{10} = 0.00225 + 0.01925 \cdot \rho; \quad (14)$$

with a determination coefficient $r^2 = 0.947$ and standard deviation $S_r = 0.00191$ W/(mK).

Basically, OPC affects the density. The higher the amount of OPC we add to the prepared mixture, the higher the LWC density we obtain. In this study, we were most interested not in the effect of OPC on density but in the effect of CEPW on both compressive strength and thermal conductivity. In Figure 3, an analysis of the effect of CEPW on the compressive strength and thermal conductivity of the specimens is presented. In the first case (Figure 3a), the compressive strength results were compared, when only EG beads were used and when part of the EG was replaced by CEPW aggregate. In the figure, and according to the ANOVA statistical model, the mean of the results is marked with a solid line and the two-sided confidence interval is marked with dashed lines. Analysis of the results shows that the difference in results is not significant.

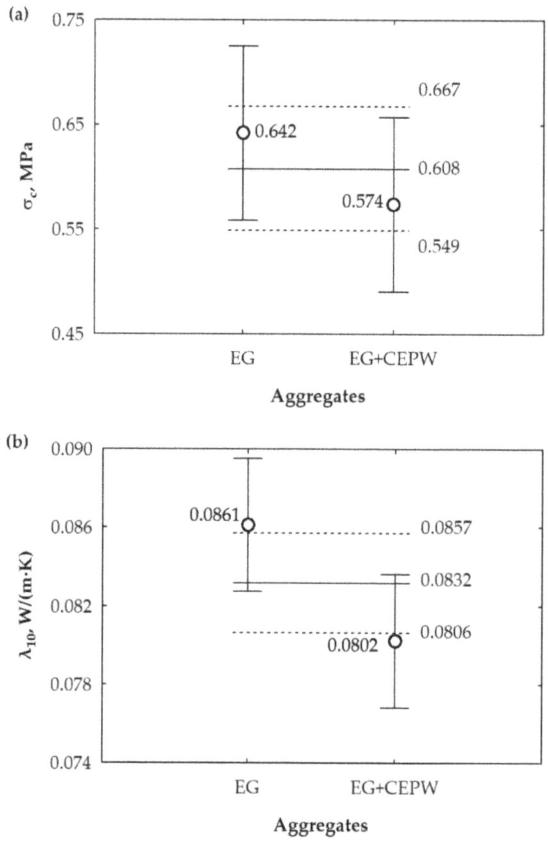

Figure 3. Analysis of the effect of CEPW on compressive strength (**a**) and thermal conductivity (**b**).

The average values obtained for compressive strength meet the 95% confidence limits, and this indicates that the replacement of part of the EG aggregate with the CEPW aggregate does not affect the compressive strength (or that the influence is insignificant). In the second case, the thermal conductivity results were evaluated (Figure 3b). The analysis of the results showed that the use of CEPW aggregate affects the reduction of thermal conductivity. The average values of the thermal conductivity measurements obtained are not within the range and this means that the results of the specimens made with only EG aggregate, and the specimens with partially replaced aggregate from CEPW, differ significantly. With CEPW, we achieved our goal of significantly reducing thermal conductivity with little or no reduction in compressive strength.

Figure 4 shows the dependence of the bending strength on the density. The bending strength values not only allow a more accurate assessment of the effects of density, binder content, and aggregate on the strength characteristics but, also, a comparison with the obtained LWC values obtained in the studies of other authors. Figure 4 shows that, as the

LWC density increases, an increase in flexural strength is also observed. To describe the dependence between density and bending strength, we obtained Equation (15):

$$\sigma_b = 1.326 \cdot \rho - 163.736; \quad (15)$$

with a coefficient of determination $r^2 = 0.869$ and standard deviation $S_r = 20.86$ MPa.

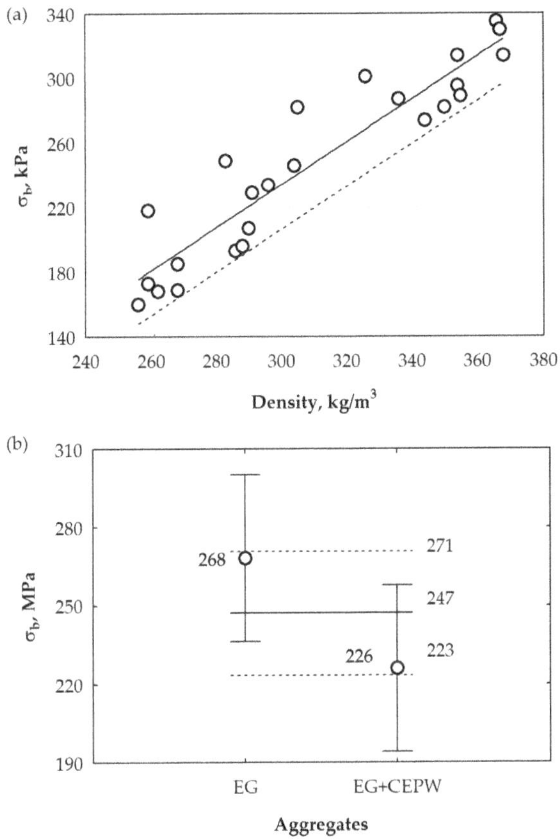

Figure 4. Dependence of the bending strength on density (**a**) and analysis of the effect of CEPW on the bending strength (**b**).

Then, an analysis of the effect of CEPW on the bending strength of the specimens was carried out. A comparison of the flexural strength results was only made when the EG aggregate was used and when part of the EG was replaced with the CEPW aggregate.

Analysis indicates that the differences in the results are not significant. The obtained average values of flexural strength meet the 95% confidence interval, and this indicates that the replacement of the EG aggregate with the CEPW aggregate does not influence flexural strength (or that the influence is insignificant). To summarise the results, it can be stated that replacing the EG aggregate with the CEPW aggregate does not affect the strength parameters.

The macrostructural studies of the LWC specimens are presented in Figure 5. The purpose of the study was to evaluate the effect of the CEPW aggregate on the macrostructure of the LWC and to explain its effects on the strength characteristics and thermal conductivity. Figure 5a shows a general view of the LWC when only EG aggregate is used, and Figure 5b shows when the 0–2 mm fraction of EG is replaced by CEPW. In the first case, we observe the formation of large voids in the LWC structure. The width of the spaces in the individual zones exceeds 5 mm. Through these voids, intensive heat transfer through the air can occur. Meanwhile, the EG aggregates themselves are well connected to each other by the touching surfaces, which allows us to ensure sufficient strength of the LWC.

Figure 5. View of the macrostructure of LWC: (**a**) with the EG aggregate; (**b**) with the EG and CEPW aggregate.

In the second case, we observe that almost all voids are filled with CEPW. Very small gaps were formed between the CEPW aggregates themselves, which are even smaller than the pores formed between the EG aggregates themselves. This good filling is due to the fact that the CEPW aggregate is very soft and fits well between the larger EG aggregates and so, in most cases, the CEPW aggregate appears slightly deformed. Deformation is likely to occur during compaction, so good void filling was observed, and there were no gaps between CEPW and EG. Since CEPW aggregates are soft and mechanically damaged during crushing, mixing, and compaction, their influence on the strength of LWC is unobservable or very insignificant. In addition, as already mentioned, EG aggregates are well connected to each other by their contact surfaces, and so the load primarily affects the EG aggregate, which has a significantly higher strength than CEPW.

The operational properties of LWC products, particularly the resistance to freezing and thawing, are closely related to the effects of water. In this work, we evaluated the effect of water on LWC by long-term immersion in water and moisture deformations.

In Figure 6a, the dependence of the long-term water absorption of the specimens on the density is presented. The figure shows that the long-term water absorption of

LWC decreases with increasing density. The test results are described by the regression Equation (16):

$$W_{lt} = 36.045 - 0.0576 \cdot \rho; \tag{16}$$

where the determination coefficient $r^2 = 0.812$ and standard deviation $S_r = 1.078\%$.

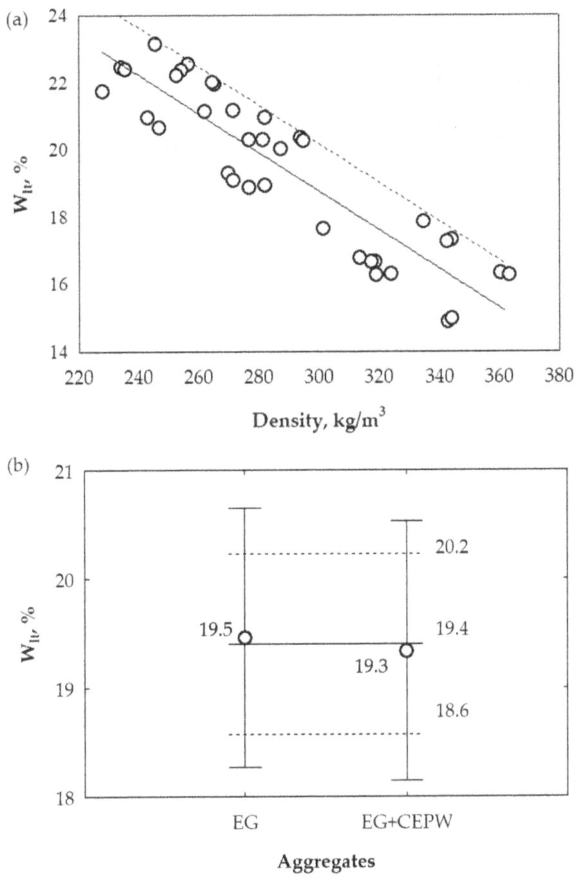

Figure 6. Dependence of the long-term water absorption of the specimens on the density (**a**) and analysis of the effect of CEPW on the water absorption (**b**).

The upper limit of the results is indicated by the dashed line in Figure 6, calculated with a 95% confidence interval.

The resulting dependence shows that, with a 1.5-times increase in LWC density, the long-term absorption decreases by approximately 1.41 times; by increasing the LWC density by 1 kg, the long-term absorption decreases by approximately 0.057%. Figure 6b shows the evaluation of the effect of the CEPW on the water absorption of the specimens. Analysis shows that there is almost no difference in the results when there is only EG aggregate or when part of the EG aggregate is replaced by CEPW aggregate. In both cases, the average values of the results are similar and close to the overall average of the results. This means that the water absorption is not affected by CEPW.

During operations, building materials become wet in the case of higher relative air humidity and dry out in dry periods. During the moisture sorption and desorption processes, most materials expand by absorbing moisture from the surrounding air and shrink by drying. Such a change in material dimensions can cause damage or collapse of structures. In Figure 7, the results of LWC expansion–shrinkage tests with light aggregate are presented.

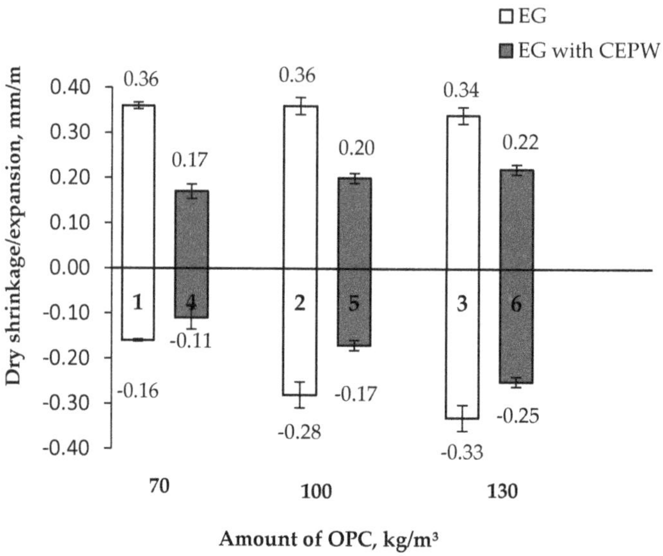

Figure 7. Dependence of LWC shrinkage/expansion on the amount of cement: 1–6—designation of the composite composition (see Table 6).

The conducted studies have shown that the use of CEPW aggregate reduces the shrinkage and expansion of the thermal insulation composite, especially at the lowest PC content (70 kg/m^3). Higher amounts of OPC are likely to result in higher internal stresses, which, in turn, lead to greater expansion or shrinkage. Additionally, it is likely that some of the moisture from the EG is taken up by the CEPW. As CEPW is sufficiently plastic, both drying shrinkage and moisture expansion have no (or negligible) influence on CEPW aggregate dimensions, and CEPW itself reduces internal stresses due to drying or wetting. Drying shrinkage is explained by the loss of moisture in the concrete, usually due to interaction with the surrounding environment, causing internal shrinkage stress, which can lead to cracking and loss of physical and mechanical properties of the concrete [37].

The average values of the structural indicators determined for LWC with EG aggregate and LWC specimens with mixed EG and CEPW aggregate are presented in Table 7.

Table 7. Average values of LWC structural indicators and calculated operational frost resistance.

Composition Codes	W_E, %	W_R, %	R, %	D, %	G_1, g/cm^2	G_2, g/cm^2	N, Cond. Units	g_2, g/cm^2·h	Operational Resistance to Cold, Conditional Cycles	
									F_{RE1}	F_{RE2}
Specimens with EG aggregate										
1	10.7	25.1	57.2	3.0	0.89	0.93	1.4	0.21	230	782
2	10.7	41.1	74.0	1.4	1.64	1.67	0.5	0.27	216	1270
3	10.8	43.6	75.2	1.3	2.26	1.89	0.7	0.25	220	1539
Specimens with EG and CEPW aggregate										
4	8.8	23.7	63.0	3.2	0.80	0.73	0.5	0.26	268	727
5	12.2	46.4	73.8	1.2	2.08	2.09	0.9	0.33	142	1027
6	8.8	45.0	80.5	1.2	1.28	1.65	1.5	0.21	162	1239

The calculated values of operational cold resistance of the specimens with EG aggregate F_{RE1} varied from 216 to 230 conditional cycles and F_{RE2} varied from 782 to 1539 conditional cycles. For specimens with EG and CEPW aggregate, the values of operational cold resistance F_{RE1} varied from 142 to 268 conditional cycles, and F_{RE2} varied from 727 to 1239 conditional cycles (see Figure 8).

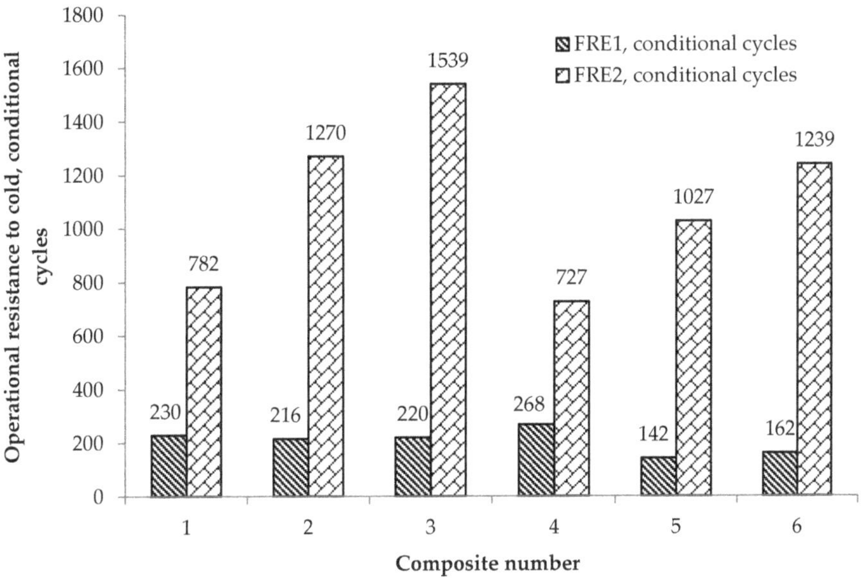

Figure 8. Predicted operational cold resistance values for the conditional cycles at the start of decomposition and the end of decomposition: 1–6—composition number of the LWC (see Table 6).

By analysing the operational frost resistance values, calculated according to the methodology adapted to predict the frost resistance of porous building materials in a one-sided method [34,35], we can see that LWC (taking into account the various aggregates used but not their composition) has a relatively high durability potential.

The minimum value of the operational cold resistance (F_{RE1}) of the composite specimens with the EG and CEPW aggregates was 142 conditional cycles and the highest F_{RE2} was 1539 conditional cycles. In our opinion, the higher durability potential was determined by the porous macrostructure formed by the specimens. The reserve of porous space, which defines the amount of reserve pores and capillaries through which water or plastic ice penetrates with great difficulty, but gradually fills up during cyclic freezing and thawing, varied from 57.2–80.5%. Effective porosity, which describes the effective pores of the specimen and the capillary potential, i.e., those pores and capillaries that are easily filled with water, varied from 8.8–12.2%, and the total open porosity index, which characterises the entire open communicating pore space of the composite specimen, varied from 23.7–46.4%.

The predicted service life of materials suitable for unprotected use under direct exposure to aggressive environmental conditions, characterised by high values of pore space reserve (54–71%), but low values of effective porosity (2–3%) and total open porosity (5–9%), varies from 722 to 1263 conditional cycles [38]. It is also stated that the values of the beginning and end of the decomposition of individual compositions are close. Therefore, when evaluating the durability potential of LWC specimens containing porous aggregates, despite the high values of pore space reserve and the high predicted in-service cold resistance based on the end of degradation (but considering the higher values of effective porosity, total open porosity, and the wide range of start and end values of the individual composition interval), it can be said that such LWC (1–6 mixture compositions) can be used in moderately aggressive environmental conditions, i.e., in structures protected from direct environmental impacts.

The direct cold resistance of the thermal insulation LWC specimens was evaluated according to changes in their compressive strength, comparing the compressive strength of the control specimens and specimens that were frozen and thawed for 25, 100, and 200 cycles by volume freezing and thawing according to the LST 1428-17 [36] methodology.

The results of compressive strength determined from the control specimens and after 25, 100, and 200 freezing and thawing cycles are presented in Figure 9; the calculated changes in compressive strength are presented in Table 8.

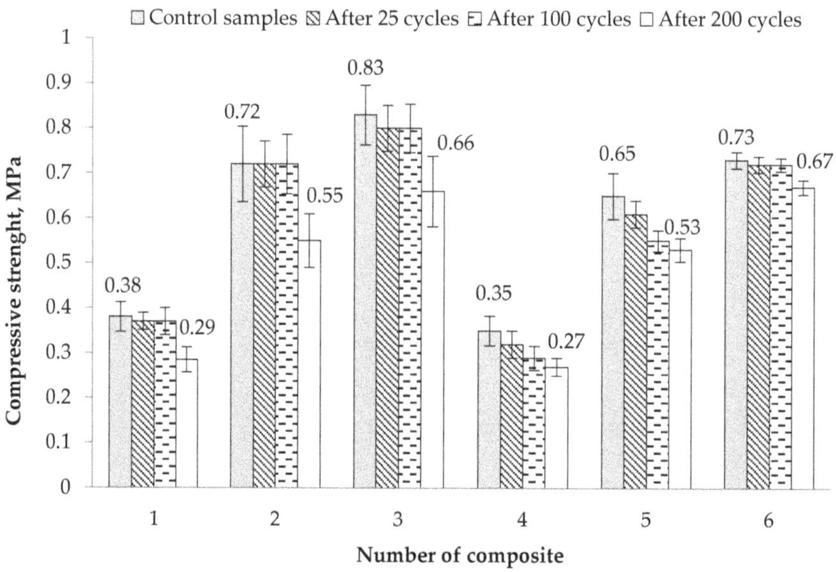

Figure 9. Compressive strength of the control specimens and after 25, 100, and 200 freeze–thaw cycles.

Table 8. Changes in compressive strength after 25, 100, and 200 freeze–thaw cycles.

Composition Codes	Change in Compressive Strength Δ*f*, %		
	After 25 Freeze–Thaw Cycles	After 100 Freeze–Thaw Cycles	After 200 Freeze–Thaw Cycles
Specimens with EG aggregate			
1	2.3	2.3	25.0
2	0	0	23.3
3	3.6	3.6	20.5
Specimens with EG and CEPW aggregates			
4	9.1	18.2	24.2
5	6.0	14.9	17.9
6	2.0	2.0	8.2

Evaluating the obtained results, Figure 9 and Table 8 show that the decrease in the compressive strength of the specimens with EG aggregate, after 25 and 100 cycles of freezing and thawing, is uniform and small, with the values of Δf varying from 0.0 to 3.6%. Here, the decrease in compressive strength of the samples after 200 cycles of freezing and thawing is large and the values of Δf vary from 20.5–25.0%.

The reduction in compressive strength of the specimens with EG and CEPW aggregate after 25 freeze–thaw cycles was greater, the maximum value of Δf being 9.1%. After 100 freeze–thaw cycles, the maximum value of Δf was 18.2% and after 200 freeze–thaw cycles and defrosting cycles, the maximum value of Δf reached 24.2%. The compressive strength of composition 6 of the specimens with the highest binder content is increased.

To summarise the research results obtained, we can say that: although LWC structures are exposed to changes in humidity and freezing and thawing, using LWC structures which are effectively protected from the direct effects of climatic conditions significantly increases their durability; they can be considered frost resistant in moderately aggressive environmental conditions.

In Figure 10a, we can see the contact zone between the bead and the binder before freezing. In this zone, the formation of a solid mass is observed. There are no cracks or other types of damage, either in the binder or in the EG bead. In Figure 10b, the view of the specimen after freezing is shown. Smaller or larger cracks are observed throughout the volume of the specimen. The largest cracks are observed in the pore walls of the EG bead, whereas individual small cracks are observed in the contact zone between the bead and the binder and in the binder layer itself. Since the pore walls of the EG beads are thin and weak, a large number of cracks form due to the freezing of water. Meanwhile, the small cracks that appeared in the contact zones between the bead and the binder and in the walls of the binder could have been caused both by the formation of small ice crystals and by the large stresses created in the EG aggregate during freezing.

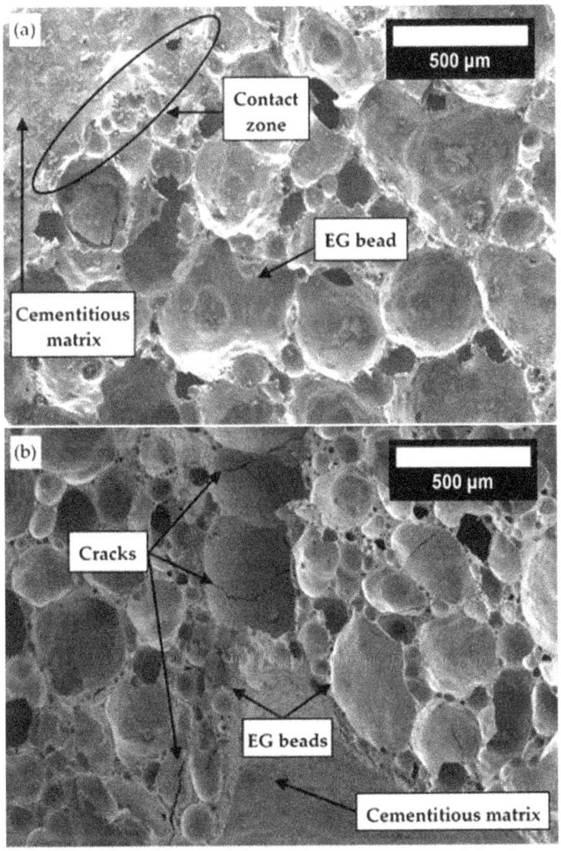

Figure 10. View of the LWC with EG aggregate: (**a**) before freezing test; (**b**) after 200 freeze–thaw cycles.

In Figure 11, a comparison of the microstructure of LWC with EG beads and CEPW aggregate, before and after freezing tests, is shown.

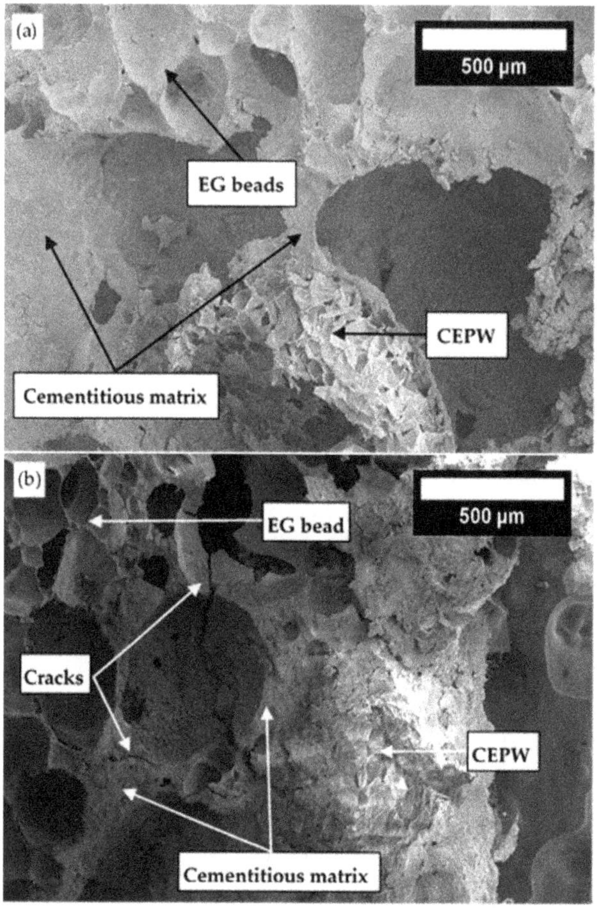

Figure 11. View of the LWC with mixed aggregate: (**a**) before freezing; (**b**) after 200 freeze–thaw cycles.

Figure 11a shows a view of the specimen before the freezing tests. Well-formed contact zones between EG, CEPW, and the cementitious matrix are observed. Several very small cracks are observed in the pore walls of EG beads, which may have appeared during the compaction of the mixture. There are no cracks in other areas, i.e., the cementitious matrix or CEPW. Figure 11b shows the view of the specimen after the freezing test. Some lager cracks are observed in the EG bead and cementitious matrix. These cracks terminate when they make contact with the CEPW. This means that CEPW has the potential to reduce cracking in LWC products. Since CEPW is sufficiently plastic, when water freezes in it, its walls do not crack but can move toward free space or expand.

The comparison of LWC is complicated by the different compositions of the mixtures, and the amount of binder, additives, fillers, densities, production technologies, etc. Density mostly reflects the main performance characteristics of LWC. In terms of density, the LWC specimens obtained are very close to modified foam concrete products or specimens prepared with polystyrene foam beads [17,23,39]. Genzel et al. [23] indicated that after using blown perlite, glass sand, and foaming agent for the production of LWC, when 1 m^3 of concrete is composed of 100 kg of foam, the compressive strength obtained is close to

1 MPa, the flexural strength ranges from 0.24–0.63 MPa, the compressive strength (after freezing) decreases to 36.2% after 25 cycles, and the bending strength decreases to 12.7%.

The authors obtained a minimum density of 594 kg/m^3, so the thermal conductivity is much higher than in our research (0.17 W/(mK)). It should be noted that the authors obtained a very high absorption in water, which reached 54.5%, and, after almost doubling the density, they reduced the absorption to 21.6%. Additionally, the amount of foam was larger. When the authors used only 50 kg/m^3 foam, the absorption ranged from 25.1–15.3%. In another work [39], at a foam concrete density of 200–300 kg/m^3, the thermal conductivity varied from 0.06–0.08 W/(mK), and the compressive stress from 0.8–1.1 MPa. Souza et al. [17] used EPS beads to replace part of the sand in the preparation of LWC. These authors used 1.54, 1.60, and 1.65 kg/m^3 polystyrene foam beads, while, in our work, 5 kg/m^3 was used. The authors obtained higher densities of 453, 482, and 510 kg/m^3, respectively. At this density, the compressive strength after 28 days of curing was 1.0–1.6 MPa and the water absorption was 60–85%.

In summary, we can say that in the work we managed to create LWC with good operational characteristics, which is suitable for operations under moderately aggressive environmental conditions.

4. Conclusions

In this study, investigation of performance characteristics of LWC with porous aggregates such as EG and CEPW was performed. During the study, LWC density, thermal conductivity, compressive strength, bending strength, water absorption, deformations, composite structure, and freezing–thawing resistance were analysed and the following results were obtained:

The modification of LWC with CEPW aggregate allows improvement of the thermal characteristics of the material without deteriorating the compressive and bending strength values. Incorporation of CEPW leads to changes in the macrostructure and results in a decrease in the number of large voids in the LWC, leading to an overall decrease in thermal conductivity from 0.0977 to 0.0720 W/(mK).

It was determined that, with a 1.5-times increase in LWC density, the long-term absorption decreases by approximately 1.41 times. The evaluation of the effect of the CEPW on the water absorption of the specimens shows that only with EG aggregate and when part of the EG aggregate is replaced by CEPW aggregate are the average values of the results similar to the overall average of the results.

The amount of LWC deformations, due to increased or decreased humidity, can be reduced by using CEPW aggregate. The lowest shrinkage value of 0.11 mm/m and an expansion value of 0.17 mm/m are observed when the LWC structure is modified with CEPW aggregate and 70 kg/m^3 of OPC is used.

The durability of LWC products increases, not only by increasing the amount of cement in the material, but also by modifying the structure with CEPW aggregate. The degradation of the compressive strength value of the specimens with EG and CEPW aggregate after 200 freeze–thaw cycles with the highest binder content of 130 kg/m^3 was about 3 times lower than of the specimens without CEPW aggregate. Microstructural studies confirmed that the CEPW contained in the LWC is not damaged during freezing and does not allow the formation of continuous cracks throughout the volume of the LWC.

In order to produce an effective thermal insulation material from LWC with sufficient strength characteristics, the influence of the CEPW size on strength characteristics and the influence of the ratio of EG and CEPW should be analysed in more detail in the future, in order to further reduce thermal conductivity, improve macrostructure, and increase durability.

Author Contributions: Conceptualisation, S.V. (Sigitas Vėjelis) and J.Š.-J.; methodology, S.V. (Sigitas Vėjelis), J.Š.-J. and V.K.; software, J.Š.-J. and S.V. (Saulius Vaitkus); validation, J.Š.-J. and V.K.; formal analysis, S.V. (Sigitas Vėjelis) and S.V. (Saulius Vaitkus); investigation, S.V. (Sigitas Vėjelis), A.K., V.K., S.V. (Saulius Vaitkus) and J.Š.-J.; resources, J.Š.-J. and S.V. (Sigitas Vėjelis); data curation, S.V. (Sigitas Vėjelis); writing—S.V. (Sigitas Vėjelis) and J.Š.-J.; writing—review and editing, S.V. (Sigitas Vėjelis) and J.Š.-J. and A.K.; visualisation, A.K.; supervision, S.V. (Sigitas Vėjelis) and J.Š.-J. All authors have read and agreed to the published version of the manuscript.

Funding: This research received no external funding.

Institutional Review Board Statement: Not applicable.

Informed Consent Statement: Not applicable.

Data Availability Statement: Not applicable.

Conflicts of Interest: The authors declare no conflict of interest. The funders had no role in the design of the study; in the collection, analyses, or interpretation of data; in the writing of the manuscript; or in the decision to publish the results.

References

1. Amran, M.; Onaizi, A.M.; Fediuk, R.; Danish, A.; Vatin, N.I.; Murali, G.; Abdelgader, H.S.; Mosaberpanah, M.A.; Cecchin, D.; Azevedo, A. An ultra-lightweight cellular concrete for geotechnical applications—A review. *Case Stud. Constr. Mater.* **2022**, *16*, e01096. [CrossRef]
2. Haller, T.; Beuntner, N.; Gutsch, H.; Thienel, K.C. Challenges on pumping infra-lightweight concrete based on highly porous aggregates. *J. Build. Eng.* **2023**, *65*, 105761. [CrossRef]
3. Shtrepi, L.; Astolfi, A.; Badino, E.; Volpatti, G.; Zampini, D. More Than Just Concrete: Acoustically Efficient Porous Concrete with Different Aggregate Shape and Gradation. *Appl. Sci.* **2021**, *11*, 4835. [CrossRef]
4. Yang, K.-H.; Kim, H.-Y.; Lee, H.-J. Mechanical Properties of Lightweight Aggregate Concrete Reinforced with Various Steel Fibers. *Int. J. Concr. Struct. Mater.* **2022**, *16*, 48. [CrossRef]
5. Lombardi, R.; Jünemann, R.; Lopez, M. Experimental assessment of the behavior of expanded glass lightweight reinforced concrete walls. *J. Build. Eng.* **2022**, *49*, 104043. [CrossRef]
6. Limbachiya, M.; Meddah, M.S.; Fotiadou, S. Performance of granulated foam glass concrete. *Constr. Build. Mater.* **2012**, *28*, 759–768. [CrossRef]
7. Macedo, P.M.B. Strong, Lower Density Composite Concrete Building Material with Foam Glass Aggregate. U.S. Patent 7695560, 13 April 2010.
8. Bumanis, G.; Bajare, D.; Korjakins, A. Mechanical and Thermal Properties of Lightweight Concrete Made from Expanded Glass. *J. Sustain. Archit. Civ.* **2013**, *2*, 26–32. [CrossRef]
9. Nemes, R. Lightweight Concrete Made with Expanded Glass Aggregate. Ph.D. Thesis, University of Technology and Economics, Budapest, Hungary, 2006; p. 13.
10. Vereshagin, V.I.; Sokolova, S.N. Granulated foam glass-ceramic material from zeolitic rocks. *Constr. Build. Mater.* **2008**, *22*, 999–1003. [CrossRef]
11. Hurley, J.A. *UK Market Survey for Foam Glass. Glass: Research and Development Final Report*; WRAP Report; The Waste and Resources Action Programme: Banbury, UK, 2003.
12. Ducman, V.; Mladenovič, A.; Šuput, J.S. Lightweight aggregate based on waste glass and its alkali–silica reactivity. *Cem. Concr. Res.* **2002**, *32*, 223–226. [CrossRef]
13. Ozguven, A.; Gunduz, L. Examination of effective parameters for the production of expanded clay aggregate. *Cem. Concr. Compos.* **2012**, *34*, 781–787. [CrossRef]
14. Jasaitiene, J.; Ivanauskas, E.; Dauksys, M. Investigation of lightweight concrete with porous aggregates. In Proceedings of the 2nd International Conference Advanced Construction, Kaunas, Lithuania, 11–12 November 2010.
15. Khatib, J.M.; Shariff, S.; Negim, E.M. Effect of Incorporating Foamed Glass on the Flexural Behaviour of Reinforced Concrete Beams. *World. Appl. Sci. J.* **2012**, *19*, 47–51. [CrossRef]
16. Kligys, M. Production Technology and Properties of Composite Material Made out of Porous Cement Paste and Crushed Expanded Polystyrene. Doctoral Dissertation, Vilnius Gediminas Technical University, Vilnius, Lithuania, 2009; p. 138.
17. Souza, T.B.; Lima, V.M.E.; Araújo, F.W.C.; Miranda, L.F.R.; Neto, A.A.M. Alkali-activated slag cellular concrete with expanded polystyrene (EPS)—Physical, mechanical, and mineralogical properties. *J. Build. Eng.* **2021**, *44*, 103387. [CrossRef]
18. Šeputytė-Jucikė, J.; Sinica, M. The effect of expanded glass and polystyrene waste on the properties of lightweight aggregate concrete. *Eng. Struct.* **2016**, *8*, 31–40. [CrossRef]
19. Rustamov, S.; Kim, S.W.; Kwon, M.; Kim, J. Mechanical behavior of fiber-reinforced lightweight concrete subjected to repeated freezing and thawing. *Constr. Build. Mater.* **2021**, *273*, 121710. [CrossRef]

20. Yuan, J.; Li, D.; Li, W.; Yang, H.; Zhang, W.; Wang, L.; Wang, J.; Xiong, Z. Study on frost resistance of EPS concrete based on EPS beads wrapping modification. *Constr. Build. Mater.* **2022**, *345*, 128400. [CrossRef]
21. Bayraktar, O.Y.; Soylemez, H.; Kaplan, G.; Benli, A.; Gencel, O.; Turkoglu, M. Effect of cement dosage and waste tire rubber on the mechanical, transport and abrasion characteristics of foam concretes subjected to H_2SO_4 and freeze–thaw. *Constr. Build. Mater.* **2021**, *302*, 124229. [CrossRef]
22. Wang, H.-T.; Zhou, Y.; Shen, J.-Y. Experimental study of dynamic biaxial compressive properties of full grade aggregate concrete after freeze thaw cycles. *Cold Reg. Sci. Technol.* **2023**, *205*, 103710. [CrossRef]
23. Gencel, O.; Bayraktar, O.Y.; Kaplan, G.; Arslan, O.; Nodehi, M.; Benli, A.; Aliakbar Gholampour, A.; Ozbakkaloglu, T. Lightweight foam concrete containing expanded perlite and glass sand: Physico-mechanical, durability, and insulation properties. *Constr. Build. Mater.* **2022**, *320*, 126187. [CrossRef]
24. Pospíchal, O.; Kucharczyková, B.; Misák, P.; Vymazal, T. Freeze-thaw resistance of concrete with porous aggregate. *Procedia Eng.* **2010**, *2*, 521–529. [CrossRef]
25. Buzhevich, G.A. *Lightweight Concretes on Porous Aggregates*; Publishing House of Literature on Construction: Moscow, Russia, 1970; 272p.
26. Komar, A.G. *Building Materials and Products*; Higher School: Moscow, Russia, 1988; p. 527.
27. Szwabowski, J.; Lazniewska-Piekarczyk, B. Air-entrainment problem in self-compacting concrete. *J. Civ. Eng. Manag.* **2009**, *15*, 137–147. [CrossRef]
28. *EN 197-1*; Cement—Part 1: Composition, Specifications and Conformity Criteria for Common Cements. European Committee for Standardization: Brussels, Belgium, 2011.
29. *EN 826*; Thermal Insulating Products for Building Applications—Determination of Compression Behaviour. European Committee for Standardization: Brussels, Belgium, 2013.
30. *EN 12664*; Thermal Performance of Building Materials and Products—Determination of Thermal Resistance by Means of Guarded Hot Plate and Heat Flow Meter Methods—Dry and Moist Products of Medium and Low Thermal Resistance. European Committee for Standardization: Brussels, Belgium, 2001.
31. *EN 12089*; Thermal Insulating Products for Building Applications—Determination of Bending Behaviour. European Committee for Standardization: Brussels, Belgium, 2013.
32. *EN ISO 16535*; Thermal Insulating Products for Building Applications—Determination of Long-Term Water Absorption by Immersion. European Committee for Standardization: Brussels, Belgium, 2019.
33. *EN 772-14*; Methods of Test for Masonry Units—Part 14: Determination of Moisture Movement of Aggregate Concrete and Manufactured Stone Masonry Units. European Committee for Standardization: Brussels, Belgium, 2001.
34. Mačiulaitis, R. *Frost Resistance and Durability of Facade Bricks [Frostwiderstand und Dauerhaftigkeit keramischer Fassadenerzeugnisse]*; Technika: Vilnius, Lithuania, 1996; 132p.
35. Mačiulaitis, R.; Malaiškienė, J.; Kičaitė, A. The regulation of physical and mechanical parameters of ceramic bricks depending on the drying regime. *J. Civ. Eng. Manag.* **2008**, *14*, 263–268. [CrossRef]
36. *LST 1428-17*; Concrete—Test Methods—Part 17: Determination of Frost Resistance to Volumetric Freezing and Thawing. Lithuanian Department of Standardization: Vilnius, Lithuania, 2016.
37. Nodehi, M.; Mohamad Taghvaee, V. Sustainable concrete for circular economy: A review on use of waste glass. *Glass Struct. Eng.* **2022**, *7*, 3–22. [CrossRef]
38. Kizinievič, V. The Effect of Technological Factors on the Cold Resistance of Ceramic Masonry Products. Doctoral Dissertation, Vilnius Gediminas Technical University, Vilnius, Lithuania, 2008; p. 139.
39. Huang, Z.; Zhang, T.; Wen, Z. Proportioning and characterization of Portland cement-based ultra-lightweight foam concretes. *Constr. Build. Mater.* **2015**, *79*, 390–396. [CrossRef]

applied sciences

MDPI

Systematic Review

Structural Performance of Fiber-Reinforced Cementitious Composite Members Reinforced with Fiber-Reinforced Polymer Bars: A Systematic Review

Helen Negash Shiferaw [1,]* and Toshiyuki Kanakubo [2]

[1] Graduate School of Science and Technology, University of Tsukuba, Tsukuba 305-8573, Japan
[2] Department of Engineering Mechanics and Energy, University of Tsukuba, Tsukuba 305-8573, Japan; kanakubo@kz.tsukuba.ac.jp
* Correspondence: s2436044@u.tsukuba.ac.jp

Abstract

The integration of fiber-reinforced cementitious composites (FRCCs) with fiber-reinforced polymer (FRP) bars represents a significant advancement in concrete technology, aimed at enhancing the structural performance of reinforced concrete elements. The incorporation of fibers into cementitious composites markedly improves their mechanical properties, including tensile strength, ductility, compressive strength, and flexural strength, by effectively bridging cracks and optimizing load distribution. Furthermore, FRP bars extend these properties with their high tensile strength, lightweight characteristics, and exceptional corrosion resistance, rendering them ideal for applications in aggressive environments. In recent years, there has been a notable increase in interest from the engineering research community regarding this topic, primarily to solve the issues of aging and deteriorating infrastructure. Researchers have conducted extensive investigations into the structural performance of FRCC and FRP composite systems. This paper presents a systematic literature review that surveys experimental and analytical studies, findings, and emerging trends in this field. A comprehensive search on the Web of Science identified 40 relevant research articles through a rigorous selection process. Key factors of structural performance, such as bond behavior, flexural behavior, ductility performance assessments, shear and torsional performance, and durability evaluations, have been documented. This review aims to provide an in-depth understanding of the structural performance of these innovative composite materials, paving the way for future research and development in construction materials technology.

Keywords: fiber-reinforced cementitious composite (FRCC); engineered cementitious composite (ECC); fiber-reinforced concrete (FRC); fiber-reinforced polymer (FRP) bar; structural performance

1. Introduction

Conventional reinforced concrete (RC) has historically served as a cornerstone in construction, providing essential strength and stability. Nevertheless, its intrinsic limitations have prompted the investigation of alternative materials that can more effectively address the requirements of contemporary engineering practices. A significant issue associated with traditional RC is the corrosion of the steel reinforcement. This problem can result in serious consequences, such as concrete spalling, reduction in load-bearing capacity, and, ultimately, structural failure [1–4]. Low tensile strength, susceptibility to cracking, and the brittle behavior of concrete are the other challenges of conventional RC members [5,6]. In response to these issues, researchers and engineers introduced an alternative material, fiber-reinforced cementitious composite (FRCC) reinforced with fiber-reinforced polymer (FRP) bars.

FRCC is an advanced construction material that mixes regular cement-based materials, like concrete, with various types of fibers to enhance their mechanical properties and durability. Some of the fibers used in FRCC are steel fiber (SF), synthetic fibers such as polyvinyl alcohol (PVA), polyethylene (PE), polypropylene (PP), aramid fiber (AF), and basalt fiber (BF) [7,8]. FRCCs such as engineered cementitious composites (ECCs), strain-hardening cementitious composites (SHCCs), and ductile fiber-reinforced cementitious composites (DFRCCs) have been broadly studied for their ability to undergo strain-hardening, deflection-hardening, and multiple cracking property. These properties make them well-suited for structural applications that demand improved durability and crack resistance [6,9]. As shown in Figure 1, FRCC and concrete display distinct differences in their tensile stress–strain responses. Concrete typically exhibits a brittle failure upon reaching its tensile strength. Fiber-reinforced concrete (FRC) demonstrates strain-softening behavior until it reaches its cracking resistance. In contrast, the engineered cementitious composite (ECC) displays strain-hardening behavior up to the rupture point in the post-cracking phase [10]. FRCC made with various raw materials or different material ratios exhibits different mechanical properties under compression and tension. For instance, to enhance the mix, mineral admixtures such as silica fume (SF) [11], fly ash (FA) [12], and ground granulated blast furnace slag (GBFS) [13] can be used as mineral additives or as a partial replacement for cement.

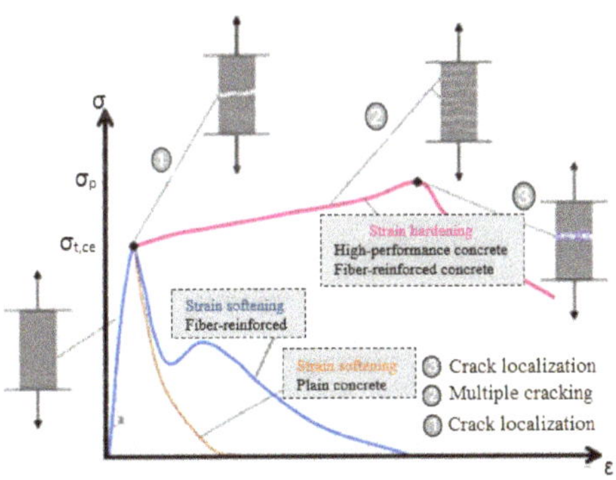

Figure 1. Behavior of fiber-reinforced cementitious composites under direct tension [10].

FRP bars were introduced in the construction industry in the 1980s as an alternative material to traditional steel reinforcement. Numerous studies [1,2,4,14] demonstrated the benefits of using FRP bars over conventional steel bars, particularly in their durability and reduced maintenance requirements. These bars are made from a polymer matrix reinforced with various fibers, including carbon (CFRP), glass (GFRP), aramid (AFRP), and basalt (BFRP). FRP bars combine strength, lightweight characteristics, and resistance to environmental degradation. These properties make them particularly advantageous in applications where high tensile strength, corrosion resistance, and reduced weight are critical. The mechanical performance of FRP bars, including their tensile strength, modulus of elasticity, and fatigue resistance, allows for innovative design solutions in various structural applications. As the demand for durable and efficient materials continues to grow, understanding the mechanical properties of FRP bars is essential for engineers and designers seeking to enhance the longevity and performance of modern structures. Each type of fiber offers properties that can be tailored to meet specific structural requirements.

The structural performance of FRCC members reinforced with FRP bars is significantly influenced by the material properties of both the composite matrix and the reinforcement bar. Due to the high tensile strength capacity of the FRP bars and the post-cracking behavior of FRCC, these members exhibit greater load-carrying capacity compared to conventional RC members. One of the main challenges of using FRP bars is their lower modulus of elasticity, resulting in higher deflection. However, the FRCC's ability to control crack width and delay crack propagation mitigates excessive deformation. The combination of FRP bars and FRCC helps maintain structural stiffness by reducing crack-induced flexibility [15]. Despite the numerous benefits of using a combination of FRP bars and FRCC materials, their application is constrained by several technical and regulatory challenges. The building codes and design standards for FRP materials are still in development [2]. Researchers are extensively examining the combination of FRCC and FRP bars to enhance structural performance, durability, and sustainability while developing comprehensive design guidelines and innovative applications.

This paper presents a systematic review of existing articles on the structural performance of FRP-reinforced FRCC members, including their bond behavior, flexural behavior, ductility performance assessments, shear and torsional performance, and durability assessment. This review aims to provide a thorough insight into the structural performance of these advanced materials, paving the way for future research and development in the field.

2. Research Methodology

A systematic literature review is a research approach employed to collect, identify, and critically assess existing studies through a structured methodology. Various methods and techniques can be employed to conduct a systematic review [16]. This research follows the guidelines outlined by the Preferred Reporting Items for Systematic reviews and Meta-Analyses (PRISMA) guidelines and is registered on the Open Science Framework (OSF) registries. The registration link can be found here [17]. The PRISMA checklist is used to ensure transparency, completeness, and accuracy of reporting in this paper. The evaluation was conducted in five stages: (1) defining objectives; (2) choosing databases; (3) identifying keywords; (4) selecting relevant papers; and (5) extracting data.

2.1. Searching Staratagy

The "Web of Science" database was used to conduct the literature search due to its extensive coverage of peer-reviewed journals, reliable citation tracking, advanced search features, and the ability to access interdisciplinary research, which collectively facilitate a thorough and efficient review of existing literature [18]. The searching criteria were "topic", which includes "title", "abstract", and "keywords". The keywords used to formulate the filtering query were "fiber-reinforced polymer", "fiber-reinforced plastic", "FRP", "FRC", "FRCC", "ECC", and "SHCC". The search link can be found here [19]. Two hundred and three articles were found in the first search, then the number of articles was refined to only peer-reviewed articles by excluding conference proceedings and pre-print articles, which reduced the number of articles to 178. Further refining was carried out to select only those related to civil engineering, material science, construction, and building technology, which reduced the number of articles to 101. Finally, the title and abstract of the articles were manually selected to assess the relevant articles that focus only on structural performance, and 40 articles were selected. Figure 2 shows the flow of the article selection methodology using the PRISMA flow chart model.

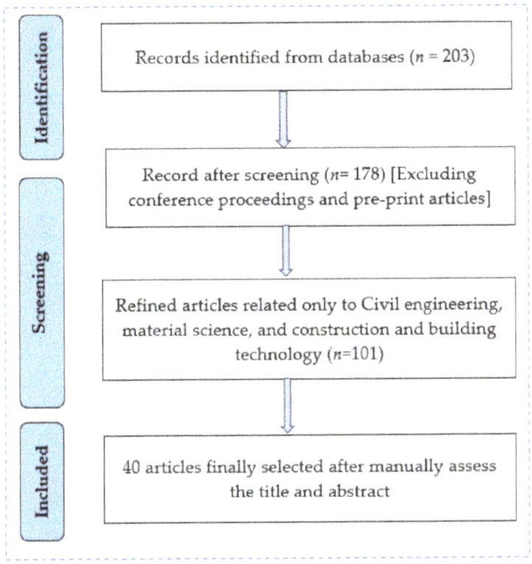

Figure 2. Article selection methodology.

2.2. Search Result

The study reviewed 40 articles published between 1999 and 2024. Recently, researchers have shown a growing interest in using fiber-reinforced polymer bars to strengthen fiber-reinforced cementitious composites. It is shown in Figure 3 that from 1999 to 2020, studies did not exceed an average of two articles per year. In 2021 and 2022, the number of articles increased to four and five per year. However, the most significant increase was in the past two years, 2023 and 2024, when there were six and nine articles published per year.

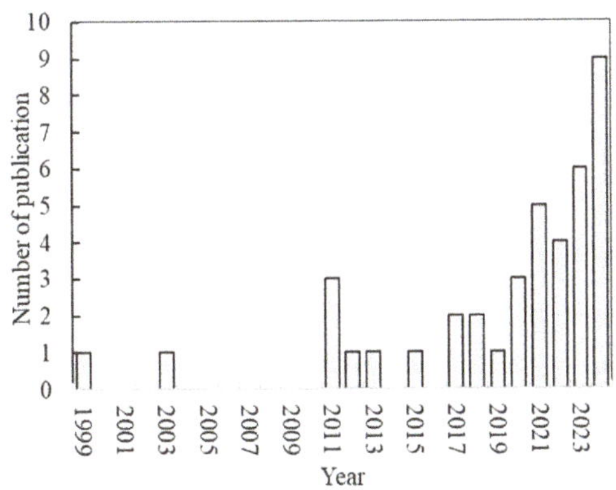

Figure 3. Time distribution of publications.

As shown in Figure 4, the selected publications are sourced from a range of academic journals. The journal "Construction and Building Materials" (nine articles) has the highest number of publications. Three journals, "Journal of Building Engineering", "Structures", and "Engineering Structures" have published five articles. The "Composite structures" journal published four articles, "Journal of Composites for Construction" published two articles, and the remaining nine journals have only one published article.

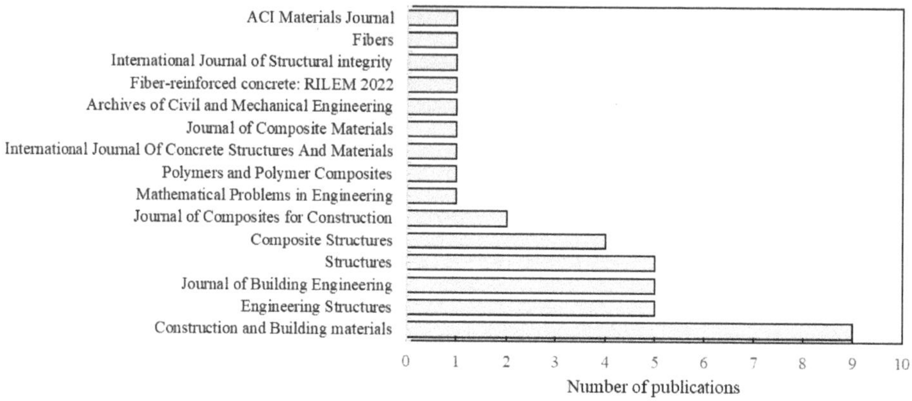

Figure 4. Sources of the selected publications.

3. Bond Behavior and Bond Strength

A comprehensive understanding of the bond characteristics between FRP bars and FRCC is crucial for the successful integration of this emerging technology within the construction industry. In contrast to the research focused on the bond behavior between FRCC and traditional steel reinforcement, there exists a notable lack of studies addressing the bond performance of FRCC in conjunction with FRP bars. Additionally, the different surface properties of FRP bars make their bond behavior much more complicated than steel bars. The mechanisms that lead to bond failure at the interface between FRP bars and FRCC are still under investigation, and important factors such as the type of bar, its diameter, and surface design have not been adequately studied. Furthermore, the absence of a reliable model for the bond strength between FRP bars and FRCC complicates the design and evaluation of FRCC-FRP bar systems, limiting the practical applications of both materials in engineering [20]. Several researchers conduct experimental, numerical, and analytical investigations to study these bond characteristics [20–29].

Li et al. [21] carried out a theoretical analysis of the debonding mechanism between FRP and ECC using a thick-walled cylinder model. The proposed model aimed to enhance the understanding of the bond behavior between the materials by examining the bond stress–slip relationship. The validity of the model is confirmed through comparison with an experimental pullout test. As shown in Figure 5, the bond stress (bond shear stress)–slip relationship was categorized into three distinct stages: an elastic stage, a softening stage, and a frictional stage. The solution to the second-order differential equation governing the bond behavior of FRP bars along the bond length was derived by incorporating the three-stage linear bond stress–slip relationship. The influence of the nonlocal parameter resulted in an increase in slip during the elastic stage and a decrease during the softening stage, while the slip remained unchanged in the frictional stage. The model's predictions for the pullout force aligned well with the experimental data. However, it was noted that the nonlocal parameters vary across different types of FRP bars.

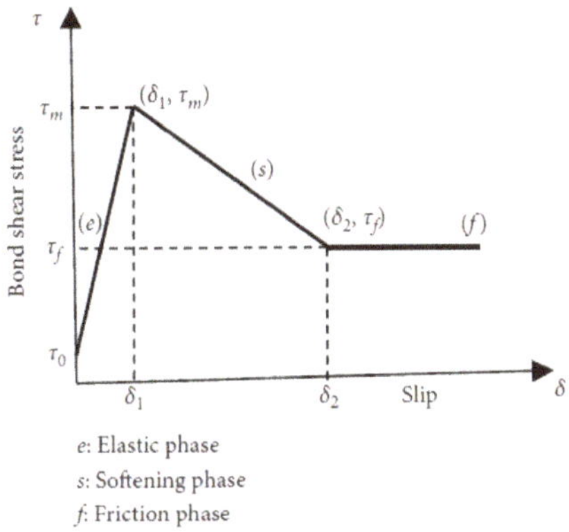

Figure 5. Bond stress–slip relationship [21].

3.1. *Influence of Bar Diameter, Modulus of Elasticity, and Surface of the Bar*

Several studies show that bond performance is highly influenced by the type of FRP bar, bar diameter, and the surface treatment and configurations.

Zhao et al. [20], Wang et al. [25] and Hossain et al. [23] reported that the average bond strength decreases when the diameter of the bar increases. During pullout testing, the bond stress is not uniformly distributed along the length of the bar; instead, it exhibits a gradual decrease from the loaded end to the free end. This stress distribution contributes to the lower bond strength observed in specimens with larger diameter bars. Additionally, the shear lag phenomenon further aggravates the reduction in bond strength as bar diameter increases. The differential movement between the core and the surface fibers under tensile loading leads to a non-uniform distribution of normal stress across the cross-section of the bar. As the diameter increases, this non-uniformity becomes more pronounced. Furthermore, the Poisson effect has also been identified as a contributing factor to the reduction in bond strength [26,30]. However, Zhao et al. reported that the finite element analysis conducted for the thick-walled cylinder demonstrated that the influence of the Poisson effect on bond strength is minimal [20]. Figure 6 shows the effect of bar diameter on bond strength. However, these general trends are not observed exclusively in combination with FRCC but have also been reported in the bond behavior between conventional concrete and FRP.

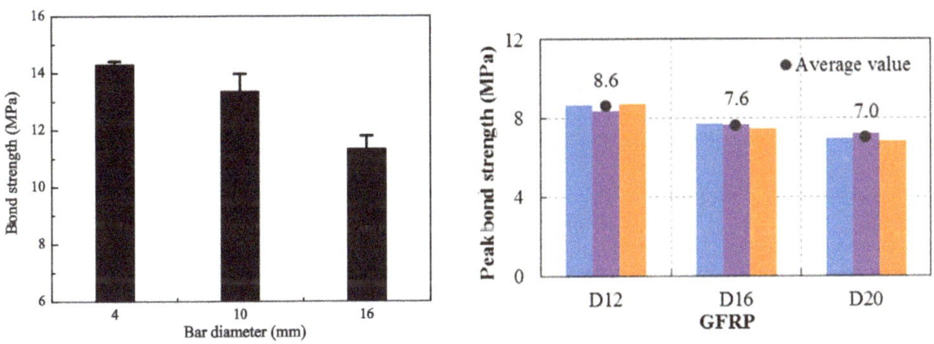

(a) (b)

Figure 6. Effect of FRP bar diameter on bond strength: (**a**) BFRP bar [25]; (**b**) GFRP bar in three different matrices [20].

Hossain et al. [23] performed a bond test using RILEM beam test method using two types of GFRP bars: low/standard modulus "LM" and high modulus "HM". The experiment results were compared against those obtained from established code-based design equations. It is reported that the bond strength of HM bars was lower than LM bars, which was primarily attributed to the premature detachment of sand coating from the rebar core. This issue is believed to be specific to the HM bar used for their study, prompting a recommendation for further investigations using commercially available HM bars. The bond strength measured from beam tests of both LM and HM bars embedded in ECC was found to exceed the prediction made by CSA S806-12 [31], CSA S6-06 [32], ACI440.1R-15 [33], and other existing predictions. Among these, ACI440.1R-15 provided a close value of bond strength prediction compared to the other equations, with a ratio of predicted to experimental values ranging from 1.18 to 1.49. Consequently, it is suggested that modifications to these codes are warranted to enhance the accuracy of bond strength predictions for ECC.

Figure 7 shows some examples of types of rebars and surface configurations. Zhao et al. [20] developed a simplified bond-strength model based on an experimental result of a direct pullout test that considered the FRP bars' surface geometry, such as the rib spacing and rib height. The model is validated using a proposed equation for bond strength. The maximum bond strength value from the proposed equations demonstrated a strong correlation with the experimental results. It is important to note that this model specifically pertains to ribbed reinforcing bars, where bond failure is predominantly governed by shear failure of the matrix. For reinforcing bars with different surface treatments, a more comprehensive investigation is required to accurately assess the calculation of maximum bond strength.

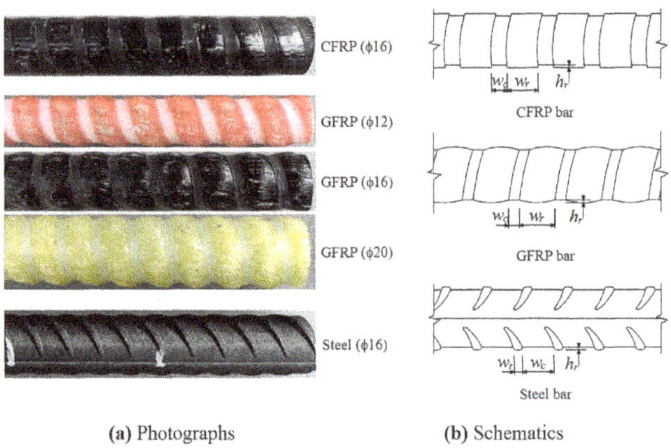

Figure 7. Examples of rebars: (a) photographs; (b) schematics [20].

Kim et al. [28] conducted an experimental investigation to study the bond–slip behavior of two types of GFRP bars, helically wrapped (HW-GFRP) and sand-coated (SC-GFRP), and a steel bar in PVA-ECC. From the experimental results, it is reported that the interfacial bond strength of HW-GFRP was much higher than that of SC-GFRP and similar to the steel bar. The overall energy absorption capacity of HW-GFRP bars exceeded that of conventional steel bars. In contrast, the SC-GFRP bars showed frictional pullout failure at the interface between the resin and the fiber reinforcement, along with reduced matrix splitting, irrespective of the properties of the matrix material.

3.2. Influence of Embedded Length

The bond strength reduced with an increase in embedded length. This reduction can be attributed to the calculation of bond strength based on the assumption of a uniform bond stress distribution. However, a more pronounced non-uniform distribution of bond stresses typically occurs with greater embedded lengths and the Poisson effect [23,25,30]. Wang et al. [25] reported that a greater load is needed to pull out the BFRP bar with increased embedment length. However, due to the Poisson effect, this extended embedment length results in a significant decrease in confinement and mechanical interlocking, subsequently reducing bond strength. Additionally, it has been reported that a linear relationship can effectively characterize the correlation between average bond strength and embedment length. Figure 8 illustrates the experimental findings regarding the influence of embedment length on the average bond strength of the BFRP bar.

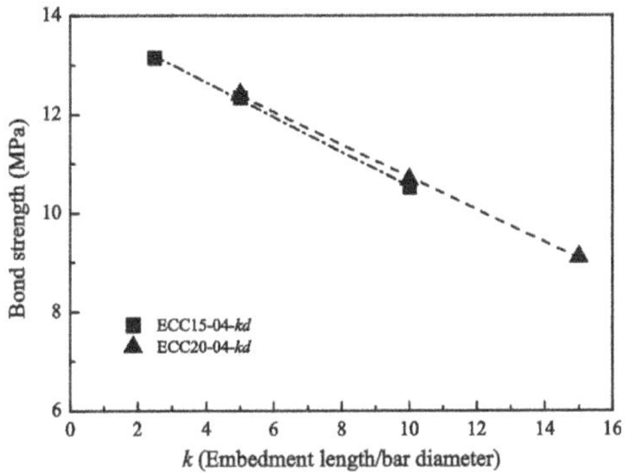

Figure 8. Effect of embedment length on average bond strength of BFRP bar [25].

Takasago et al. [22] studied the bond characteristics between braided AFRP and PVA FRCC and proposed a trilinear model for the bond stress–slip relationship, as shown in Figure 9. Pullout tests with varying bond lengths, fiber-volume fraction in FRCC, and cross-sectional dimensions were performed. The bond strength improved as both the amount of fiber and the size of the cross-section of FRCC increased, due to the fiber's bridging effect along the cracks. The proposed model was checked using numerical analysis for longer bond lengths, and the results showed that the average bond stress–slip relationship matched well with the experimental results.

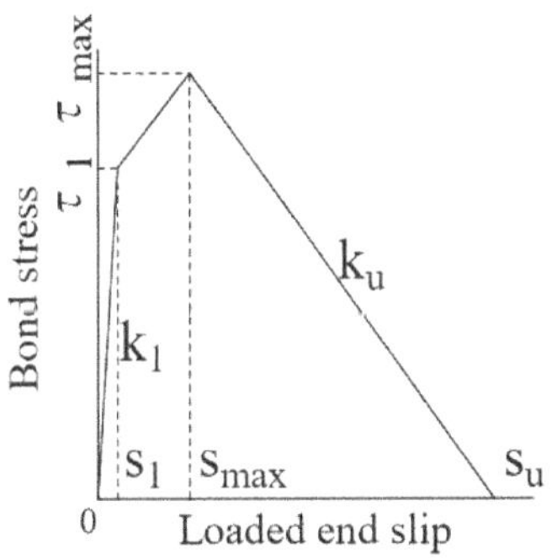

Figure 9. Trilinear model [22].

3.3. Influence of FRCC or ECC Characteristics

The FRCC or ECC characteristics are affected by many factors such as type of fiber, fiber-volume fraction, mix design, and curing age. The bond characteristics results of Takasago et al. [22] showed that the bond strength increased with the increase in fiber-volume fraction.

Wei et al. [27] evaluated the bond performance of GFRP bars embedded in normal-strength sea-water sea-sand ECC (SS-ECC) with PVA fibers and high-strength SS-ECC with PE fiber. A parallel test was also performed on GFRP bars in ECC made with fresh water (FR-ECC). In this study, it is reported that the influence of the saline content enhances the bold performance, and it refines the microstructures of ECC. This paper provided a more comprehensive understanding of the bond performance relationship with chemical adhesion, friction resistance, and mechanical interaction.

Table 1 provides an overview of the study design and findings reported by various researchers regarding the bond behavior between FRCC and FRP bars. The researchers employed different experimental setups to examine the bond strength. Figure 10 schematically shows the different setups used in the studies. From Table 1, it can be concluded that the bond performance of the FRP with FRCC is affected by variables such as FRP type, diameter of the bar, type of surface of the bar, modulus of elasticity of the bar, fiber type in the FRCC, fiber-volume fraction, bond length, grade of the matrix, environment, and so on.

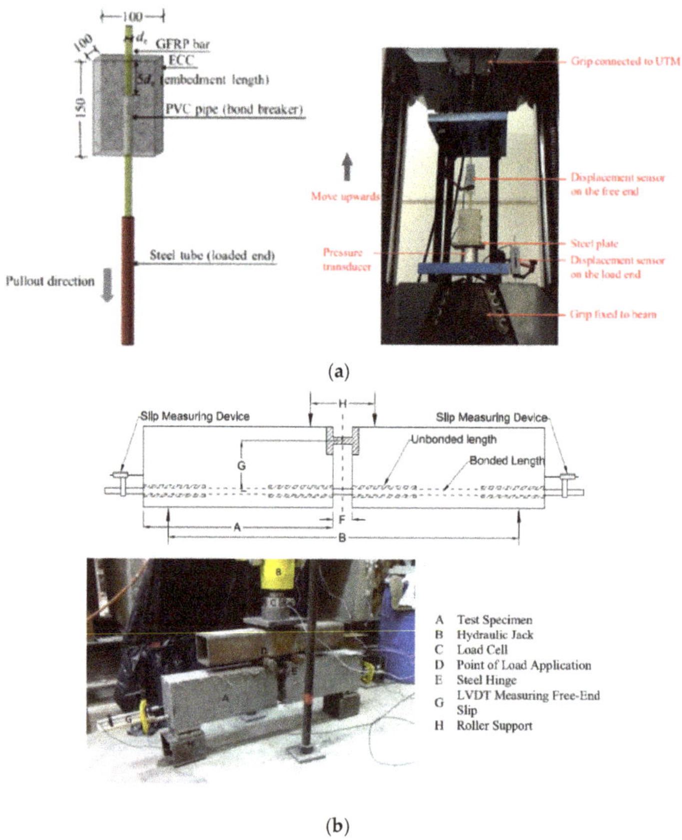

Figure 10. Test specimen geometry and test setup: (**a**) direct pullout method [26]; (**b**) RILEM beam method [23].

Table 1. The research plan and result summary of bond behavior.

Ref.	FRP Type	Fiber Type in FRCC	Study Type	Type of Test	Bond Strength (MPa)	Variables
Zhao et al. [20]	CFRP GFRP	PE	Experimental	Direct pull out	4–15	• Bar diameter
Li et al. [21]	GFRP CFRP		Analytical			• FRP type
Takasago et al. [22]	AFRP	PVA	Experimental and analytical	Direct pull out	6–9	• Bond length • Fiber-volume fraction
Hossain et al. [23]	GFRP	PVA	Experimental and analytical	RILEM beam method	9–22	• Bar diameter • Modulus of elasticity • Embedded length
Cao et al. [24]	BFRP	PVA	Experimental and analytical	Direct pull out	10–20	• Grade of ECC • Surface of the bar
Wang et al. [25]	BFRP	PVA	Experimental	Direct pull out	9–15	• Bar diameter • Embedded length • Cover thickness
Wang et al. [26]	GFRP	PVA	Experimental and analytical	Direct pull out	2–23	• Grade of ECC • Freeze–thaw cycle
Wei et al. [27]	GFRP	PVA PE	Experimental	Direct pull out	11–25	• Fiber type • Grade of ECC • Testing age (curing time) • Water type (fresh/sea water) • Sand type (river/sea sand)
Kim et al. [28]	GFRP	PVA	Experimental	Direct pull out	17–27	• Fiber-volume fraction • FRP surface treatment (sand coated/helically wrapped)
Wu et al. [29]	GFRP	PVA	Experimental and analytical	Direct pull out	3–18	• Grade of ECC • Testing age (curing time)

4. Flexural Behavior

The flexural behavior of structural elements provides insight into their load-carrying capacity, failure mode, ductility, and overall performance under bending loads, which are essential for safe and effective structural design. The corrosion resistance of FRP combined with the superior mechanical properties of FRCC contributes to the durability and longevity of structures, reducing maintenance costs and ensuring safety. Additionally, insights gained from studying flexural behavior can inform innovative design solutions and optimize structural systems, ultimately contributing to the development of design guidelines and standards that ensure the safe and effective use of these advanced materials in construction. Several investigations have examined the flexural behavior of FRCC members reinforced with FRP bars [22,34–43].

Yuan et al. [34] investigated ECC beams reinforced with BFRP bars through an experimental program and proposed an analytical model for the moment–curvature relationship based on the conventional strip method developed for conventional concrete by Wu et al. [44]. It is reported that the deformation capacity of BFRP-reinforced ECC beams with reduced reinforcement ratios showed an improved deformation capacity, resulting from the higher ultimate compressive strain characteristics of ECC. When comparing beams with the same size and BFRP reinforcement configurations, ECC beams exhibited significantly greater ultimate load, ultimate deformation, and energy dissipation compared to conventional concrete beams. Notably, the removal of shear stirrups from the ECC beam resulted in only a marginal decrease in load capacity and deformation when compared to a concrete beam with dense stirrups, underscoring the effectiveness of ECC in enhancing shear capacity. In a parametric study of BFRP-reinforced ECC beams, it was observed that the maximum ultimate curvature was achieved at the balanced reinforcement ratio. An increase in the compressive strength of ECC markedly improved both the ultimate moment capacity and the ultimate curvature of the beam, particularly when failure was governed by ECC crushing in the compression zone. However, the ductility, defined as the ratio of inelastic energy to total energy, remained relatively constant with variations in ECC compressive strength, irrespective of the failure mode. The moment–curvature response of the BFRP-reinforced ECC beam was predicted using a strip model, which showed a strong correlation between experimental and analytical findings.

Cai et al. [35] analyzed the flexural behavior of ECC beams reinforced with BFRP bars using a nonlinear finite element model. The simulation outcomes were validated against experimental data published by Yuan et al. [34]. Through finite element analysis, the influence of various parameters on the mechanical behavior of BFRP-reinforced ECC beams was investigated. These parameters included compressive strength (CS) values of 30, 45, 60, and 75 MPa, ultimate tensile strain (TA) levels of 0.5, 1.5, 2.5, 3.5, and 4.5%, tensile strength of ECC (TE) at 2, 4, 6, and 8 Mpa, and longitudinal reinforcement ratios (LR) of 0.2, 0.4, 0.6, 0.8, and 1%. The findings, illustrated in Figure 11, indicate that the ultimate strength of the beams increased with all parameters, except for the tensile strain of ECC when it exceeded 1.5%. Additionally, the ultimate deflection was observed to rise with increasing compressive strength and tensile strain of ECC, while the ultimate tensile stress exhibited minimal influence on the ultimate deflection of BFRP-reinforced ECC beams. Furthermore, variations in the longitudinal reinforcement ratio were found to alter the failure mode of the ECC beams. Notably, for over-reinforced beams, an increase in the longitudinal reinforcement ratio resulted in a decrease in ultimate deflection. The simulation results showed good agreement with the experimental results.

Fischer et al. [36] examined the influence of the stress–strain characteristics of FRP bars on the load–deformation behavior of FRP-reinforced ECC members subjected to reversed cyclic loading conditions. An analytical model was formulated to represent the load–deflection envelope, which was derived from the moment–curvature relationship and the sectional stiffness of the flexural members as a function of the applied load. The findings indicated that the load–deformation response of FRP-reinforced ECC is predominantly governed by flexural deformation, even at substantial drift levels, with crack initiation appearing to be largely independent of the interfacial bond properties. The inelastic deformation of ECC under compressive loads results in a reduction in flexural stiffness and ultimately leads to a gradual failure mode. However, it concurrently induces compressive strain and reduces the tensile strength of the FRP reinforcement. The analytical model demonstrated reasonable agreement with the experimental results; nevertheless, it did not account for rotational sliding, which can significantly affect the flexural stiffness and strength in certain scenarios.

Zhou et al. [37] proposed a modified model that aimed at predicting the maximum crack width in ECC beams reinforced with CFRP bars, specifically within the service limit state. The calculation methodology for this model was based on the principles of the Chinese standard GB50608 [45]. However, certain empirical equations of the critical parameters were modified to better suit different types of matrix materials. Furthermore, the average crack spacing and coefficient for the strain distribution of the FRP bars were established through regression analysis of experimental data. As a result, the applicability of this modified model is confined exclusively to ECC beams reinforced with sand-coated CFRP bars under service limit conditions.

Takasago et al. [22] conducted a four-point bending test to investigate the correlation between reinforcement strain and crack width. The assessment of crack width was performed using a predictive formula that integrated the proposed bond model along with a fiber-bridging law. The findings indicated that the number of cracks increased with an increase in fiber-volume fraction in FRCC. Furthermore, the relationship between reinforcement strain and crack width derived from the bending test was consistent with the predictions made by the formula.

Al Marahla et al. [39,46] examined how synthetic fibers affect the short- and long-term flexural performance of GFRP beams, comparing their findings with established guidelines. Their study revealed that the addition of fibers led to reductions in tensile creep, compressive creep, and shrinkage strain by 15%, 22%, and 26%, respectively. Additionally, deflection decreased by 25% to 43% depending on the fiber-volume fraction used. The experimental outcomes were assessed against prediction models for both short-term and long-term deflection as outlined in ACI 440 [33], ISIS [47], and CSA [31], which were found to be inaccurate for FRC beams. While the creep and shrinkage strain prediction models from Eurocode 2 [48] and ACI 209 [49] aligned well with the compressive strain, they tended to overestimate shrinkage strains. The authors suggested that further research and adjustments to current methodologies are necessary to make them suitable for FRC beams reinforced with FRP bars.

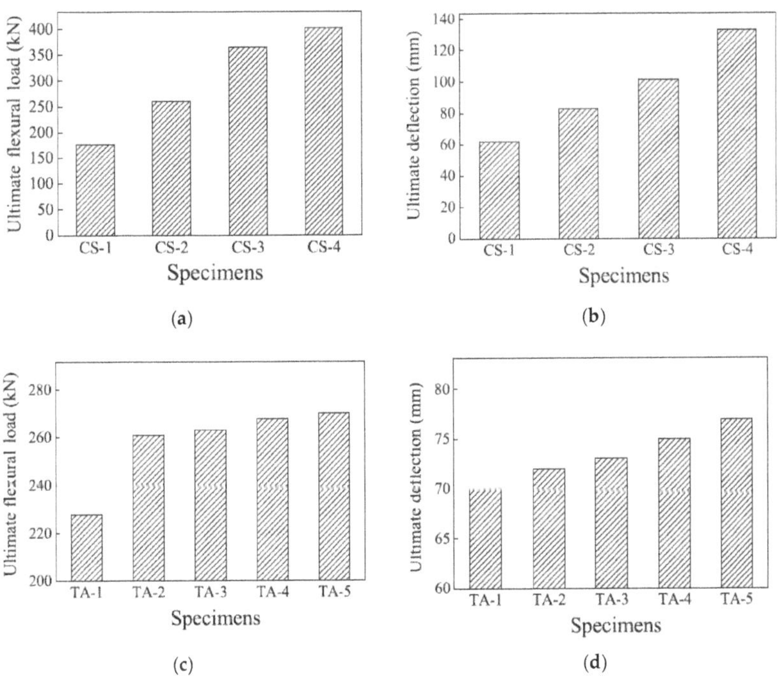

Figure 11. *Cont.*

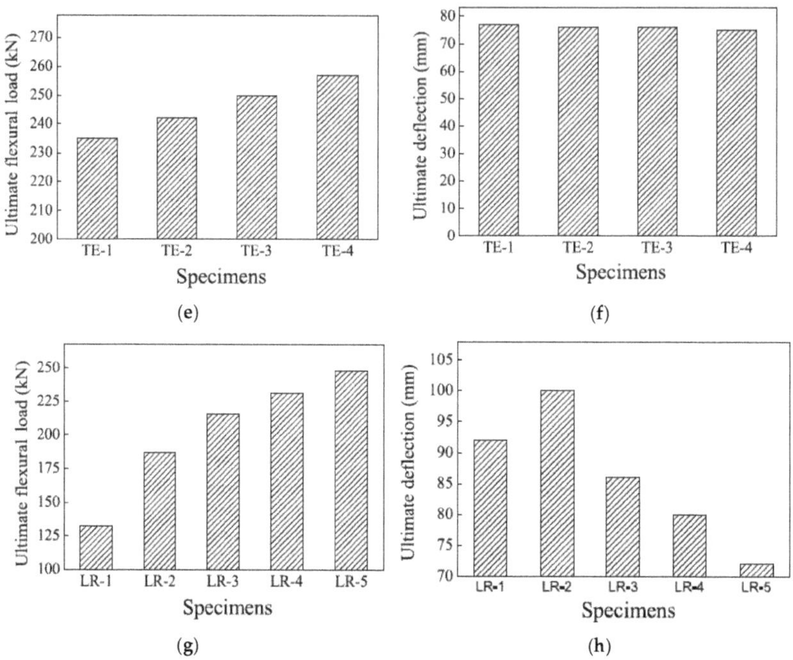

Figure 11. Analyzed flexural behavior of ECC beams reinforced with BFRP bars: (**a**,**b**) effect of compressive strength of ECC on the ultimate flexural load and ultimate deflection; (**c**,**d**) effect of tensile strain of ECC on the ultimate flexural load and ultimate deflection; (**e**,**f**) effect of tensile strength of ECC on the ultimate flexural load and ultimate deflection; (**g**,**h**) effect of longitudinal reinforcement ratio on the ultimate flexural load and ultimate deflection [35].

Behnam et al. [42] proposed a model for estimating the flexural resistance factors for FRC beams that are reinforced with GFRP bars. To assess the reliability of this model, a Monte Carlo simulation was employed for calibration. The analysis utilized data from three previously published studies, Issa et al. [50], Wang et al. [51], and Alsayed et al. [52], which were based on experimentally obtained results for FRC reinforced with GFRP bars. The target reliability index used in this assessment aligned with that of the current design code specifications. The estimated flexural resistance factors were found to range between 0.72 and 0.95.

Attia et al. [40] and Abushanab et al. [41] studied basalt FRC members reinforced with BFRP bars and reported that basalt microfibers (BMF) enhanced the flexural performance of BFRP reinforced members in terms of load carrying capacities, cracking, and ductility.

Reinforced concrete structures are designed and maintained in adverse environmental conditions, where they are subjected to cycles of drying and wetting, freeze–thaw phenomena, carbonation, and chloride ingress. These factors contribute to the corrosion of the steel reinforcement and the deterioration of the concrete matrix. Consequently, these structures may not fulfill the requirements of the ultimate limit state and the serviceability limit state when exposed to corrosive environments [53]. As discussed in the introduction, FRP-reinforced FRCC structures can resist this harsh environment. Wang et al. [38] conducted experimental and analytical investigations on the flexural behavior of FRP-reinforced ECC specimens subjected to freeze–thaw cycles. They developed formulas for the balanced reinforcement ratio, moment capacity, and stiffness of FRC-reinforced ECC specimens, based on ACI 440 [33] and GB 50608 [45]. These formulas account for the tensile capacity provided by ECC materials. The proposed model demonstrated good agreement with the experimental findings.

Despite the numerous advantages associated with FRP bars, their limited fire resistance remains a significant limitation for their utilization in the construction industry. Investigating the post-fire performance of FRP-reinforced FRCC members is crucial for assessing their serviceability and structural integrity. Jafarzadeh et al. [43] conducted both experimental and analytical investigations to evaluate the flexural behavior of FRC beams reinforced with GFRP bars following exposure to elevated temperatures of 20 °C, 250 °C, 400 °C, and 600 °C. The findings indicated that the residual flexural performance of GFRP-FRC beams is significantly influenced by the temperature of exposure. It was determined that the deflection values predicted by the equations proposed by ACI 440.1R-06 [54] and ACI 440.1R-15 [33] closely aligned with the experimental results, whereas the equations proposed by CSA S806-12 [31] and Faza et al. [55] showed very distinct values. The authors recommended further research that considers the manufacturing processes of the FRP bars, the specific exposure conditions, and the duration of exposure to heat, as well as the cooling procedures applied to the specimens.

Table 2 provides a summary of the research and findings reported by various researchers regarding the flexural performance of FRCC reinforced with FRP bars. From this table, it can be concluded that the flexural performance of FRP in combination with FRCC is influenced by several factors, and efforts have been made to understand the structural response under various loading conditions and environmental factors, including the effects caused by differences in materials used for FRP and FRCC.

Table 2. Summary of the research for flexural behavior.

Ref.	FRP Type	Fiber Type in FRCC	Study Type	Type of Test	Variables
Takasago et al. [22]	AFRP	PVA	Experimental and analytical	Bending test	• Fiber-volume fraction
Yuan et al. [34]	BFRP	PVA	Experimental and analytical	Bending test	• Bar diameter • Reinforcement ratio • Presence/absence of shear reinforcement
Cai et al. [35]	BFRP	PVA	Analytical	Bending test	• ECC characteristics • Reinforcement ratio
Fischer et al. [36]	AFRP	PE	Experimental and analytical	Reversed cyclic loading	• Reinforcement ratio • FRP surface treatment
Zhou et al. [37]	CFRP and GFRP	PVA	Experimental and analytical	Bending test	• Elastic modulus of FRP bar • Reinforcement ratio • FRP bar type
Wang et al. [38]	BFRP	PVA	Experimental and analytical	Bending test	• Environmental conditions
Al Marahla [39,46]	GFRP	PP	Experimental	Bending test	• Fiber-volume fraction • FRP bar diameter
Attia et al. [40]	BFRP	BF	Experimental and analytical	Bending test	• Fiber-volume fraction
Abushanab et al. [41]	BFRP	BF	Numerical		• Fiber-volume fraction • Stirrup spacing

Table 2. *Cont.*

Ref.	FRP Type	Fiber Type in FRCC	Study Type	Type of Test	Variables
Bahnam et al. [42]	GFRP	PP, GF, and SF	Analytical		• Fiber-volume fraction • Fiber type
Jafarzadeh et al. [43]	GFRP	SF	Experimental and analytical	Bending test	• Temperature • Reinforcement ratio

5. Ductility Evaluation

To assess the ductility of members reinforced with FRP bars, various methods have been developed. These methods include the energy-based approach [56] and the deformation-based approach [57]. In the energy-based approach, ductility is defined as the ability to absorb energy, represented by the ratio of total energy to elastic energy. Conversely, the deformation-based approach evaluates ductility by examining the deformability margin between the ultimate and service stages, considering both strength and deflection. Figure 12 illustrates the ductility index definition.

Jafarazdeh et al. [43] used the deformation approach to determine the ductility factor of FRC beams reinforced with GFRP bars. It is reported that all the beams have failed due to the rupture of the bar. For this reason, the effect of the fiber had no noticeable effect on the ductility factor. They also reported that specimens with a higher reinforcement ratio showed a higher deformability factor. In this regard, increasing the reinforcement ratio and using discrete fibers indicated a positive effect on the concrete-crushing mode at higher temperatures when steel fiber was used.

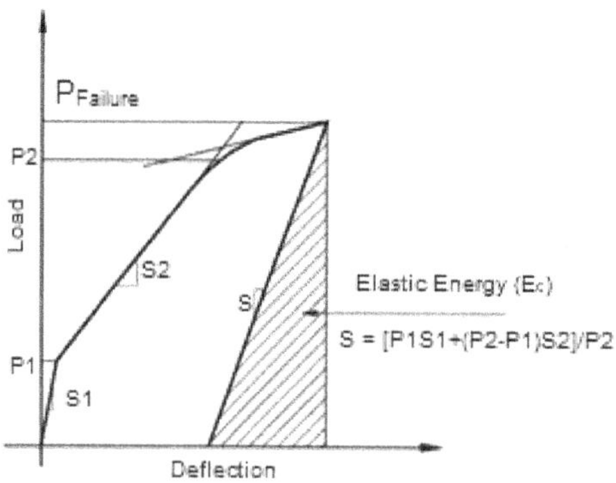

Figure 12. Definition of ductility index [51].

Liao et al. [58,59] employed a modified deformation-based method to determine the ductility indices of BFRP bars in SS-ECC beams. As illustrated in Figure 13, the study examined how the longitudinal reinforcement ratio, shear span ratio, and stirrup ratio influenced the ductility indices. The findings indicated that the ductility index decreased with an increase in the longitudinal reinforcement ratio, while it increased with higher shear span and stirrup ratios.

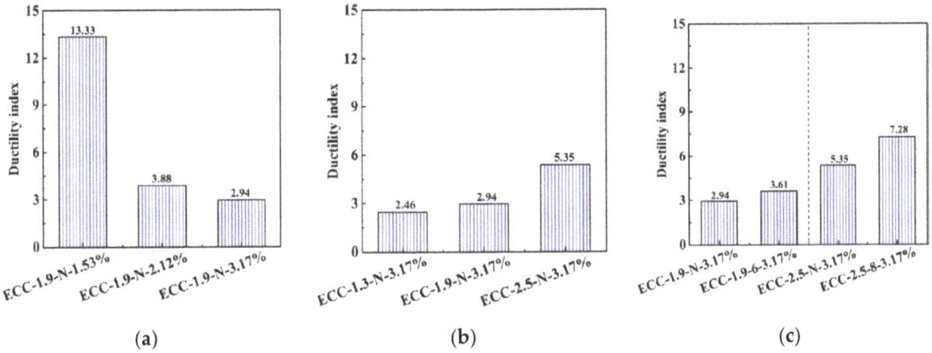

Figure 13. Influencing factors on ductility index: (**a**) effect of longitudinal reinforcement ratio; (**b**) effect of shear span ratio; (**c**) effect of stirrup ratio [59].

Wang et al. [51] employed both energy-based and deformation-based methods to assess the ductility indices of GFRP-reinforced FRC beams. They found that the energy-based approach inadequately accounts for the influence of discrete fibers in the concrete, particularly regarding the improvement in ultimate moment capacity and the corresponding increase in deflection. On the other hand, the ductility index derived from the deformation-based method effectively describes factors like load capacity and the impact of deformation on ductility.

6. Shear and Torsional Performance

Liao et al. [59–61] developed a modified calculation model based on ACI 440.1R-06 [54] to predict the ultimate shear capacity of SS-ECC beams reinforced with BFRP bars. The ratio of the predicted values to the experimental results was 0.97, indicating a good agreement between the model and the experimental findings. As illustrated in Figure 14, the experimental data revealed that SS-ECC beams with a low longitudinal reinforcement ratio confirmed greater shear capacity, which increased with a higher stirrup ratio or a lower shear span ratio.

Yu et al. [62] studied the torsional performance of BFRP-reinforced PE SS-ECC beams without stirrups through experiment and analysis. It is reported that the SS-ECC beams with a high fiber-volume fraction failed due to compression damage. It was found that rather than the longitudinal reinforcement ratio, the strength of SS-ECC and the height/width ratio of the beam are the critical factors to determine the ultimate torque. The effectiveness of developed FEM for prediction of torsional performance was verified and the torque–angle of twist relationship achieved a good agreement with the experimental results, as shown in Figure 15. The change in stress distribution of the cross-section was marked with three points in the typical curve. Point A represents the section subjected to pure torsional loading, where the longitudinal BFRP bars experience tensile stress. At Point B, with an increase in the angle of twist, the cross-section is subjected to shear stress. The shear stress within the SS-ECC is comparatively high, while the tensile stress in the longitudinal BFRP bars remains significantly lower than their tensile strength. At Point C, although the ultimate torque was attained, the tensile stress in the BFRP bars continued to be minimal. Nevertheless, the SS-ECC reached its maximum shear stress, leading to the failure of the SS-ECC.

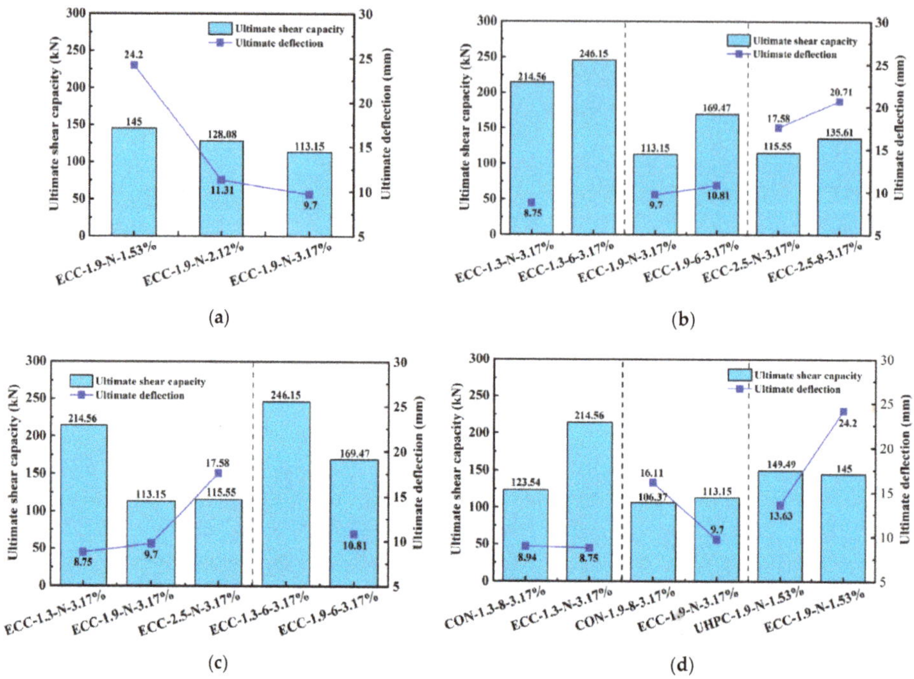

Figure 14. Ultimate shear capacity of tested beams: (**a**) effect of longitudinal reinforcement ratio; (**b**) effect of stirrup ratio; (**c**) effect of shear span ratio; (**d**) effect of concrete type [59].

Zhou et al. [63] proposed an empirical formula to estimate the ultimate torsional moment for FRC and ECC beams that are reinforced with GFRP bars. They analyzed the individual contributions of the concrete, PP fibers, and GFRP bars to torsional strength. The predicted outcomes showed good agreement with the experimental results.

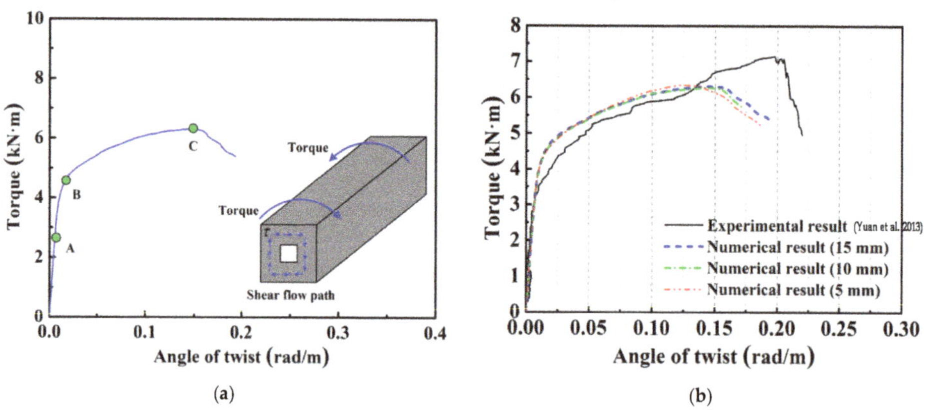

Figure 15. Torque–angle of twist curve: (**a**) typical curve; (**b**) example of curve of the experiment and analysis for various mesh sizes [34,62].

7. Durability Performance

The study conducted by Yuan et al. [30], Hao et al. [64], and Zhou et al. [65] examined the durability of GFRP bars made from different resin types, epoxy, vinyl ester, and polyester, when subjected to sustained loads. These bars were embedded in ultra-high performance engineered cementitious composites (UHP-ECC) and immersed in an alkaline solution at an elevated temperature of 40 °C for six months. The findings indicated that GFRP bars made from polyester-based resin experienced greater degradation compared to those made from vinyl ester and epoxy resins. Furthermore, the GFRP bars embedded in UHP-ECC exhibited less degradation than bare GFRP bars, although this protective effect was less effective under sustained loading conditions. Analyses using scanning electron microscopy (SEM) and X-ray computed tomography (CT) revealed that the primary cause of degradation in the GFRP bars was matrix hydrolysis. This process negatively impacted the stress transfer efficiency between the fibers, ultimately leading to a reduction in the tensile strength of the GFRP bars.

8. Conclusions

The integration of FRP and FRCC represents a significant advancement in concrete technology, addressing the limitations of traditional RC, especially issues like steel corrosion, low tensile strength, and brittleness. A systematic overview and discussion regarding the structural performance of FRCC members reinforced with FRP bars has been presented here. Based on the findings from this review, the following conclusions can be drawn:

1. The bond behavior between FRP bars and FRCC is complex and influenced by factors such as bar diameter, surface treatment, and the characteristics of the FRCC matrix. The bond strength decreases when the bar diameter and embedded length increase. It improves as both the amount of fiber and the size of the cross-section of FRCC increases. Further research is needed to develop reliable models for predicting bond strength and performance, as existing studies show variability in results based on different experimental setups and material properties.
2. FRCC members reinforced with FRP bars exhibit superior load-carrying capacities and ductility compared to conventional RC, highlighting their potential for improved structural performance. Type of FRP, reinforcement ratio, and environmental conditions are some of the factors that highly affect the flexural and ductility performance of these members.
3. The shear capacity of FRCC members reinforced with FRP bars increases with low longitudinal reinforcement ratio, higher shear reinforcement ratio, or low shear span ratio.
4. It was found that the torsional resistance of FRCC members reinforced with FRP bars is more dependent on the height/width ratio than the reinforcement ratio.

In conclusion, the findings aim to inform the development of design guidelines and standards for the effective use of these innovative materials in construction, ultimately contributing to enhanced structural performance and sustainability in civil engineering applications.

9. Recommendations for Future Studies

Based on the discussion in this paper, it can be concluded that additional experimental and numerical research is necessary to fill various research gaps concerning the structural performance of FRCC elements reinforced with FRP bars. The following gaps have been identified:

1. Lack of comprehensive experimental studies that cover a wide range of FRCC and FRP combinations has led to insufficient data on their performance under various loading conditions, such as dynamic, cyclic, or impact loads.
2. There are inconsistencies in findings across different studies, which could arise from variations in testing methods, material properties, and environmental conditions. This validity makes it challenging to draw a definitive conclusion.
3. There are fewer studies on bond behavior under different loading scenarios and environmental conditions. More research is needed to develop reliable predictive models.

4. More research on the environmental impact of producing and using FRCC and FRP materials, including life cycle assessment, is needed.
5. There are insufficient design guidelines or standards specifically made for the use of FRCC and FRP in structural applications, which could limit their adoption in engineering practice.

As the search for articles was restricted to peer-reviewed publications in English indexed on the Web of Science, the results of the study may not fully represent the existing literature. Future research could enhance the data by exploring publications from a range of databases for both quantitative and qualitative studies, such as Google Scholar, Scopus, and Science Direct.

Author Contributions: Conceptualization, H.N.S. and T.K.; methodology, H.N.S.; validation, T.K.; investigation, H.N.S.; resources, H.N.S.; writing—original draft preparation, H.N.S.; writing—review and editing, T.K.; supervision, T.K. All authors have read and agreed to the published version of the manuscript.

Funding: This study was supported by the Japan Society for the Promotion of Science KAKENHI Grant Number JP25K01362.

Conflicts of Interest: The authors declare no conflicts of interest.

Abbreviations

The following abbreviations are used in this manuscript:

FRCC	Fiber-reinforced cementitious composite
FRP	Fiber-reinforced polymer
FRC	Fiber-reinforced concrete
ECC	Engineered cementitious composite
GFRP	Glass fiber-reinforced polymer
CFRP	Carbon fiber-reinforced polymer
AFRP	Aramid fiber-reinforced polymer
BFRP	Basalt fiber-reinforced polymer
SF	Steel fiber
PE	Polyethylene
PVA	Polyvinyl alcohol
PP	Polypropylene
SHCC	Strain hardening cementitious composite
DFRCC	Ductile fiber-reinforced cementitious composite
SF	Silica fume
FA	Fly ash
GBFS	Ground granulated blast furnace
HM	High modulus of elasticity
LM	Low/standard modulus of elasticity
SC-GFRP	Sand-coated GFRP
HW-GFRP	Helically wrapped GFRP
SS-ECC	Sea-water sea-sand ECC
FR-ECC	Fresh water ECC
UHP-ECC	Ultra-high performance ECC
SEM	Scanning electron microscope
CT	X-ray computed tomography

References

1. Soric, Z.; Kisicek, T.; Galic, J. Deflections of Concrete Beams Reinforced with FRP bars. *Mater. Struct.* **2010**, *43*, 73–90. [CrossRef]
2. Nanni, A.; De Luca, A.; Jawaheri, Z.H. *Reinforced Concrete with FRP Bars*, 1st ed.; CRC Press: New York, NY, USA, 2014. [CrossRef]

3. *ACI 440.1R-16*; Guide for the Design and Construction of Structural Concrete Reinforced with FRP Bars. American Concrete Institute: Farmington Hills, MI, USA, 2016.
4. Said, H. Deflection Prediction for FRP-Strengthened Concrete Beams. *J. Compos. Constr.* **2010**, *14*, 244–248. [CrossRef]
5. Ge, W.J.; Ashour, A.; Ji, X.; Cai, C.; Cao, D.F. Flexural Behavior of ECC-Concrete Composite Beams Reinforced with Steel Bars. *Constr. Build. Mater.* **2018**, *159*, 175–188. [CrossRef]
6. Labib, W. Fibre Reinforced Cement Composites. In *Cement-Based Materials*; Saleh, H.E.-D.M., Rahman, R.O.A., Eds.; InTech: London, UK, 2018. [CrossRef]
7. Kim, D.; Naaman, A.; El-Tawil, S. Comparative Flexural Behavior of Four Fiber Reinforced Cementitious Composites. *Cem. Concr. Compos.* **2008**, *30*, 917–928. [CrossRef]
8. Lin, C.; Kayali, O.; Morozov, E.; Sharp, D. Influence of Fibre Type on Flexural Behaviour of Self-Compacting Fibre Reinforced Cementitious Composites. *Cem. Concr. Compos.* **2014**, *51*, 27–37. [CrossRef]
9. Bentur, A.; Mindess, S. *Fibre Reinforced Cementitious Composites*, 2nd ed.; Taylor & Francis: London, UK, 2007.
10. Tahmouresi, B.; Nemati, P.; Asadi, M.; Saradar, A.; Mohtasham, M. Mechanical Strength and Microstructure of Engineered Cementitious Composites: A New Configuration for Direct Tensile Strength, Experimental and Numerical Analysis. *Constr. Build. Mater.* **2021**, *269*, 121–136. [CrossRef]
11. Zhu, Y.; Zhang, Z.; Yang, Y.; Yao, Y. Measurement and Correlation of Ductility and Compressive Strength for Engineered Cementitious Composites (ECC) Produced by Binary and Ternary Systems of Binder Materials: Fly Ash, Slag, Silica Fume, and Cement. *Constr. Build. Mater.* **2014**, *68*, 192–198. [CrossRef]
12. Shuxin, W.; Victor, C.L. Engineered Cementitious Composites with High-Volume Fly Ash. *Mater. J.* **2007**, *104*, 233–241. [CrossRef]
13. Lim, I.; Chern, J.C.; Liu, T.; Chan, Y.W. Effect of Ground Granulated Blast Furnace Slag on Mechanical Behavior of PVA-ECC. *J. Mar. Sci. Technol.* **2012**, *20*, 11. [CrossRef]
14. Wu, Z.; Yin, J. Fracturing Behaviors of FRP-Strengthened Concrete Structures. *Eng. Fract. Mech.* **2003**, *70*, 1339–1355. [CrossRef]
15. Ge, W.; Ashour, A.F.; Cao, D.; Lu, W.; Gao, P.; Yu, J. Experimental Study on Flexural Behavior of ECC-Concrete Composite Beams Reinforced with FRP Bars. *Compos. Struct.* **2019**, *208*, 454–465. [CrossRef]
16. Carrera, A.; Ochoa, W.; Larrinaga, F.; Lasa, G. How to conduct a systematic literature review: A quick guide for computer science research. *MethodsX* **2022**, *9*, 101895. [CrossRef]
17. Available online: https://osf.io/gkt5f (accessed on 4 July 2025).
18. Falagas, M.; Pitsouni, E.; Malietzis, G.; Pappas, G. Comparison of PubMed, Scopus, Web of Science, and Google Scholar: Strengths and Weaknesses. *FASEB J.* **2008**, *22*, 338–342. [CrossRef] [PubMed]
19. Available online: https://www.webofscience.com/wos/woscc/summary/b14ae2ac-ec76-4797-b0a0-ff400c383dc0-015cbcbed7/relevance/1 (accessed on 3 April 2025).
20. Zhao, D.; Zhou, Y.; Xing, F.; Sui, L.; Ye, Z.; Fu, H. Bond Behavior and Failure Mechanism of Fiber Reinforced Polymer Bar–Engineered Cementitious Composite Interface. *Eng. Struct.* **2021**, *243*, 112520. [CrossRef]
21. Li, Q.; Fu, M.; Xie, B. Analyzing the Bond Behavior of Fiber-Reinforced Polymer (FRP) Bars Embedded in Engineered Cementitious Composites (ECCs) with the Nonlocal Continuum Rod Model. *Math. Probl. Eng.* **2020**, *2020*, 1710364. [CrossRef]
22. Takasago, S.; Kanakubo, T.; Kobayashi, H.; Sasaki, H. Bond and Cracking Characteristics of PVA-Fiber-Reinforced Cementitious Composite Reinforced with Braided AFRP Bars. *Fibers* **2023**, *11*, 107. [CrossRef]
23. Hossain, K. Bond Strength of GFRP Bars Embedded in Engineered Cementitious Composite using RILEM Beam Testing. *Int. J. Concr. Struct. Mater.* **2018**, *12*, 6. [CrossRef]
24. Cao, J.; Wu, F.; Zhao, B.; Lei, H.; Li, Z.; Chen, A. Experimental and Analytical Investigations for the Bond Slip and Anchorage Design Between BFRP Bar and Engineered Cementitious Composites. *Structures* **2024**, *68*, 107075. [CrossRef]
25. Wang, H.; Sun, X.; Peng, G.; Luo, Y.; Ying, Q. Experimental Study on Bond Behaviour between BFRP Bar and Engineered Cementitious Composite. *Constr. Build. Mater.* **2015**, *95*, 448–456. [CrossRef]
26. Wang, H.; Wu, L.; Xu, X.; Lin, Z. Experimental Study on Bond Durability of GFRP Bar/Engineered Cementitious Composite Exposed to Freeze-Thaw Environments. *J. Build. Eng.* **2024**, *84*, 108592. [CrossRef]
27. Wei, J.; Ke, L.; Wang, P.; Li, W.; Leung, C. Microstructure, Mechanical Properties and Interaction Mechanism of Seawater Sea-Sand Engineered Cementitious Composite (SS-ECC) with Glass Fiber Reinforced Polymer (GFRP) bar. *Compos. Struct.* **2024**, *343*, 118302. [CrossRef]
28. Kim, B.; Lee, J.Y. Polyvinyl Alcohol Engineered Cementitious Composite (PVA-ECC) for the Interfacial Bond Behaviour of Glass Fibre Reinforced Polymer Bars (GFRP). *Polym. Polym. Compos.* **2012**, *20*, 545–558. [CrossRef]
29. Wu, L.; Wang, H.; Xu, X.; Lin, Z. Investigation on Bond Durability of GFRP bar/Engineered Cementitious Composite Under Alkaline-Saline Environments. *J. Build. Eng.* **2023**, *77*, 107343. [CrossRef]

30. Godat, A.; Aldaweela, S.; Aljaberi, H.; Al Tamimi, N.; Alghafri, E. Bond Strength of FRP Bars in Recycled Aggregate Concrete. *Constr. Build. Mater.* **2021**, *267*, 120919. [CrossRef]
31. *CSA S806-12*; Design and Construction of Building Components with Fiber-Reinforced Polymers. Canadian Standards Association: Toronto, ON, Canada, 2012.
32. *CSA S6-06*; Canadian Highway Bridge Design Code (CHBDC). Canadian Standards Association: Toronto, ON, Canada, 2006.
33. *ACI 440.1R-15*; ACI Committee 440. Guide for the Design and Construction of Structural Concrete Reinforced with FRP Bars. American Concrete Institute: Farmington Hills, MI, USA, 2015.
34. Yuan, F.; Pan, J.; Leung, C. Flexural Behaviors of ECC and Concrete/ECC Composite Beams Reinforced with Basalt Fiber-Reinforced Polymer. *J. Compos. Constr.* **2013**, *17*, 591–602. [CrossRef]
35. Cai, J.; Pan, J.; Zhou, X. Flexural Behavior of Basalt FRP Reinforced ECC and Concrete Beams. *Constr. Build. Mater.* **2017**, *142*, 423–430. [CrossRef]
36. Fischer, G.; Victor, C.L. Deformation Behavior of Fiber-Reinforced Polymer Reinforced Engineered Cementitious Composite (ECC) Flexural Members under Reversed Cyclic Loading Conditions. *Struct. J.* **2003**, *100*, 25–35. [CrossRef]
37. Zhou, J.; Xia, G.; Wang, P.; Zhao, J.; Xu, L.; Pan, J. Investigation of Flexural Cracks in Engineered Cementitious Composites Beams Reinforced with CFRP Bars. *Constr. Build. Mater.* **2024**, *438*, 136900. [CrossRef]
38. Wang, Y.; Wang, G.; Guan, Z.; Ashour, A.; Ge, W.; Zhang, P. The Effect of Freeze–Thaw Cycles on Flexural Behaviour of FRP-reinforced ECC beams. *Archiv. Civ. Mech. Eng.* **2021**, *21*, 101. [CrossRef]
39. Al Marahla, R.; Shehzad, M.; Garcia, E. Flexural and Deflection Behaviour of Synthetic Fibre Reinforced Concrete Beams Reinforced with Glass Fibre Reinforced Polymer Bars under Sustained Service Load. *Structures* **2023**, *54*, 946–955. [CrossRef]
40. Attia, K.; Alnahhal, W.; Elrefai, A.; Rihan, Y. Flexural Behavior of Basalt Fiber-Reinforced Concrete Slab Strips Reinforced with BFRP and GFRP Bars. *Compos. Struct.* **2019**, *211*, 1–12. [CrossRef]
41. Abushanab, A.; Alnahhal, W. Numerical Parametric Investigation on the Moment Redistribution of Basalt FRC Continuous Beams with Basalt FRP Bars. *Compos. Struct.* **2021**, *277*, 114618. [CrossRef]
42. Behnam, B.; Al-Iessa, M. Effect of Randomly Dispersed Short Fibers on the Flexural Resistance Factor of Concrete Beams Reinforced with GFRP Bars. *IJSI* **2022**, *13*, 226–236. [CrossRef]
43. Jafarzadeh, H.; Nematzadeh, M. Evaluation of Post-Heating Flexural Behavior of Steel Fiber-Reinforced High-Strength Concrete Beams Reinforced with FRP Bars: Experimental and Analytical Results. *Eng. Struct.* **2020**, *225*, 111292. [CrossRef]
44. Wu, Y.; Oehlers, D.; Griffith, M. Rational Definition of the Flexural Deformation Capacity of RC Column Sections. *Eng. Struct.* **2004**, *26*, 641–650. [CrossRef]
45. *GB50608-2010*; Technical Code for Infrastructure Application of FRP Composites. Central Research Institute of Building and Construction Co., Ltd., MCC Group; China Planning Press: Beijing, China, 2011.
46. Al Marahla, R.; Garcia, E. Cracking Behaviour of FRC Members Reinforced with GFRP Bars under Sustained Loads. In *Fibre Reinforced Concrete: Improvements and Innovations II*; Serna, P., Llano-Torre, A., Martí-Vargas, J.R., Navarro-Gregori, J., Eds.; Springer International Publishing: Cham, Switzerland, 2022; Volume 36, pp. 319–330. [CrossRef]
47. ISIS Manual, Manual no.3. Strengthening Reinforced Concrete Structures with Externally Bonded Fiber Reinforced Polymers. In *The Canadian Network of Centers of Excellence on Intelligent Sensing for Innovative Structures*; ISIS Canada, University of Winnipeg: Manitoba, MB, Canada, 2001; p. 151.
48. *BS EN 1992-1-1*; Eurocode 2: Design of Concrete Structures–Part 1-1: General Rules and Rules for Buildings. European Committee for Standardization: Geneva, Switzerland, 2004. [CrossRef]
49. ACI Committee 209. *Prediction of Creep, Shrinkage, and Temperature Effects in Concrete Structures*; Creep Volume Changes in Concrete; American Concrete Institute: Farmington Hills, MI, USA, 1997.
50. Issa, M.; Metwally, I.; Elzeiny, S. Influence of Fibers on Flexural Behavior and Ductility of Concrete Beams Reinforced with GFRP rebars. *Eng. Struct.* **2011**, *33*, 1754–1763. [CrossRef]
51. Wang, H.; Belarbi, A. Ductility Characteristics of Fiber Reinforced Concrete Beams Reinforced with FRP Rebars. *Constr. Build. Mater.* **2011**, *25*, 2391–2401. [CrossRef]
52. Alsayed, S.; Alhozaimy, A. Ductility of Concrete Beams Reinforced with FRP Bars and Steel Fibers. *J. Compos. Mater.* **1999**, *33*, 1792–1806. [CrossRef]
53. Kumar, M.P. Concrete Durability-Fifty Years Progress. In Proceedings of the 2nd International Conference on Concrete Durability, Montral, QC, Canada, 4–9 August 1991; American Concrete Institute: Farmington Hills, MI, USA, 1991; pp. 1–31.
54. *ACI 440.1R-06*; Guide for the Design and Construction of Structural Concrete Reinforced with FRP Bars. American Concrete Institute: Farmington Hills, MI, USA, 2006.

55. Faza, S.; Ganga, H. Pre- and Post-cracking Deflection Behavior of Concrete Beams Reinforced with Fiber-Reinforced Plastic Rebars. In Proceedings of the First International Conference, Use Advanced Composite Materials in Bridges and Structures, Sherbrooke, QC, Canada, October 1992.
56. Naaman, A.; Jeong, S. Structural Ductility of Concrete Beams Prestressed with FRP Tendons. In Proceedings of the Second International RILEM Symposium (FRPRCS-2) on Reinforcement for Concrete Structures, Ghent, Belgium, 23–25 August 1995; pp. 379–386.
57. Jaeger, L.; Mufti, A.; Tadros, G. The Concept of the Overall Performance Factor in Rectangular-Section Reinforced Concrete Members. In Proceedings of the 3rd International Symposium on Non-Metallic (FRP) Reinforcement for Concrete Structures, Sapporo, Japan, 14–16 October 1997; Volume 2, pp. 551–559.
58. Liao, Q.; Su, Y.; Yu, K.; Yu, J. An Innovative Strategy for Improving Ductility of Seawater Sea-Sand Engineered Cementitious Composites Beam Reinforced with GFRP Bar. *J. Build. Eng.* **2024**, *98*, 111063. [CrossRef]
59. Liao, Q.; Yu, J.; Shi, T.; Yu, K.; Su, Y. Seawater Sea-Sand Engineered Cementitious Composites Contribution to Shear Performance of Beams Reinforced with BFRP Bars. *Compos. Struct.* **2023**, *320*, 117183. [CrossRef]
60. Liao, Q.; Yu, J.; Dong, F.; Su, Y.; Yu, K. FRP bars Reinforced Seawater Sea-Sand Engineered Cementitious Composites Beams with Various Salinities: Shear Behaviors and Cost Effectiveness. *J. Build. Eng.* **2024**, *83*, 108452. [CrossRef]
61. Liao, Q.; Yu, J.; Shi, T.; Su, Y. Mechanical Behaviors and Failure Criteria of Seawater Sea-Sand Engineered Cementitious Composites under Combined Tension and Shear. *J. Build. Eng.* **2022**, *54*, 104552. [CrossRef]
62. Yu, J.; Su, Y.; Liao, Q.; Yu, K. Torsional Performance of FRP Bars Reinforced Seawater Sea-Sand Engineered Cementitious Composites (SS-ECC) Beams Without Stirrup: A numerical study. *Constr. Build. Mater.* **2023**, *397*, 132286. [CrossRef]
63. Zhou, J.; Shen, W.; Wang, S. Experimental Study on Torsional Behavior of FRC and ECC Beams Reinforced with GFRP bars. *Constr. Build. Mater.* **2017**, *152*, 74–81. [CrossRef]
64. Hao, Z.; Zeng, J. Effect of Sustained Loads on the Durability of GFRP Bars with and without UHP-ECC Cover. *Eng. Struct.* **2024**, *309*, 118050. [CrossRef]
65. Zhou, J.; Hao, Z.; Zeng, J.; Feng, S.; Liang, Q.; Zhao, B. Durability Assessment of GFRP Bars Embedded in UHP-ECCs Subjected to an Accelerated Aging Environment with Sustained Loading. *Constr. Build. Mater.* **2024**, *419*, 135364. [CrossRef]

applied sciences

MDPI

Article

Mechanical Behavior of Carbon Fiber-Reinforced Concrete Structures After Structural Failure

Tomas Gómez Prieto *, María Isabel Prieto , Alberto Leal and Daniel Ferrández

Departamento de Tecnología de la Edificación, Universidad Politécnica de Madrid, Avda. Juan de Herrera, 6, 28040 Madrid, Spain; mariaisabel.prieto@upm.es (M.I.P.); alberto.leal.matilla@upm.es (A.L.); daniel.fvega@upm.es (D.F.)

* Correspondence: tomas.gomez.prieto@upm.es

Abstract: The research presented below examines the use of carbon fiber-reinforced concrete specimens that have been previously tested under compression. For this study, two types of specimens were used: half without the addition of polypropylene fibers and the other half with the addition of 3 kg/m^3 of polypropylene fibers. These specimens were tested in two stress groups, one above 15 N/mm^2 and the other below this value. The confinement was carried out using four types of carbon fibers, two unidirectional and two bidirectional, analyzing their stress and maximum strain and comparing them with reference tests. The results indicate that confinement with carbon fibers is highly effective for stresses both above and below 15 N/mm^2. In fact, the confinement increased the compressive strength compared to the original specimens, which had very low strengths, reaching values up to 15 N/mm^2. Finally, the polypropylene fibers demonstrated a greater energy absorption capacity, achieving a non-explosive failure.

Keywords: FRP; concrete; compression; low-strength concrete; polypropylene

1. Introduction

Over the past 30 years, fiber-reinforced polymers (FRPs) have been continuously used in the repair, strengthening, reconstruction, and new construction of structures [1–3]. The use of this material offers advantages such as high strength, durability, corrosion resistance, lightweight properties, ease of installation, and fatigue resistance, making it an attractive option for improving the capacity and performance of structures [4,5]. The use of fiber-reinforced polymers (FRPs) to strengthen concrete columns through confinement is an effective solution to address the deterioration of mechanical properties of columns over time or to enhance their mechanical capacity to support higher loads [6].

The scientific literature has demonstrated the benefits of confining concrete columns with FRPs. Columns confined with FRPs show significant improvements in compressive strength, ductility, stiffness, and durability compared to unrein-forced concrete columns [7–10]. These improvements are particularly relevant in the context of rehabilitating existing structures, where the aim is to enhance the load-bearing capacity and structural behavior of structural elements without the need for major interventions [11–13].

There are some concerns and limitations associated with the strengthening of fully confined concrete columns with FRPs, such as the adhesion between the FRP and the concrete surface and the high associated costs [14–16]. Moreover, in certain cases, full reinforcement of the columns may not be necessary in terms of compressive strength. Instead, they can be partially confined with FRP bands, achieving sufficient performance with a reduced amount of reinforcement fiber [17–20].

Another important aspect of this type of reinforcement is the minimum requirements that the substrate must meet. The surfaces must be clean and free of anything that could compromise adhesion, must have sufficient flatness, and must have a moisture content below 4%. Regarding mechanical requirements, the minimum compressive strength of the concrete must exceed 15 N/mm^2 and the substrate's pull-off strength must exceed 15 N/mm^2 [21].

In recent years, there has been an increase in the incorporation of fibers into concrete. Among the fibers used, synthetic and polymeric fibers have been widely employed to improve the crack resistance of concrete. In particular, polypropylene macrofibers have gained significant importance due to their strength capacity and good adhesion to the concrete matrix [22–24]. The incorporation of polypropylene (PP) fibers, or their hybridization with other types of fibers, provides concrete with various benefits, such as toughness, lightness, improved fire performance by reducing the "spalling" effect, and enhanced chemical resistance in alkaline environments [25–27]. Studies have also examined the parameters that determine the behavior of fiber-reinforced concretes, with the most important being fiber geometry, fiber content, and the orientation and distribution of the fibers [28–32].

In the proposed research context, some studies have investigated the influence of structural synthetic discontinuous polypropylene fibers (DSSPFs) on the adhesion between fiber-reinforced concrete and carbon fiber-reinforced polymers (CFRPs). These studies have shown that the addition of fibers slightly improves adhesion strength with CFRP composites [33].

In concrete elements subjected to compression and reinforced with CFRPs, with the addition of fibers, it was observed that lightweight concretes achieved lower strengths than normal-weight concrete [33,34]. Furthermore, the addition of carbon fibers or hybridization with polypropylene fibers barely changed the axial compression behavior of lightweight concrete confined with CFRPs [34].

The compressive strength of the concrete to which the reinforcement is applied was investigated in concretes with hybrid fibers, focusing on low-strength concretes and considering the number of reinforcement layers. The study concluded that CFRP confinement is more effective for fiber-reinforced low-strength concrete compared to high-strength concrete, in terms of both strength and de-formation [35]. Similarly, Sathwik, M. C. et al. [36] analyzed the influence of defects in reinforced concrete by considering defects as a preloading process. Specimens were subjected to different axial loads to simulate varying degrees of damage and were subsequently reinforced with CFRPs. It was observed that the reinforcement significantly increased the strength of the specimens, even when they had been preloaded up to 70% of their ultimate load capacity.

Building on these premises, the objective of this work is to evaluate the effectiveness of carbon fiber (CFRP) reinforcements in the confinement of concrete specimens subjected to compression when the reinforcement is applied to low-strength concretes, with and without polypropylene fibers. Additionally, the minimum compressive strength requirement was assessed by analyzing the behavior of concretes with tensile strength below and above 15 N/mm^2.

2. Materials and Methods

Based on the objective of the study, a mix design was developed to produce low-performance concrete [37,38], resulting in the preparation of 24 reference concrete specimens and 24 specimens with polypropylene fibers. The 48 specimens were tested to failure, separating each type into two groups based on whether their compressive strength was greater or less than 15 N/mm^2, the threshold recommended by manufacturers of this type of reinforcement and the reference value for the study.

Within each group, the specimens were randomly reinforced with four types of reinforcements (unidirectional and bidirectional, used for compression confinement and flexural-compression reinforcements, respectively, with two fabric weights for each type to evaluate the improvement in structural performance). The reinforced specimens were then tested to failure under compression to evaluate the increase in stress and strain that reinforced concrete can achieve.

The materials used for the experimental development of the research are as follows:

- Cement type CEM II/B—L 32.5 N, in accordance with UNE-EN 197-1:2011 (Cement. Part 1: Composition, specifications, and conformity criteria for common cements) [39] and RC-16(Code for the design and execution of plain or reinforced concrete structures) [40].
- Coarse aggregate: Rolled aggregate with a granulometric fraction of 4–12 mm and a maximum aggregate size of 12 mm, siliceous in nature, in compliance with UNE-EN 12620:2003+A1:2009 (Aggregates for concrete) [41].
- Fine sand: Washed sand with a granulometric fraction of 0–4 mm, in accordance with standard UNE-EN 13139/AC:2004 [42].
- Potable water supplied by the Canal de Isabel II in the community of Madrid, meeting the technical requirements established for structural concrete.
- Plasticizer additive Sikament-165 ES, commonly used for all types of concrete.
- Polypropylene fibers: Extruded monofilaments with a diameter < 0.30 mm and a length of 12 mm (Sikafiber M-12).
- Unidirectional carbon fiber fabrics (MapeWrap C UNI-AX) with weights of 300 and 600 g/m^2, equivalent dry fabric thickness of 0.164 mm, tensile strength > 4900 N/mm^2, tensile modulus of 252 GPa, and elongation at break of 2%.
- Bidirectional carbon fiber fabrics (MapeWrap C BI-AX), with weights of 230 and 360 g/m^2, equivalent dry fabric thicknesses of 0.064 mm and 0.1 mm, respectively, tensile strength > 4800 N/mm^2, tensile modulus of 230 GPa, and elongation at break of 2.1%.
- Epoxy resin MapeWrap 31, a two-component resin used to bond the carbon fiber fabrics to concrete [43].

A total of 48 cylindrical specimens with a diameter of 100 mm and a height of 200 mm were prepared, in accordance with UNE-EN 12390-1:2022 [44]. Half of the specimens were produced without any additions (SA), while the other half were produced with the addition of 3 kg/m^3 of polypropylene fibers (PP), with a cement–sand–gravel–water ratio of 1:2:3:0.5.

All specimens were subjected to compressive failure testing in accordance with UNE-EN 12390-3:2020 [45] and were divided into two groups based on the results of the compressive test, using the following nomenclature: Type I specimens, those with a compressive strength greater than 15 N/mm^2, and Type II specimens, those with a compressive strength less than 15 N/mm^2. Once tested to failure, the specimens were reinforced with the four types of carbon fiber fabric.

The experimental phase was conducted in accordance with applicable international standards (CNR-DT 200 R1/2013; EN 14889-2:2006) [46,47].

The nomenclature for specimens with and without polypropylene fibers, reinforced with the various types of fabric, is presented in Table 1.

Table 1. The nomenclature of concrete test specimens with and without polypropylene fibers, with the four types of reinforcement fabrics.

Specimens/ Type of Fibers	UNI AX-300	UNI AX-600	Specimens/ Type of Fibers	UNI AX-300	UNI AX-600
Type I Stresses	SA-1-U 300 SA-2-U 300 SA-3-U 300	SA-1-U 600 SA-2-U 600 SA-3-U 600	Type I Stresses	PP-1-U 300 PP-2-U 300 PP-3-U 300	PP-1-U 600 PP-2-U 600 PP-3-U 600
Type II Stresses	SA-4-U 300 SA-5-U 300 SA-6-U 300	SA-4-U 600 SA-5-U 600 SA-6-U 600	Type II Stresses	PP-4-U 300 PP-5-U 300 PP-6-U 300	PP-4-U 600 PP-5-U 600 PP-6-U 600
Specimens/ Type of Fibers	BI AX-230	BI AX-360	Specimens/ Type of Fibers	BI AX-230	BI AX-360
Type I Stresses	SA-1-BI 230 SA-2-BI 230 SA-3-BI 230	SA-1-BI 360 SA-2-BI 360 SA-3- BI 360	Type I Stresses	PP-1-BI 230 PP-2-BI 230 PP-3-BI 230	PP-1-BI 360 PP-2-BI 360 PP-3-BI 360
Type II Stresses	SA-4-BI 230 SA-5-BI 230 SA-6-BI 230	SA-4-BI 360 SA-5- BI 360 SA-6-BI 360	Type II Stresses	PP-4-BI 230 PP-5-BI 230 PP-6-BI 230	PP-4-BI 360 PP-5-BI 360 PP-6-BI 360

SA: concrete specimens without the addition of polypropylene fibers; PP: concrete specimens with the addition of polypropylene fibers; BI: type of bidirectional carbon fiber fabric with balanced weight; U: type of bidirectional carbon fiber fabric with balanced weight; 300/600/230/360: weight of the carbon fiber.

Once the materials were prepared in the laboratory of the Escuela Técnica Superior de Edificación de Madrid, they were mixed in the IBERTEST planetary vertical-axis mixer, model CIB-701, updated to IB32-040V0. The cement, gravel, sand, and fibers were dry-mixed for two minutes to achieve material homogenization. Subsequently, water and the admixture were added, and the mixture was processed for five minutes until a plastic consistency of the concrete was achieved.

The concrete was then poured into the molds in accordance with the standard for fresh concrete testing (UNE-EN 12390-3) [45]. The molds were leveled and kept at laboratory ambient temperature, approximately 22 ± 3 °C, and a relative humidity of approximately 60% for 24 h before demolding. After 24 h, the specimens were removed from the molds and stored in a curing chamber at a temperature of 20 ± 2 °C and a relative humidity of ≥95% for 28 days for curing and hardening.

The fibers used for the study, supplied by the company MAPEI SPAIN S.A, have different morphologies. One type of fiber is unidirectional, and the other type is bidirectional. For each type of fiber, strips were cut with dimensions of 180 × 420 mm, with an overlap of 100 mm, meaning that each strip has a length of 180 mm and a width of 420 mm, with a 100 mm overlap at the end.

For the application of the carbon fiber on the concrete specimen, it was ensured that the surface of the specimens was clean, free of dust or foreign elements. Next, a first thin layer of resin was applied to the entire surface of the specimen using a roller. Immediately after, the carbon fiber was placed over the resin, and a second layer of resin was applied using a trowel, thereby confining the entire specimen. Finally, pressure was applied to the fiber to remove any excess resin that had been applied Figure 1a.

Once the specimens had cured, compression testing was carried out on reference specimens (with and without polypropylene fibers), which were then reinforced and retested under compression.

The process of reinforcing the specimens with carbon fiber fabric and their subsequent compression testing is illustrated in Figure 1b.

Figure 1. Reinforcement with CFRP (**a**); compression test (**b**).

3. Results y Analysis

For each concrete specimen, a stress–strain diagram is created, which allows us to study in a more graphical manner the most characteristic and significant values of the test and analyze the results analytically.

From the data obtained in the compression test using the MIB 60/AM universal press, a graph of force (F) in kN / stroke (δ) in mm is generated, containing all the results. Figure 2 shows the typical graph used.

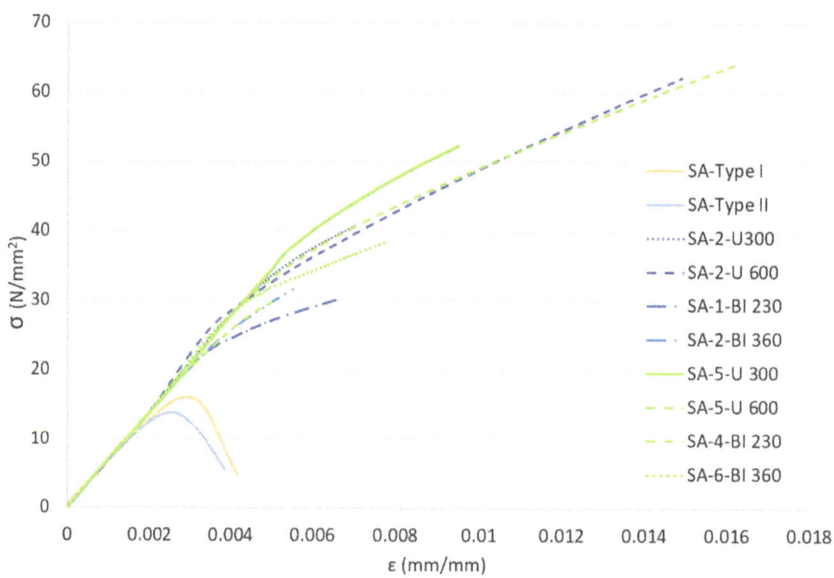

Figure 2. Tensile strength–strain curves in concretes without the addition of polypropylene fibers.

Since what we are really interested in is the concrete's compressive strength, the results for the compressive stress (σc) in N/mm^2 and the unit longitudinal strain (εc) are obtained from the force "F" in kN, the cross-sectional area "A" of the cylindrical specimen in mm^2, and its length (1) and (2):

$$\sigma_c = F/A \quad (1)$$

$$\varepsilon_c = \Delta L/L \quad (2)$$

σ_c: compressive stress (N/mm^2);
ε_c: dimensionless;
ΔL: length variation(mm);
L: specimen length(mm).

With these data, stress–strain graphs are generated and the results are analyzed to extract the key findings from the test.

First, the average values of the most representative results obtained from the compression tests on concretes with and without the addition of polypropylene fibers before reinforcement are presented. The concretes are divided into groups based on whether they correspond to Type I or Type II concretes (Table 2). These values include maximum stresses (σ_{max}), ultimate stresses (σ_{ult}), strains corresponding to the maximum stresses (ε_{max}), and strains corresponding to the ultimate stresses (ε_{ult}).

Table 2. Most representative average values from the compression test on concretes with and without polypropylene fibers prior to reinforcement.

	Specimens	σ_{max} (N/mm^2)	σ_{ult} (N/mm^2)	ε_{max}	ε_{ult}
Type I ≥ 15 N/mm^2	SA	16.24 ± 0.67	5.23 ± 0.43	0.0027 ± 0.003	0.0042 ± 0.006
	PP	18.93 ± 0.78	6.09 ± 0.35	0.0032 ± 0.002	0.0047 ± 0.001
Type II ≤ 15 N/mm^2	SA	12.62 ± 0.52	4.11 ± 0.50	0.0023 ± 0.004	0.0040 ± 0.008
	PP	13.60 ± 0.52	4.09 ± 0.18	0.0025 ± 0.001	0.0045 ± 0.004

The results obtained from the compression test are presented below, using the most representative graph for each case. These graphs show the evolution of stress relative to strain in Type I and Type II concretes (Figure 2), without polypropylene fibers and with the four types of reinforcement investigated. The highest values are achieved in Type II concretes, with strengths lower than 15 N/mm^2 [36]. In all types of reinforcement, the stresses significantly improve after reinforcement, with the greatest increases in strength observed for U 600 reinforcements [36].

Figure 3 shows the results obtained from the compression test, through the most representative graph for each case, illustrating the stress–strain evolution in Type I and Type II concretes with polypropylene fibers and the four types of reinforcement investigated. As observed in concretes without polypropylene fibers, the highest strengths are achieved in Type II concretes, whose strengths are below 15 N/mm^2. For all types of reinforcement, the stresses significantly improve when reinforced, with the highest strength increases observed for U 600 reinforcements.

Although the strain energy density (area under the stress–strain curve) was not calculated, it can be clearly observed in Figures 2 and 3 that the strain energy density is significantly higher in reinforced concretes compared to unreinforced concretes. Additionally, the highest strain energy densities are observed in concretes reinforced with unidirectional fibers of the highest weight.

The most representative values from the compression test on concretes without polypropylene fibers are shown in Table 3, based on the type of concrete. The table illustrates the increase in tensile strength and strain for each type of reinforcement compared to its unreinforced specimen. As can be observed, for confinement reinforcement, the unidirectional reinforcement performs better, significantly enhancing the ductility of the concrete.

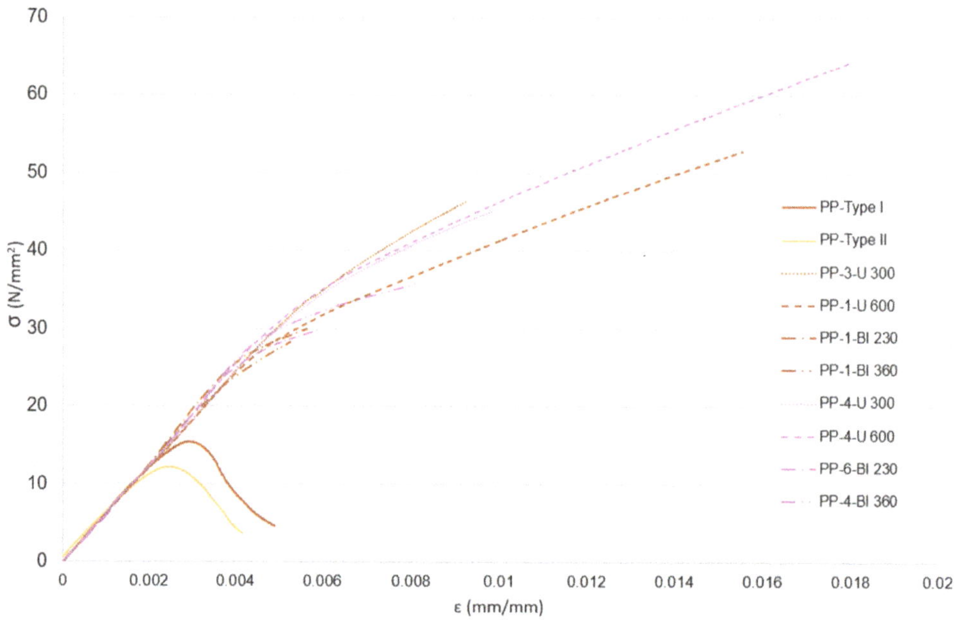

Figure 3. Tensile strength–strain curves in concretes with the addition of polypropylene fibers.

Table 3. Test results of reinforced specimens without the addition of polypropylene fibers.

	Specimens	σ_{max} (N/mm^2)	(%)	σ_{ult} (N/mm^2)	(%)	ε_{max}	(%)	ε_{ult}	(%)
Type I	SA-2-U 300	40.81 ± 1.17	195	40.81 ± 1.17	636	0.007 ± 0.04	192	0.007 ± 0.04	90
	SA-2-U 600	62.21 ± 0.27	288	62.21 ± 0.27	1187	0.015 ± 0.003	415	0.015 ± 0.003	259
	SA-1-BI 230	30.05 ± 0.41	101	30.05 ± 0.41	570	0.007 ± 0.01	163	0.007 ± 0.01	48
	SA-2-BI 360	31.91 ± 0.85	66	31.91 ± 0.85	450	0.006 ± 0.01	70	0.006 ± 0.01	15
Type II	SA-5-U 300	52.39 ± 0.38	328	52.39 ± 0.38	964	0.009 ± 0.09	327	0.009 ± 0.09	145
	SA-5-U 600	64.41 ± 0.89	464	64.41 ± 0.89	1778	0.016 ± 0.02	670	0.016 ± 0.02	603
	SA-4-BI-230	30.21 ± 0.61	97	30.21 ± 0.61	557	0.005 ± 0.09	104	0.005 ± 0.09	40
	SA-6-BI 360	38.35 ± 0.23	214	38.35 ± 0.23	944	0.008 ± 0.007	244	0.008 ± 0.007	106

Table 4 shows the results of the most representative graphs from the compression test on concrete with polypropylene fibers, based on the type of concrete. It can be observed that reinforcements with unidirectional fabrics achieve the highest strengths and the greatest strain increments, which implies an improvement in the ductility of the concrete.

As observed in Figure 4, all failures of the reinforced specimens are brittle, yet they exhibit significantly higher strength and energy absorption capacity compared to the unreinforced specimens [36]. The unreinforced specimens strengthened with unidirectional fiber experience a very violent brittle failure, whereas those reinforced with bidirectional fiber fail in a brittle manner but retain their structure. In concretes with polypropylene fibers, despite the brittle failure, the fragments remained connected, maintaining the cohesion of the specimen.

Table 4. Test results of specimens reinforced with the addition of polypropylene fibers.

	Specimens	σ_{max} (N/mm^2)	(%)	σ_{ult} (N/mm^2)	(%)	ε_{max}	(%)	ε_{ult}	(%)
Type I	PP-3-U 300	46.45 ± 0.26	229	46.45 ± 0.26	992	0.009 ± 0.014	120	0.009 ± 0.014	120
	PP-1-U 600	52.86 ± 1.57	211	52.86 ± 1.57	937	0.016 ± 0.028	187	0.016 ± 0.028	187
	PP-1-BI 230	29.98 ± 0.59	31	29.98 ± 0.59	328	0.006 ± 0.017	27	0.006 ± 0.017	27
	PP-1-BI 360	28.28 ± 0.7	34	28.28 ± 0.7	345	0.005 ± 0.018	−13	0.005 ± 0.018	−13
Type II	PP-4-U 300	45.15 ± 0.68	214	45.15 ± 0.68	942	0.010 ± 0.027	103	0.010 ± 0.027	103
	PP-4-U 600	64.24 ± 0.66	393	64.24 ± 0.66	1539	0.018 ± 0.035	270	0.018 ± 0.035	270
	PP-6-BI 230	29.95 ± 0.22	125	29.95 ± 0.22	645	0.006 ± 0.001	51	0.006 ± 0.001	51
	PP-4-BI 360	35.88 ± 0.61	160	35.88 ± 0.61	762	0.008 ± 0.017	86	0.008 ± 0.017	86

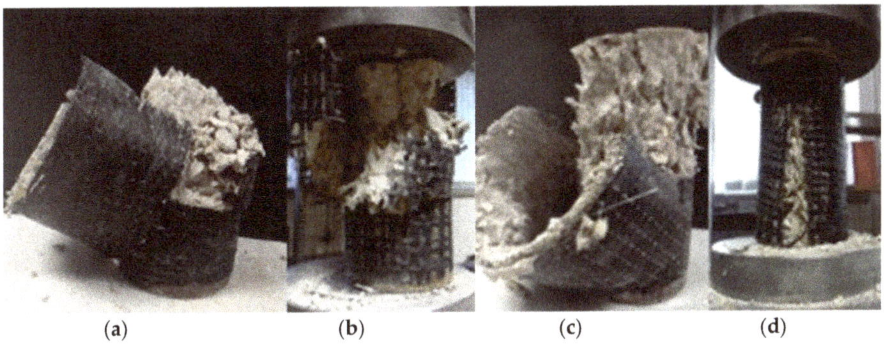

Figure 4. Specimen failures. (**a**) SA-2B-U 300; (**b**) SA-1L-BI 230; (**c**) PP-3C-U 300; (**d**) PP-1L-BI 230.

In Figure 5, it can be observed that the greatest strength increases in confined specimens compared to unconfined specimens occur in concretes with stresses below 15 N/mm^2 (Type II), both in concretes with and without polypropylene fibers and in concretes reinforced with higher weight unidirectional fabrics. A compressive strength increase of 464% was achieved in concretes without polypropylene fibers (SA-5-U600) and 393% in concretes with polypropylene fibers (PP-4-U600).

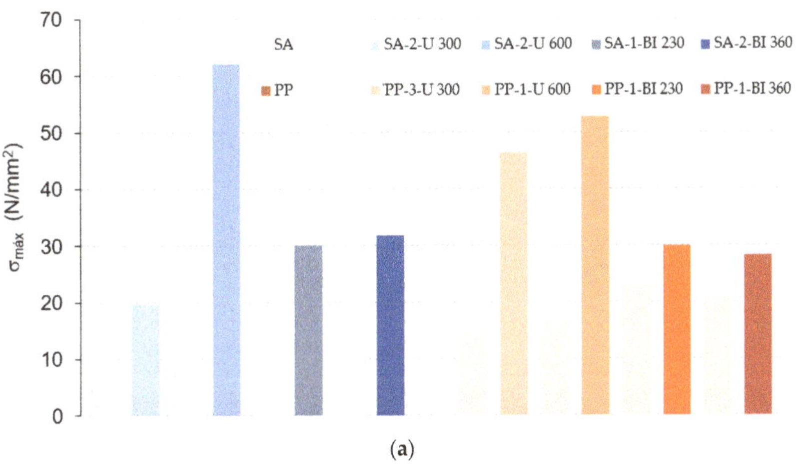

Figure 5. *Cont.*

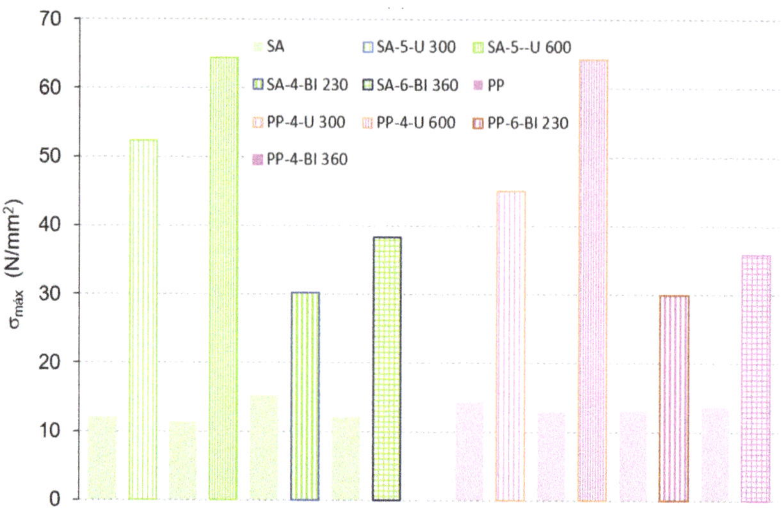

(**b**)

Figure 5. Maximum compressive stresses. (**a**) Type I specimens; (**b**) Type II specimens.

Figure 6 shows the maximum stresses reached in concretes with and without polypropylene fibers for each type of reinforcement. For bidirectional reinforcements, the behavior is similar regardless of whether the concrete contains polypropylene fibers or not. In the case of unidirectional reinforcements, the best stress performance is observed in concretes without polypropylene fibers, except for the U-300 reinforcement in Type I concretes.

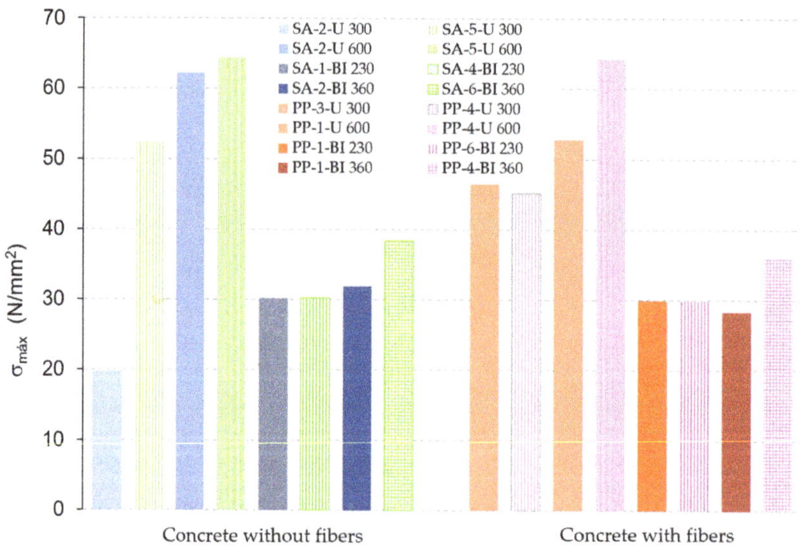

Figure 6. Maximum compressive stress in concretes with and without polypropylene fibers and different types of reinforcement.

Table 5 shows the percentage variation in stresses and strains among the different reinforced specimens, depending on whether or not they contain polypropylene fibers. As can be observed, the addition of fibers does not generally improve performance in terms of stress. However, an improvement is evident in the increase in maximum and ultimate strains when the concrete contains polypropylene fibers, which enhances ductility.

Table 5. Comparison between Type I and Type II stress specimens for specimens with and without polypropylene fibers.

Specimens	Stresses	σ_{max}, σ_{ult} (N/mm^2)	(%)	ε_{max}, ε_{ult}	(%)
SA-2-U 300 PP-3-U 300	Type I Type I	40.81 ± 1.17 46.45 ± 0.26	14	0.007 ± 0.04 0.009 ± 0.014	33
SA-5-U 300 PP-4-U 300	Type II Type II	52.39 ± 0.89 45.15 ± 0.68	−14	0.009 ± 0.009 0.010 ± 0.027	4
SA-2-U 600 PP-1-U 600	Type I Type I	62.21 ± 0.27 52.86 ± 1.57	−15	0.015 ± 0.03 0.016 ± 0.028	4
SA-5-U 600 PP-4-U 600	Type II Type II	64.41 ± 0.89 64.24 ± 0.66	−0.3	0.016 ± 0.02 0.018 ± 0.035	10
SA-1-BI 230 PP-1-BI 230	Type I Type I	30.05 ± 0.41 29.98 ± 0.59	−0.2	0.007 ± 0.01 0.006 ± 0.017	−14
SA-4-BI 230 PP-6-BI 230	Type II Type II	30.21 ± 0.61 29.95 ± 0.22	−1	0.005 ± 0.02 0.006 ± 0.001	15
SA-2-BI 360 PP-1-BI 360	Type I Type I	31.91 ± 0.85 28.28 ± 0.70	−11	0.006 ± 0.01 0.005 ± 0.018	−6
SA-6-BI 360 PP-4-BI 360	Type II Type II	38.35 ± 0.23 35.88 ± 0.61	−6	0.008 ± 0.007 0.008 ± 0.017	5

Figure 7 shows the maximum stresses achieved in concretes with and without polypropylene fibers as a function of the type of reinforcement. As can be observed, the highest strengths were achieved with unidirectional U-600 reinforcements in concretes with strengths below 15 N/mm^2, reaching strength increases of over 450% [36]. Reinforcements with unidirectional U-300 fabrics also achieved significant strength increases on the order of 220%. Although lower than those achieved by the fabric with a higher weight, reinforcing concrete with the unidirectional fabric of 600 g/m^2 results in a 22% increase in strength compared to the 300 g/m^2 fabric in Type I reference concretes and a 52% increase in Type II reference concretes. In concretes with polypropylene fibers, the increase in fabric weight leads to a 13% increase in strength in Type I concretes and a 42% increase in Type II concretes. The bidirectional fabrics studied achieved much smaller strength increases compared to the unidirectional fabrics. In the case of the BI-230 reinforcement, the strength increase was similar across all specimens, regardless of their initial strength or whether the concrete contained polypropylene fibers. For the BI-360 reinforcement, the behavior was similar to the previous case, but better performance was observed in Type II concretes.

In summary, the confined specimens, both with and without the addition of fibers, show improvements in maximum stress and ultimate strain compared to the unconfined specimens [36]. This demonstrates that the jointly confined specimen possesses good mechanical properties, with the U 600 fiber yielding the best results in this regard. This suggests that combining multiple confinement methods can be beneficial for improving the mechanical properties of structures subjected to loading.

Next, the existing theoretical models will be validated using the results obtained from the experimental tests, comparing the values with the formulations of ACI 440-17 (ACI PRC-440.2-17: *Guide for the Design and Construction of Externally Bonded FRP Systems for Strengthening Concrete Structures*) [48] and FIB (*Externally bonded FRP reinforcement for*

RC structures. International federation for Structural Concrete, Laussanne) [49]. Figure 3 illustrates the two stages of the stress–strain curve of a jointly confined specimen during a uniaxial compression test. As can be observed, the stress–strain curve in this study has two well-differentiated zones:

Elastic zone: In this stage, the stress–strain curve is a straight line. As the load increases, the line gradually and slightly curves. This curve is similar to the constitutive curve of unconfined concrete. That is, the elastic behavior of confined concrete is similar to that of unconfined concrete, although there may be some differences due to the confinement provided by the CFRP.

Plastic stress stage: In this stage, the presence of CFRP prevents the rapid expansion caused by the internal cracks in the concrete. This results in a triaxial stress state in the core area of the concrete. During this stage, the stress–strain curve is practically a straight line, allowing the confinement to enable the concrete to reach higher stress levels before failure compared to unconfined concrete.

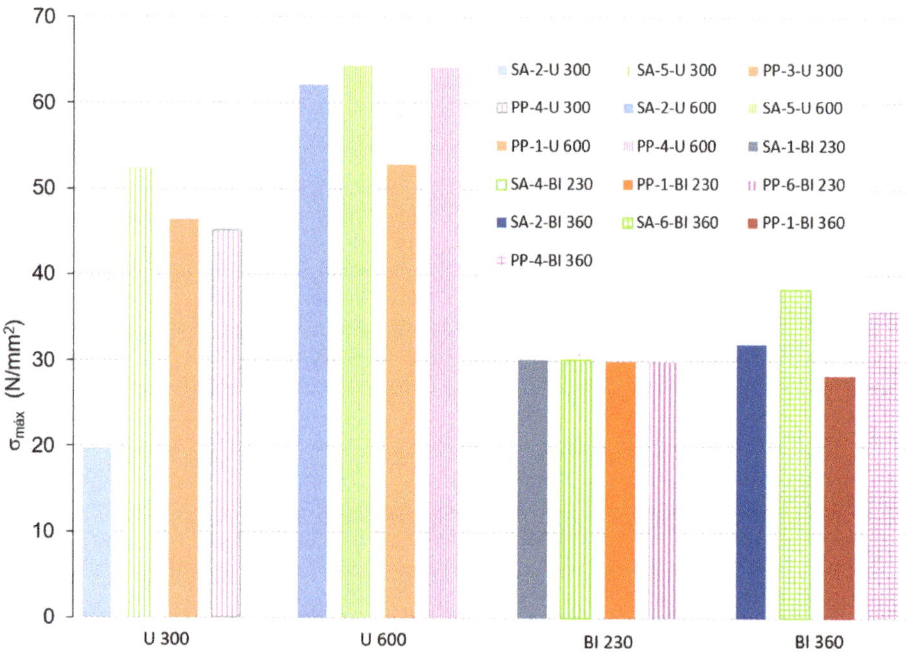

Figure 7. Maximum compressive stress with different types of reinforcement.

Table 6 shows the models to be validated. Using the ACI model [48]. the calculation of the maximum strength of confined concrete can be determined using Equations (3) and (4) proposed by Lam and Teng. To validate the model indicated by the FIB [49]. Equations (5)–(9) will be used.

Table 6. Summary of calculation models.

ACI440-17	FIB
$f_{cc} = f_c + \varphi_f * 3.3 * f_l$ (4)	$f_{cc} = f_c * (2.254\ x \sqrt{1 + 7.94 * \frac{f_l}{f_c}} - 2 * \frac{f_l}{f_c} - 1.254)$ (7)
$f_l = \frac{2 * E_f * n * t_f * \varepsilon_{fe}}{D}$ (5)	$f_l = \frac{2 * E_f * n * t_f * \varepsilon_{fe}}{D}$ (8)
$\varepsilon_{fe} = K_e * \varepsilon_{fu} = 0.55 * \varepsilon_{fu}$ (6)	ε_{fe} (9)

The above can be explained as follows:

f_{cc}: Compressive strength of confined concrete. (N/mm^2).
f_c: Compressive strength of unconfined concrete. (N/mm^2).
f_l: Maximum confinement pressure due to FRP reinforcement. (N/mm^2).
E_f: Tensile modulus of elasticity of FRP. (N/mm^2).
n: Number of layers.
t_f: Thickness of the FRP material (mm).
ε_{fe}: Effective strain in the FRP reinforcement at failure (mm/mm).
ε_{fu}: Design tensile strain of the FRP reinforcement (mm/mm).
K_e, φ_f: Coefficients.

From the above Equations (1)–(6), the theoretical values of confinement stresses are calculated and compared with the experimentally obtained values. The analysis begins with the specimens without fiber addition. For this purpose, the average failure stress values in confined concrete (both theoretical and experimental) and the average failure stress values in unconfined concrete for Type I and Type II stresses are calculated (Figure 8).

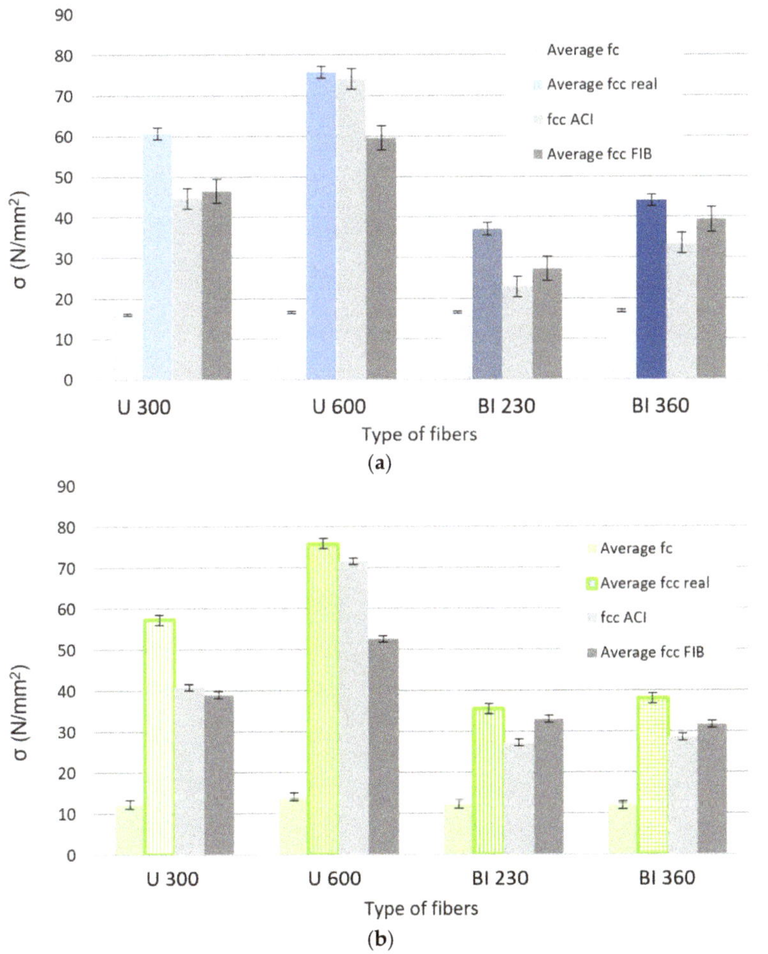

Figure 8. Experimental and theoretical confinement values in concretes without polypropylene fibers. (**a**) Type I stresses; (**b**) Type II stresses.

Both the theoretical and experimental results demonstrate a significant improvement in the strength of reinforced concretes for Type I and Type II stresses. The experimental results are on the safe side for all types of reinforcement when the concrete does not contain polypropylene fibers, as the confinement stress values are higher than the theoretical values calculated using the ACI [48] and FIB [49]. The best theoretical results align with the experimental ones, with the highest stresses achieved when reinforced with the unidirectional fabric of the highest weight.

When concretes with polypropylene fibers are reinforced, the experimentally obtained stresses are not always greater than the theoretical values, meaning they are not on the safe side. Therefore, additional tests should be conducted to better correlate the theoretical and experimental behavior of confined concretes with polypropylene fiber additions (Figure 9).

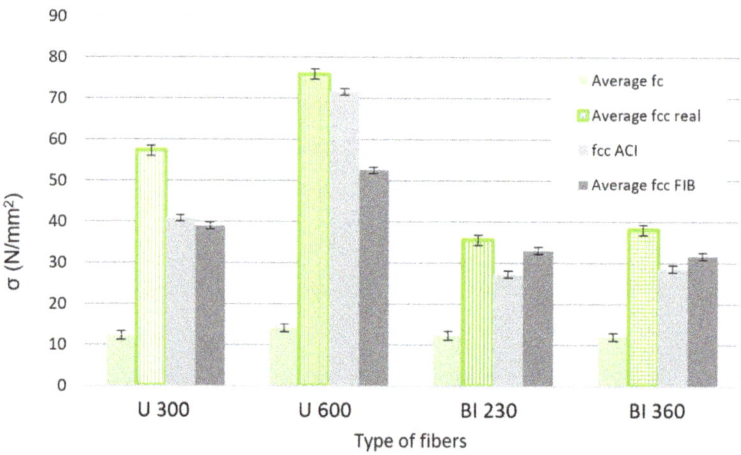

(**a**)

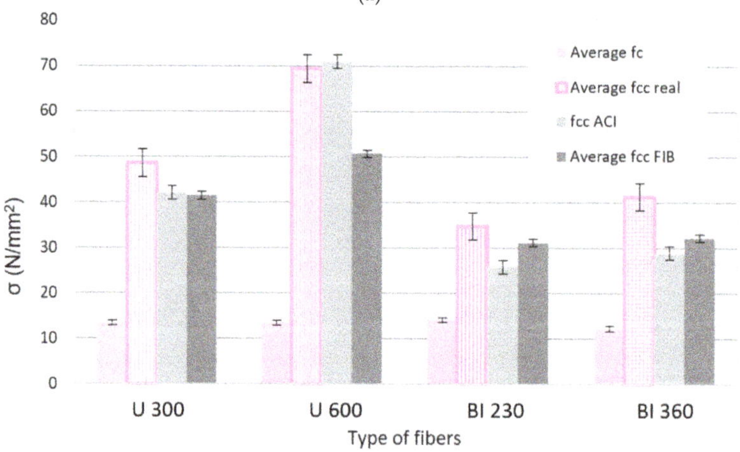

(**b**)

Figure 9. Experimental and theoretical confinement values in concretes with polypropylene fibers. (**a**) Type I stresses; (**b**) Type II stresses.

Conversely, for the specimens with polypropylene fiber addition (Figure 9) subjected to Type I stresses, the U 300 fiber exceeds the theoretical values by a large margin, and although the U 600 fiber remains above the theoretical value from FIB, it is much closer to the theoretical value from ACI. For the values corresponding to the bidirectional fabrics BI 230 and BI 360, both actual values fall below the theoretical values.

When examining the Type II stress values, all theoretical values are lower than the actual resistances, except in the case of the U 600 reinforcement, where the theoretical value exceeds the experimental value.

4. Conclusions

This article presents an experimental study on the compressive behavior of cylindrical concrete specimens with and without polypropylene fibers, reinforced with CFRP. Forty-eight specimens were tested under axial compression load with stresses below 15 N/mm^2 and above 15 N/mm^2. From the analysis of the results, the following conclusions can be drawn:

- The greatest compressive strength increases obtained by confining concrete were achieved in concretes with unconfined strengths below 15 N/mm^2 (Type II), reinforced with unidirectional fabrics of 600 g/m^2 weight. Specifically, a 464% increase was obtained in reference concretes (SA-5-U600) (Specimen 5 confined with MapeWrap C UNI-AX 600 carbon fibers without the addition of polypropylene fibers) and a 393% increase in concretes with polypropylene fibers (PP-4-U600) (Specimen 4 confined with MapeWrap C UNI-AX 600 carbon fibers with the addition of polypropylene fibers).
- The reinforcement of CFRP with the unidirectional fabric of 600 g/m^2 results in a greater increase in strength compared to unreinforced concrete than when using a unidirectional fabric reinforcement of 300 g/m^2. When reinforcement is carried out with bidirectional fabrics of 230 and 360 g/m^2, the strength increases are significantly lower than those achieved with unidirectional fabrics of 300 and 600 g/m^2, regardless of whether the concrete contains polypropylene fibers or not.
- As shown in Figures 2 and 3, the strain energy density (area under the stress–strain curve) increases substantially when the concrete is reinforced with carbon fiber fabrics, reaching the highest values when unidirectional reinforcements with a weight of 600 g/m^2 are used. This is observed both in reference concretes and in concretes with polypropylene fibers, significantly enhancing the ductility of the concrete.
- The confinement reinforcement of concrete with bidirectional carbon fibers increases the strength compared to unreinforced concrete, but in a similar manner regardless of the fabric weight used (230 and 360 g/m^2) or whether the concrete contains polypropylene fibers or not. Higher stress levels were only achieved when the unreinforced concrete had compressive strengths below 15 N/mm^2 (Type II), using the reinforcement with the higher fabric weight.
- The addition of polypropylene fibers to the concrete does not improve its compressive strength but enhances the cohesion of the concrete, thereby preventing explosive failure of the specimens.
- Type II stresses ($\sigma < 15$ N/mm^2) have achieved the highest values compared to Type I stresses. Additionally, there is a closer alignment of stress values for bidirectional fibers compared to unidirectional fibers.
- The mechanical behavior of confined reference concretes reinforced with carbon fiber is on the safe side compared to the theoretical values formulated by specialized standards (ACI [48] and FIB [49]). However, the same cannot be said for concretes with polypropylene fibers.
- Low-performance concretes with strengths below 15 N/mm^2 can achieve significant increases in compressive strength when confined with unidirectional carbon fiber fabrics of 600 g/m^2 weight. The same type of reinforcement in concretes with the addition of polypropylene fibers also generates significant improvements in compressive

strength. However, the values obtained are not always on the safe side concerning the existing standards (ACI [48] and FIB [49]).

Author Contributions: Conceptualization, T.G.P. and M.I.P.; methodology, T.G.P. and M.I.P.; software, A.L.; validation, D.F., M.I.P. and T.G.P.; formal analysis, A.L.; investigation, M.I.P.; resources, D.F.; data curation, A.L.; writing—original draft preparation, T.G.P.; writing—review and editing, M.I.P.; visualization, D.F.; supervision, A.L.; project administration, D.F.; funding acquisition, D.F. All authors have read and agreed to the published version of the manuscript.

Funding: This research received no external funding.

Institutional Review Board Statement: Not applicable.

Informed Consent Statement: Not applicable.

Data Availability Statement: Data available on request.

Acknowledgments: The authors thank the company Mapei Spain S.A.U. for the supplied materials.

Conflicts of Interest: The authors declare no conflict of interest.

References

1. Bakis, C.E.; Bank, L.C.; Brown, V.L.; Cosenza, E.; Davalos, J.F.; Lesko, J.J.; Machida, A.; Rizkalla, S.H.; Triantafillou, T.C. Fiber-reinforced polymer composites for construction—State-of-the-art review. *J. Compos. Constr.* **2002**, *6*, 73–87. [CrossRef]
2. De, B.; Bera, M.; Bhattacharjee, D.; Ray, B.C.; Mukherjee, S. A comprehensive review on fiber-reinforced polymer composites: Raw materials to applications, recycling, and waste management. *Prog. Mater. Sci.* **2024**, *146*, 101326. [CrossRef]
3. Hammad, M.; Bahrami, A.; Khokhar, S.A.; Khushnood, R.A. A State-of-the-Art Review on Structural Strengthening Techniques with FRPs: Effectiveness, Shortcomings, and Future Research Directions. *Materials* **2024**, *17*, 1408. [CrossRef] [PubMed]
4. Arabshahi, A.; Tavakkolizadeh, M. Predictive model for slenderness limit of circular RC columns confined with FRP wraps. *Struct. Concr.* **2021**, *23*, 849–875. [CrossRef]
5. Jiang, T.; Wang, X.F.; Zhang, W.; Chen, G.; Lin, Z. Behavior of FRP-Confined Recycled Brick Aggregate Concrete under Monotonic Compression. *J. Compos. Constr.* **2020**, *24*, 04020067. [CrossRef]
6. Feng, C.; Yu, F.; Fang, Y. Mechanical behavior of PVC tube confined concrete and PVC-FRP confined concrete: A review. *Structures* **2021**, *31*, 613–635. [CrossRef]
7. Bai, Y.; Dai, J.; Yaseen, Z.M. Cyclic stress-strain model incorporating buckling effect for steel reinforcing bars embedded in FRP-confined concrete. *Compos. Struct.* **2017**, *182*, 54–66. [CrossRef]
8. Del Rey Castillo, E.; Griffith, M.C.; Ingham, J. Seismic behavior of RC columns flexurally strengthened with FRP sheets and FRP anchors. *Compos. Struct.* **2018**, *203*, 382–395. [CrossRef]
9. Ouyang, L.; Zeng, J.; Zhen, B.; Lu, Z. Seismic retrofit of square reinforced concrete columns using basalt and carbon fiber-reinforced polymer sheets: A comparative study. *Compos. Struct.* **2017**, *162*, 294–307. [CrossRef]
10. Alabduljabbar, H.; Mamun, A.A.; Alyousef, R.; Amran, Y.M. Strengthening of reinforced concrete beams by using fiber-reinforced polymer composites: A review. *J. Build. Eng.* **2019**, *25*, 100798. [CrossRef]
11. Soudki, K.; Alkhrdaji, T. Guide for the design and construction of externally bonded FRP systems for strengthening concrete structures (ACI 440.2 R-02). In *Structures Congress 2005: Metropolis and Beyond*; American Society of Civil Engineers: Reston, VA, USA, 2005; pp. 1–8.
12. Abdalla, J.A.; Hraib, F.; Hawileh, R.A.; Mirghani, A.M. Experimental investigation of bond-slip behavior of aluminum plates adhesively bonded to concrete. *J. Adhes. Sci. Technol.* **2017**, *31*, 82–99. [CrossRef]
13. Smith, S.T.; Teng, J.G. FRP-strengthened RC beams. I: Review of debonding strength models. *Eng. Struct.* **2002**, *24*, 385–395. [CrossRef]
14. Haroun, M.A.; Feng, M.Q.; Bhatia, H.; Baird, K.; Elsanadedy, H. *Structural Qualification Testing of Composite-Jacketed Circular and Rectangular Bridge Columns*; No. FHWA/CA/UCI-99-01; The National Academies of Sciences, Engineering, and Medicine: Washington, DC, USA, 1999.
15. Park, T.; Na, U.J.; Chung, L.; Feng, M.Q. Compressive behavior of concrete cylinders confined by narrow strips of CFRP with spacing. *Compos. Part B-Eng.* **2008**, *39*, 1093–1103. [CrossRef]
16. Chen, D.; Sun, G.; Meng, M.; Jin, X.; Li, Q. Flexural performance, and cost efficiency of carbon/basalt/glass hybrid FRP composite laminates. *Thin-Walled Struct.* **2019**, *142*, 516–531. [CrossRef]

17. Zeng, J.; Guo, Y.; Zeng, J.; Chen, W. Stress-strain behavior of concrete in circular concrete columns partially wrapped with FRP strips. *Compos. Struct.* **2018**, *200*, 810–828. [CrossRef]
18. Hao, H.; Hadi, M.N.S.; Youssef, J.A. Optimized FRP Wrapping Schemes for Circular Concrete Columns under Axial Compression. *J. Compos. Constr.* **2015**, *19*, 04015015. [CrossRef]
19. Zeng, J.; Guo, Y.; Zeng, J.; Li, J.; Xie, J. Behavior of partially and fully FRP-confined circularized square columns under axial compression. *Constr. Build. Mater.* **2017**, *152*, 319–332. [CrossRef]
20. Triantafyllou, G.G.; Rousakis, T.C.; Karabinis, A.I. Axially Loaded Reinforced Concrete Columns with a Square Section Partially Confined by Light GFRP Straps. *J. Compos. Constr.* **2015**, *19*, 04014035. [CrossRef]
21. Baquer Sistach, J. *La Fibra de Carbono en Refuerzo de Estructuras de Hormigón*; Institut d´Estudis Estructurals: Barcelona, Spain, 2021.
22. Akeed, M.H.; Qaidi, S.M.A.; Ahmed, H.U.; Faraj, R.H.; Mohammed, A.; Emad, W.; Tayeh, B.A.; Azevedo, A.R. Ultra-high-performance fiber-reinforced concrete. Part IV: Durability properties, cost assessment, applications, and challenges. *Case Stud. Constr. Mater.* **2022**, *17*, e01271. [CrossRef]
23. Cao, Q.; Gao, Q.; Jia, J.; Gao, R. Early-Age Cracking Resistance of Fiber-Reinforced Expansive Self-Consolidating Concrete. *Am. Concr. Inst. J.* **2019**, *116*, 15–26. [CrossRef]
24. Yousefieh, N.; Joshaghani, A.; Hajibandeh, E.; Shekarchi, M. Influence of fibers on drying shrinkage in restrained concrete. *Constr. Build. Mater.* **2017**, *148*, 833–845. [CrossRef]
25. Malchiodi, B.; Pelaccia, R.; Pozzi, P.; Siligardi, C. Three sustainable polypropylene surface treatments for the compatibility optimization of PP fibers and cement matrix in fiber-reinforced concrete. *Ceram. Int.* **2023**, *49*, 24611–24619. [CrossRef]
26. Mao, J.; Liang, N.; Liu, X.; Zhong, Z.; Zhou, C. Investigation on early-age cracking resistance of basalt-polypropylene fiber reinforced concrete in restrained ring tests. *J. Build. Eng.* **2023**, *70*, 106155. [CrossRef]
27. Esakki, A.D.K.; Dev, S.K.A.; Gomathy, T.; Gifta, C.C. Influence of adding steel–glass–polypropylene fibers on the strength and flexural behaviour of hybrid fiber reinforced concrete. *Mater. Today Proc.* 2023, *in press*. [CrossRef]
28. Taghipoor, H.; Sadeghian, A.G. Experimental investigation of single and hybrid-fiber reinforced concrete under drop weight test. *Structures* **2022**, *43*, 1073–1083. [CrossRef]
29. Zhai, K.; Fang, H.; Li, B.; Guo, C.; Yang, K.; Du, X.; Du, M.; Wang, N. Failure experiment on CFRP-strengthened prestressed concrete cylinder pipe with broken wires. *Tunn. Undergr. Space Technol.* **2023**, *135*, 105032. [CrossRef]
30. Khan, M.S.; Hashmi, A.F.; Shariq, M.; Ibrahim, S.M. Effects of incorporating fibres on mechanical properties of fibre-reinforced concrete: A review. *Mater. Today Proc.* 2023, *in press*. [CrossRef]
31. Hosseinzadeh, H.; Salehi, A.M.; Mehraein, M.; Asadollahfardi, G. The effects of steel, polypropylene, and high-performance macro polypropylene fibers on mechanical properties and durability of high-strength concrete. *Constr. Build. Mater.* **2023**, *386*, 131589. [CrossRef]
32. Su, Q.; Xu, J. Durability and mechanical properties of rubber concrete incorporating basalt and polypropylene fibers: Experimental evaluation at elevated temperatures. *Constr. Build. Mater.* **2023**, *368*, 130445. [CrossRef]
33. Alhassan, M.A.; Al Rousan, R.Z.; Al Shuqari, E.A. Bond-slip behavior between fiber reinforced concrete and CFRP composites. *Ain Shams Eng. J.* **2019**, *10*, 359–367. [CrossRef]
34. Liu, X.; Wu, T.; Chen, H.; Liu, Y. Compressive stress-strain behavior of CFRP-confined lightweight aggregate concrete reinforced with hybrid fibers Author links open overlay panel. *Compos. Struct.* **2020**, *244*, 112288. [CrossRef]
35. Raza, A.; Rafique, U.; Masood, B.; Ali, B.; ul Haq, F.; Nawaz, M.A. Performance evaluation of hybrid fiber reinforced low strength concrete cylinders confined with CFRP wraps Ali Raza. *Structures* **2021**, *31*, 182–189. [CrossRef]
36. Sathwik, M.C.; Prashanth, M.H.; Naik, S.C.; Satish, A. Experimental and numerical studies on compressive behaviour of CFRP wrapped cylindrical concrete specimens subjected to different pre-loading conditions. *Mater. Today Proc.* **2020**, *27*, 327–335. [CrossRef]
37. Serrano, R.; Cobo, A.; Prieto, M.I.; González, M.N. Analysis of fire resistance of concrete with polypropylene or steel fibers. *Constr. Build. Mater.* **2016**, *122*, 302–309. [CrossRef]
38. Fernández-Cánovas, M.; González-García, M.N.; Piñero, J.Á.; Cobo, A. Compressive strength behaviour of low- and medium-strength concrete specimens confined with carbon fibres in defective implementation conditions: An experimental study. *Mater. De Constr.* **2016**, *66*, e103. [CrossRef]
39. *UNE-EN 197-1:2011*; Cement. Part 1: Composition, Specifications, and Conformity Criteria for Common Cements. Spanish Association for Standardization: Madrid, Spain, 2011.
40. Ministry of Public Works. RC-16: Instruction for the Reception of Cements. Technical General Secretariat of the Ministry of Public Works. 2016. Available online: https://www.mitma.gob.es/ (accessed on 15 June 2024).
41. *UNE-EN 12620:2003+A1:2009*; Aggregates for Concrete. Spanish Association for Standardization: Madrid, Spain, 2009. Available online: https://www.une.org/encuentra-tu-norma/busca-tu-norma/norma?c=N0043155 (accessed on 3 June 2024).
42. *UNE-EN 13139/AC:2004*; Aggregates for Mortar. Spanish Association for Standardization: Madrid, Spain, 2004. Available online: https://www.une.org/encuentra-tu-norma/busca-tu-norma/norma?c=N0032204 (accessed on 15 May 2024).

43. Mapei Adhesivos, Selladores y Productos Químicos Para la Construcción (No Date) Spain. Available online: https://www.mapei.com/es/es/pagina-de-inicio (accessed on 13 June 2023).
44. *UNE-EN 12390-1:2022*; Testing Hardened Concrete. Part 1: Shape, Dimensions, and Other Requirements for Specimens and Moulds. Spanish Association for Standardization: Madrid, Spain, 2022. Available online: https://www.une.org/encuentra-tu-norma/busca-tu-norma/norma?c=N0069630 (accessed on 18 April 2024).
45. *UNE-EN 12390-3:2020*; Testing Hardened Concrete. Part 3: Compressive Strength of Test Specimens. Spanish Association for Standardization: Madrid, Spain, 2020. Available online: https://www.une.org/encuentra-tu-norma/busca-tu-norma/norma?c=N0063272 (accessed on 13 May 2024).
46. *CNR-DT 200 R1/2013*; Guide for the Design and Construction of Externally Bonded FRP Systems for Strengthening Existing Structures. Consiglio Nazionale Delle Ricerche: Roma, Italy, 2013.
47. *Norma Europea EN 14889-2:2006*; Fibras Para Hormigón. Parte 2: Fibras Poliméricas. Definiciones, Especificaciones y Conformidad. Asociación Española de Normalización: Madrid, Spain, 2006.
48. *ACI PRC-440.2-17*; Guide for the Design and Construction of Externally Bonded FRP Systems for Strengthening Concrete Structures. American Concrete Institute (ACI): Toronto, ON, Canada, 2017. Available online: https://www.concrete.org/store/productdetail.aspx?ItemID=440217&Language=English&Units=US_AND_METRIC (accessed on 6 May 2023).
49. Fédération Internationale du Béton (FIB). *Externally Bonded FRP Reinforcement for RC Structures*; International Federation for Structural Concrete: Lausanne, Switzerland, 2001.

applied sciences

MDPI

Article

Efficiency of Flange-Bonded CFRP Sheets in Relocation of Plastic Hinge in RC Beam–Column Joints

Farzad Hejazi [1,*], Ramin Azarm [2] and Ali Akbar Firoozi [3]

1 School of Engineering, University of the West of England, Bristol BS16 1QY, UK
2 Department of Civil Engineering, University Putra Malaysia, Serdang 43400, Malaysia; ramin.azarm@gmail.com
3 Department of Civil Engineering, University of Botswana, Gaborone UB0061, Botswana; a.firoozi@gmail.com
* Correspondence: farzad.hejazi@uwe.ac.uk

Abstract: Beam–column connection zones are high regions of interest in reinforced concrete (RC) structures, which are expected to respond elastically to seismic loads. Using carbon fiber-reinforced polymers (CFRP) to improve these connections, performance is critical in retrofitting deficient RC frames because existing slabs may pose numerous limitations in the design and wrapping of CFRP sheets in joints. The main aim of this research is to develop a new design for flange-bonded CFRP retrofit of frames, including slabs, for the relocation of plastic hinges of the connection area toward the beam and to develop beam–column joint capacity and building stability in cases of subjection to dynamic loads. The performance of these proposed retrofittings was explored both experimentally and numerically. Two full-scale fabricated interior RC joints of a real moment-resisting frame with moderate ductility were subjected to monotonic loads before and after retrofitting, and the results were used to detail the numerical progress and verify of the beam–column connection. Moreover, a parametric study was conducted on CFRP sheets' optimal thickness to examine its influence on plastic hinge relocation in the connection region. Results show that the retrofitting method can efficiently relocate the plastic hinge to the mid-span of the beam, which, in turn, leads to improved capacity and achievement of the RC frame and guarantees better structural safety a lower cost.

Keywords: seismic retrofitting; RC joints; CFRP sheets; non-linear static analysis; flange-bonded plastic hinge; safety

1. Introduction

The use of carbon fiber-reinforced polymers (CFRPs) is to repair and retrofit reinforced concrete members, and this has lately given rise to an extreme eagerness and interest expressed by researchers and engineers alike [1–3]. The rehabilitation of such frames can be carried out locally by strengthening individual elements such as beams, columns, and beam–column connections, or globally by including shear-resisting elements such as shear walls or braces. Various strengthening methods for RC frames, which were not designed for seismic loads, were proposed by employing carbon fiber-reinforced polymer (CFRP) to enhance the collapse-resistant capacity of the joints. The experimental test results showed that the proposed reinforcement designs could effectively increase the load resistance capacity of the joint but failed to reduce the severe softening behavior [4]. The response of non-ductile, heat-damaged reinforced concrete (RC) beam–column joints was investigated through some experimental research works. The findings proved a considerable improvement in the cyclic performance of the joints strengthened with a hybrid CFRP system [5]. In an experimental study, the behavior of three-dimensional reinforced concrete joints under combined lateral and gravity loads was tested via the strength method with CFRP composite sheets on two reinforced samples, which showed an increase in strength of 9% to 23% [6]. Many studies were carried out on the strengthening of 3D external RC beam–column joints without transverse bars in their common core using CFRP. The results indicated that strengthening prevents the diagonal shear failure at the column in the transverse beam connection point, and also increases the lateral load capacity by 11–15%, energy absorption by 10–20%, and energy loss by 2.2 times [7]. Thereafter, the seismic response of a beam–column connection with a shear deficiency (BCJ) reinforced with UHPFRC and/or CFRP was investigated. The results revealed that the synergistic action of UHPFRC and CFRP can increase the structural stiffness, strength, and energy dissipation capacity of the BCJ compared to other strengthening techniques [8]. In general, beam–column connections are a vital component of the structure not only in terms of stability but also in terms of seismic performance. By strengthening a basic building model, one can witness the bending effects of improving accuracy and reducing computational effort, and this method is effective for the earthquake-resistant design of frames [9]. In particular, shifting the formation location of plastic hinges off the column toward the beam interval is a useful method geared toward increasing the beam–column connection capacity, which can further enhance the overall seismic structural performance. To effectively relocate a plastic hinge, different retrofitting plans can be positioned, for example, web- and flange-bonded approaches and simultaneous web–flange-bonded approaches [10,11]. Literature-wise, much of the bulk of research in this area in recent years has focused on the former.

The experimental assessment of RC beam–column connections, either internally with steel bars or externally via CFRP sheets, has proven its capacity for noticeable joint enhancement through this strengthening technique [12]. In particular, the work has emphasized joint behavior under reverse cyclic loading to assess deformation capacity and strength. As a result, the findings have indicated that the bearing capacity and ductility of the connections are closely related to the original element condition and strengthening application. Furthermore, several rehabilitation schemes concerning joints with CFRP composites have been evaluated, whereby their configurations in the forms of U and X can be exemplified accordingly [13]. Therefore, it is reported that all schemes selected can successfully modify the mode from joint non-ductile shear failure to ductile-flexural. The results of experimental and numerical studies revealed that when the joints in an RC frame are reinforced by CFRP, at that moment, the capacity increases by 37%, whereas a maximum of 18% decrement of joint rotation is observed [14,15].

The results of the test for the RC joints retrofitted by hybrid CFRP sheets consequently suggest that this scheme may efficiently ameliorate the stiffness and rehabilitate the joint capacity efficiently [16]. The effectiveness of near-surface-mounted carbon-reinforced polymer ropes for enhancing interior-reinforced concrete (RC) beam–column joints was investigated. The X-shaped rope, implemented with two ropes on each side of the beam and two on each side of the column, imparted the best effectiveness on stiffness and yielding load, and the stiffness increased by 236.7% compared to the control beam–column joint [17]. Thereafter, a full-scale 3D exterior beam–column joint, lacking shear reinforcement but externally shear-strengthened with CFRP sheets, was investigated with respect to its behavior under both gravity and lateral cyclic loadings. The results revealed enhancements in lateral load-bearing capacity, energy dissipation, and secant stiffness [18]. Meanwhile, the effects of externally strengthened connections by utilizing several kinds of composites have been experimentally studied, and the results revealed that the simultaneous use of CFRP and glass fiber-reinforced polymer (GFRP) is capable of resolving the issues of lacking ductility and shear strength of problematic joints [19]. The effect of a set of parameters related to the contribution of CFRP sheets that were externally attached to the joint was numerically investigated. The obtained results confirmed the effectiveness of the CFRP-strengthening technique with respect to a relatively wide range of considered parameters [20]. An investigation detailing RC beam–column joints strengthened with steel-reinforced polymer (SRP) systems and their seismic performance has been presented by Alessandro De Vita, and the results of an experimental campaign performed at the Laboratory of Materials and Structural Testing of the University of Salerno, Italy, have been published [21]. In the research, the outcomes of cyclic tests performed on SRP-strengthened joints have been examined via a comparison against those obtained from a prior experimental program. They include companion specimens otherwise not provided with transverse beam stubs and strengthened by CFRP systems.

Parallel to the above work, another scholarly attempt by Antonopoulos and Triantafillou has tested some scaled joints modified by using CFRP [22]. In particular, either partial or full debonding of composites has been identified as the dominating factor across the test results. Nonetheless, the work has shown that a 20% increment in strength and energy dissipation and a 100% increase in stiffness are both possible. Additionally, the analytical prediction of shear strength carried out was in decent agreement with the experimental results. Thereafter, to evaluate the joint performance, CFRP-retrofit methods were applied to fortify RC joints and results revealed that the use of appropriately increasing composites can considerably enhance the ductility and strength of joint samples laterally [23]. Moreover, another numerical study concluded that the extent of the effectiveness of the CFRP web-bonding of joints toward the relocation of a hinge is used in order to yield a seismically enhanced performance of an RC structure [24]. Here, comparisons have been made accordingly between the results of frame retrofitting at the joints' webs using CFRP sheets and steel braces [25]. The numerical model for RC joints with CFRP-cover has been developed and nonlinear analyses have been conducted. As a result, the numerical model proposed in the work is capable of predicting the experimental work with precision [26].

The investigation on the method of web-bonded strengthening in the case of deficient RC joints with CFRP confirmed its ability to restore and further upgrade the strength of the system [27]. Moreover, the proposed scheme has been demonstrated as possibly preventing brittle failure from occurring in the core of a joint, which is attributable to the formation of plastic deformations in the form of hinges relocated off the column. Moreover, the study has revealed that the schemes are capable of improving the behavior factor of the frame in the CFRP, as well as influencing the ductility while the bracing approach influenced the strength. As a further matter, the authors have highlighted the limitations of the web-bonded approach in the context of plastic hinge relocation in large joints, suggesting that the combined web–flange-bonding method could be more effective. A diamond wire saw-cutting technology was adopted to achieve annular wrapping of the joint core area with the CFRP sheets. It was indicated that the proposed method improved the shear bearing capacity of the joint core area, inhibited the development of diagonal cracks in the joint core area, and transferred the failure location mode [28]. Furthermore, similar studies have yielded the same results with regard to the joint web-bonded scheme in which the practical limitations of web-bonded and web–flange-bonded schemes are both considered for retrofitting real 3D RC joints [29,30].

Through the same application, another study utilized cementitious mortar to enhance the bond between CFRP rod panels [31], whereas the use of embedded CFRP rods in the beam–column joints in areas of high humidity has also been proposed [32]. The latter study opted for this method to prevent steel bar corrosion from occurring, therefore conducting the experimental study on a prototype under applied cyclic loads. In another numerical investigation, the efficiency of RC joints retrofitted with CFRP laminates has been investigated in detail [33]. For this purpose, three practical CFRP configurations were proposed, and the joint behavior was assessed via a detailed numerical model in FE software (i.e., ANSYS, 2019 R1) by subjecting them to monotonic non-linear analyses. The results revealed that the grooving method can shift the plastic hinge location off the column vicinity. Similarly, another numerical investigation aimed at studying the effects of the flange-bonded approach, which is geared at having the location of plastic hinges changed off the column face [34]. Scholars have conclusively stated that the flange-bonded approach leads to better results compared to the use of the web-bonded method, whereby their differences encompass several aspects including ductility, cost, capacity, and performance level. Moreover, they have given a report of the excellent compliance shown by the CFRP flange-bonded scheme performance compared with that of the steel-bracing method [35,36].

More recently, another study has presented a complete summary of the results obtained following the analytical and numerical simulations of the structural behavior possessed by RC beam–column joints retrofitted with different types of fiber-reinforced polymeric (CFRP) composite laminates and hybrid connectors. In the process, non-linear numerical simulations were performed to evaluate the behavior of the beam–column joints, of which the behavior of eight full-scale interior RC beam–column specimens was numerically evaluated [37]. As the beam–column joint is considered as the initial point to introduce plastic hinges when the frame is subjected to lateral load [38], many efforts have been made to relocate the plastic hinge formation from the joint to the beam section. This is achieved by using CFRP strengthening to sustain the structure stability [39]. In line with this, many numerical and analytical studies have been conducted to predict concrete section deformation retrofitted using CFRP plates and rods [40], while increasing the bonding between CFRP and concrete.

Based on the extensive literature review in this research, beam–column joints have been identified as the main causes of vulnerability in the RC frame in forming plastic hinges during structure loading, which may lead to its collapse. However, most of the research in the literature has focused on increasing the strength of beams or columns in the joint area; the location of the plastic hinge in the retrofitted joint has not been considered, even though the occurrence of a plastic hinge in the beam–column connection is the most critical threat to the functioning of the joint when the structure is subjected to catastrophic excitations. Hence, implementing an appropriate retrofitting method is required to prevent plastic hinge formation in the beam–column joint area. Therefore, the main aim of this research is proposing a new retrofitting scheme for beam–column connections in the RC frame using CFRP to relocate and shift plastic hinge from near the face of the column to the further length of the beam. Then, the joint is protected against applied excitations and able to function properly to provide structural safety and integrity.

For this purpose, in the present study, the newly proposed retrofitting scheme has been applied to the RC joint (beam–column connection), and experimental and numerical investigations have been carried out. Accordingly, two full-scale RC interior joints from a moment-resisting frame with moderate ductility have been cast and experimentally tested, and the results were used to validate the numerical solutions and to evaluate the performance of the CFRP flange-bonded retrofitting scheme. Furthermore, a parametric study has been conducted to find the optimal thicknesses of the CFRP sheets for relocating the plastic hinges from near the face of the column to further along length of the beam.

2. Seismic Retrofitting of Reinforced RC Structures Using CFRP Sheets

Recent earthquakes have underscored the vulnerability of existing structures to seismic loads, emphasizing the urgent need to address the structural integrity and seismic resilience in older buildings. These events have highlighted the shortcomings of many existing structures, particularly those built before modern seismic codes and standards were established. Consequently, there has been a growing focus on seismic retrofitting methods to enhance the resistance and performance of reinforced concrete structures against severe ground motion.

One widely employed method in seismic retrofitting is the use of carbon fiber-reinforced polymer (CFRP) sheets to strengthen beams and columns. Civil engineers frequently turn to this technique due to its effectiveness in enhancing the seismic performance of structures. However, the critical challenge lies in the proper design of the retrofitting scheme. Achieving the desired level of performance while maintaining the structural integrity and stability of the existing building is of paramount importance. A well-executed retrofitting plan not only bolsters the structure's ability to withstand seismic forces but also ensures that the retrofit elements integrate seamlessly with the original construction, minimizing the risk of unforeseen issues. In the ongoing effort to mitigate earthquake-related risks, special attention to retrofit design and execution remains a vital aspect of safeguarding vulnerable structures and, ultimately, the safety of the people who inhabit them. The interest in the use of CFRP materials in the retrofitting of defective RC frame beam and column connections is growing remarkably, as the methods have advantages, for example:

(i) Improved seismic performance: Beam–column connections are critical areas in a structure, and defects in these areas can severely deteriorate seismic performance of buildings. CFRP materials can be effectively used to strengthen these connections to increase the ductility, which led to better the withstanding of earthquake forces. This increase in ductility helps to prevent easy failures and ensures the structure's resilience in the event of an earthquake by reducing catastrophic damage.

(ii) Lightweight and corrosion resistant: CFRP composites are lightweight as compared to the conventional steel jacketing method. This advantage makes them particularly suitable for retrofitting and strengthening work as they add less weight to the existing structure. Furthermore, CFRP materials do not corrode, which is a significant advantage over steel reinforcement. Steel corrosion can weaken connections over time, while CFRP retains its strength and integrity, providing long-term durability and reliability.

(iii) Ease of installation: CFRP materials are relatively easy to install and can be used for retrofitting a structure to address specific defects in beam–column connections. This ease of installation can reduce maintenance work during the renovation.

(iv) Tailored retrofitting solution: CFRP retrofitting can be conducted according to the specific configuration of structural elements. Engineers can target specific defects and weaknesses in a given RC frame by designing a CFRP reinforcement system. This tailored approach ensures repeatability.

Therefore, these advantages make CFRP a valuable and versatile tool in enhancing the safety and longevity of existing structures, particularly in earthquake-prone regions. In this study also, CFRP sheets have been selected for retrofitting the joints among all other fibers. CFRP is preferred because it possesses higher tensile strength and stiffness compared to other types of fibers thanks to the exceptional strength of carbon fibers. Additionally, it is lighter, more corrosion-resistant, and better equipped to withstand high temperatures when compared to other fiber types. Therefore, CFRP materials are better suited for retrofitting RC joints, given their critical function.

Although using carbon fiber-reinforced polymer (CFRP) sheets in retrofitting existing RC frames with slabs presents specific challenges that impact the performance of the retrofitted structure. The primary challenge when using CFRP is achieving a strong and durable bond between the CFRP sheets and the existing concrete surface to transfer the load. Proper surface preparation of the concrete is highly effective in ensuring a strong bond between CFRP and the existing concrete. Inappropriate adhesive selection and inadequate surface finishing can lead to delamination of CFRP sheets during seismic events, compromising the effectiveness of the retrofitting.

Furthermore, the type of adhesive and its thickness are highly effective in bonding CFRP to the concrete and consequently in the strength of the RC beam–column joint. Implementing the appropriate adhesive is able to influence the durability of the bonding and the load-bearing capacity significantly. Special types of adhesives are formulated to provide high-strength bonding; other types provide flexible bonding but are more resistant against environmental threats. Similarly, adhesive thickness is highly impactful on the quality of bonding and the strength and performance of the retrofitted RC joint using CFRP.

An insufficient adhesive layer may lead to a reduction in the integrity of CFRP sheets with the concrete surface, while excess thickness can create voids or uneven stress distribution, potentially compromising the structural integrity. Therefore, implementing the appropriate adhesive type and suitable thickness can be highly contributed to achieving a reliable binding between CFRP sheets and concrete to perform together to enhance the strength and functionality of the retrofitted RC joint.

This bonding is especially critical for RC frames with slabs due to the limited accessibility for creating a full loop of CFRP layers around the beams. Proper anchorage and confinement can be achieved by applying CFRP sheets to the surrounding structural elements, which is not feasible for beams in frames with slabs due to their geometry.

The beam–column joint plays a crucial role in the integrity of reinforced concrete structures when subjected to lateral loads, such as earthquakes. Due to the rigidity of these connections in RC structures, the first plastic hinges often appear in the vicinity of these joints as the lateral load is transferred from the beams to the columns. This can result in a reduction in the strength of the beam–column joint, leading to instability and compromising the overall structural integrity. Therefore, the aim of this research is to propose a retrofitting scheme using CFRP sheets to relocate the occurrence of plastic hinges in the RC beam–column connections from near the column face to further within the beam. This approach is designed to enhance the safety of the joint, providing stability and preserving the structural integrity when exposed to severe earthquake excitations.

3. Experimental Setup

The experimental setup consisted of two full-scale RC interior joints of a moment-resisting frame with moderate ductility (Figure 1) prepared according to ACI codes (440.2R-17) [41]. A plain specimen (CS1) was ordered as the control specimen, and another section that was later retrofitted with a flange-bonded scheme (RS1) was constructed and tested under monotonic loading until sample failure. The joints were chosen from a typical four-story RC structure and consisted of half-length lower and upper floor columns that were 2700 mm (2.7 m) in height with half-length left and right beams with a length of 4000 mm (4 m) and dimensions of column joint cross-sections of 400 × 400 mm^2. Additionally, 8Φ25 bars were used to longitudinally reinforce the columns, complying with a 2.4% reinforcement ratio. Rectangular Φ10 ties were used as transverse reinforcements. Each two-column ties were 170 mm apart inside the joint, 600 mm above and below the joint, and 200 mm in the general length of the column. All beams had a length of 1800 mm (1.8 m) from the column to the free end, and the cross-section selected was 300 × 300 mm^2. Two bars (i.e., Φ25) were employed for reinforcement of the beams at the top and two bars (i.e., Φ20) at the bottom. Φ8 rectangular stirrups were used as transverse reinforcements of the beams (Figure 1). The stirrups were 50 mm from one another in the first 600 mm. The yield strengths of the reinforcing steel were 400 MPa for the bars and ties. However, the concrete compressive strength reached 21.5 MPa for the standard cylinder samples. The ultimate stress and strain and elasticity modulus of the unit-directional CFRP used in the models were, respectively, f_FRP = 3900 MPa, 0.0155 mm/mm, and 240 GPA. In the retrofitted specimen with the flange-bonded scheme (RS1), FRP laminates covered the top and bottom flanges of the ends of the beams, which joined the connection, extending so as to cover the adjoining column faces. The joint was covered by 5 FRP layers. The total thickness for 5 layers of FRP is 0.825 mm. The length of CFRP sheets was kept constant at 300 mm for the joint based on recommendations by Pauley and Priestly [31] for desirable plastic hinge relocation. In the design of this experiment, an attempt has been made to investigate the relocation of plastic hinges when the load is applied to the end of beams. Therefore, to maintain the same strength for the beam against both in-plane and out-of-plane loads, the square section has been employed to extend the results of this study for the frame exposed to lateral loads in future studies.

3.1. Test Setup and Loading Sequence

The laboratory experiment details are shown in Figure 2. The samples were placed under a constant axial load on the column and two monotonic loads on the beam end. The applied load was measured by a load cell. As to simulation of the inflection points at the upper and lower centers of the columns, each end of the column is specially designed to support it, ensuring free end rotation. For this purpose, a pinned support was placed at the column base, and a roller support was exerted on the top. A vertical slot with a length of 2 cm was provided for the roller to allow vertical deformations in the column to suitably transmit the axial load in the column to the hinge support (Figure 2c). The test was conducted step-by-step, and monotonic "pushover" loads were applied until destruction. In the design of the experimental test, to provide a realistic RC joint condition, an attempt was made to apply the axial load equivalent to 10% of the column strength, which is 150 kN, to subject the column to a compressive condition. The column load, equaling 150 kN, was kept constant, whereas the loads on the beam's tips were simultaneously and gradually increased, starting from zero, while corresponding displacements and strains were recorded.

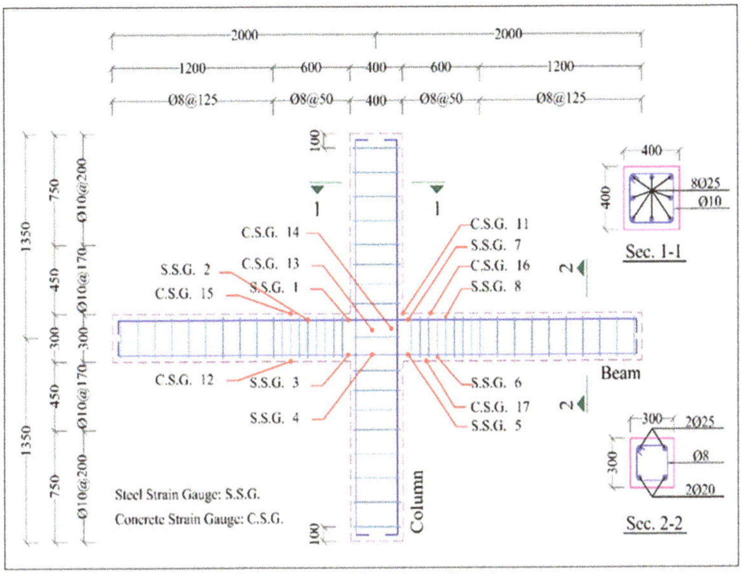

(**a**) Reinforcements and locations of strain gauges.

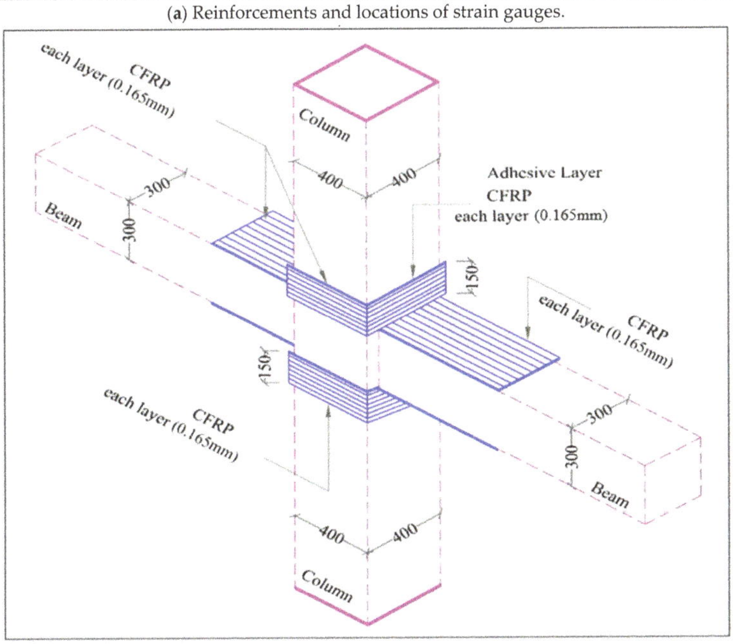

(**b**) Retrofit configuration for plastic hinge relocation off the column face.

Figure 1. Details of sections and retrofitting configuration.

(**a**) Test setup (CS1).

(**b**) Test setup (RS1).

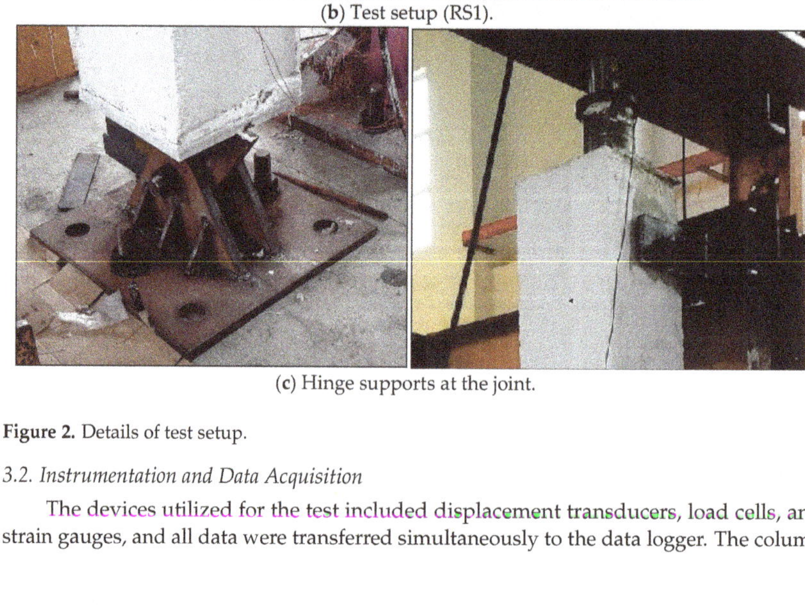

(**c**) Hinge supports at the joint.

Figure 2. Details of test setup.

3.2. Instrumentation and Data Acquisition

The devices utilized for the test included displacement transducers, load cells, and strain gauges, and all data were transferred simultaneously to the data logger. The column axial load and the loads applied to the beams' tips were controlled by load cells. For this purpose, in the experimental test setup, three Kyowa load cells (force transducers) with a capacity of 300 kN and 150 kN have been employed to measure the applied force on the column and at the ends of two beams. The positions of these load cells are depicted in Figure 3.

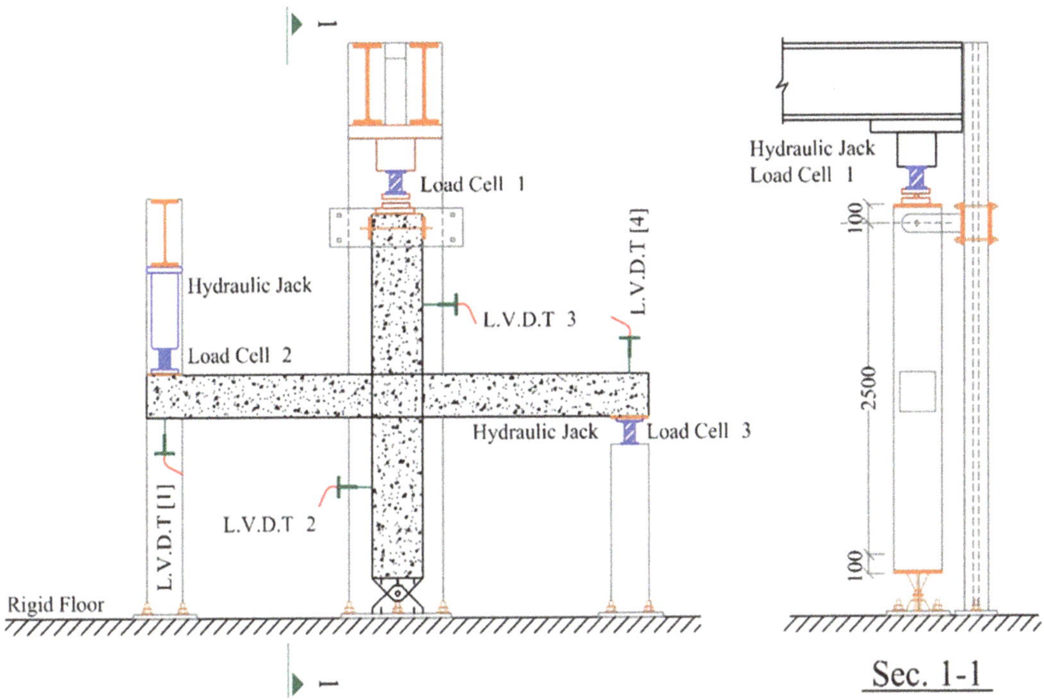

Figure 3. Schematic view of the LVDT sensor positions in experimental test.

Additionally, four units of KYOWA LVDT sensors (displacement transducers) capable of measuring deformations up to 100 mm in the test setup have been utilized. Two of the LVDTs were installed at the midpoints of the columns, both at the top and bottom, to measure the deformation of the column under applied axial pressure at the top. The other two LVDTs were placed at the ends of the beams to measure the displacements of the beams when pushing loads were applied to both ends (Figure 3).

Strains of the reinforcing steel and concrete were gained at different sections by means of electric resistance gauges, bonded to the bars at pre-ground locations for smooth contact.

3.3. Results and Discussion

The control sample CS1, which bore no FRP, was tested monotonically to failure. The tip load versus displacement of the CS1 and RS1 is shown in Figure 4. The capability of the flange-bonded retrofit in enhancing capacity can be deduced from these curves. It is obvious that the response is almost linear up to 9 KN, at which load the first crack develops and the joint enters non-linear responses with a gradually decreasing stiffness as the load increases. The joint cracking is not visible at that load; it is only assumed from the shape of the force–displacement curve. The first flexural cracks of the beam initiate with the tip load of 20 KN, corresponding to a displacement of about 9 mm. Moreover, beam flexural cracks were raised at different locations off the column face as loading progressed. The longitudinal reinforcing of the beam yielded an ordinary load equal to 39 KN with an average yield displacement of 32 mm. The test was stopped at the load reached the joint's peak capacity, which was almost 45 KN, corresponding to a 43.6 mm displacement. As the column was constantly subjected to a 150 kN load, the experimental test was halted when the beam's resistance significantly dropped, primarily due to the development of severe cracks and the crushing of the compression area of the beam concrete, posing a safety concern.

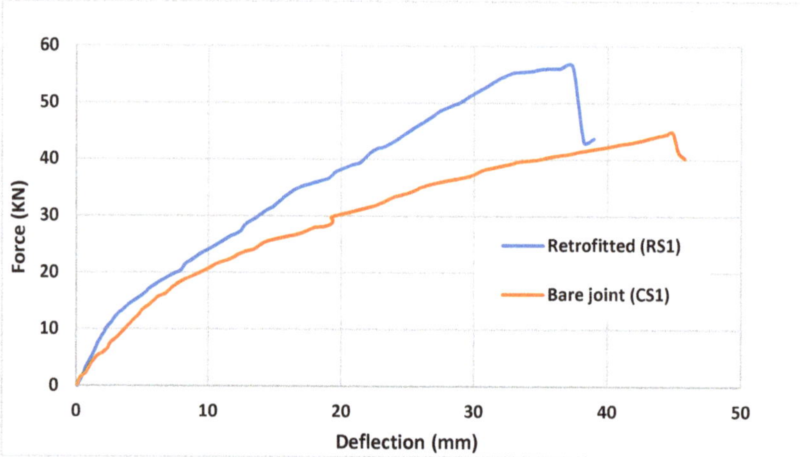

Figure 4. Beam tip load versus tip displacement of the specimens (left side).

Consequently, the force–displacement graph has been plotted up to the point where the beam's resistance reached its maximum, i.e., before experiencing a noticeable reduction in strength capacity. This peak resistance force is then used to determine the load-bearing capacity of the beam. Though loading was over, some flexural cracks were noticed to have been formed near the beam, close to the column face, and others had progressed into the beam (Figure 5a).

The test for case RS1 (retrofitted and strengthened by five 300 mm CFRP layers stretching from the column) was stopped when the load reached the joint's peak capacity at 56.58 KN, which corresponded to a 38.4 mm displacement. As shown in Figure 5b, the beam failed, demonstrating ductile behavior right where FRP was cut off and sending the plastic hinge off the joint core.

No considerable debonding was experimentally observed in this research, which presents moderate ductility. Furthermore, Table 1 shows that the flange-bonded retrofit had increased the original frame capacity by 26%, while decreasing the displacement capacity by 17%. It is seen that the maximum bar strains at the end point of FRP in the longitudinal reinforcements are greater than the rebar yield strain.

Table 1. Steel yield stress–plastic strain.

Model		Max Bar Strain		Max Concrete Strain		Pmax (kN)	Δmax (mm)
		Column Face	End of CFRP	Column Face	End of CFRP		
Original (CS1)	Left	2.24×10^{-2}	-	-3.48×10^{-3}		43.65	45
	Right	2.15×10^{-2}		-3.21×10^{-3}		30.12	43.4
Retrofitted (RS1)	Left		1.8×10^{-2}		-4.44×10^{-3}	53.58	38.4
	Right		1.6×10^{-2}		-3.56×10^{-3}	39.75	37.1

(**a**) Crack pattern in CS1

(**b**) Crack pattern in RS1 (Overall view).

Figure 5. Crack pattern after conducting an experimental test.

4. Numerical Model of the Retrofitted Joints

In this study a designed RC frame has been considered which has been used in the literature to investigate the effects of retrofitting joints via the CFRP web-bonded scheme [24,25,42]. The ANSYS 2019 R1 finite element software was used for modeling, as shown in Figure 6. In the finite element modeling process, the contact between the CFRP sheets and concrete is defined as the tie bonding since it is assumed that both concrete and CFRP sheets are properly bonded to each other using an adhesive. At each node of the elements, three degrees of freedom were considered, comprising eight-noded solid elements (possible translations in the nodal X-, Y-, and Z-directions). This element renders possible cracks in three orthogonal directions, plastic deformations, and crushing. Link180 element is used to model the steel reinforcements. A bi-linear curve with 2% post-yield strain hardening was used to represent the steel stress–strain relation. The steel plates were modeled at the hinge support by Solid185, which is a solid element with eight nodes.

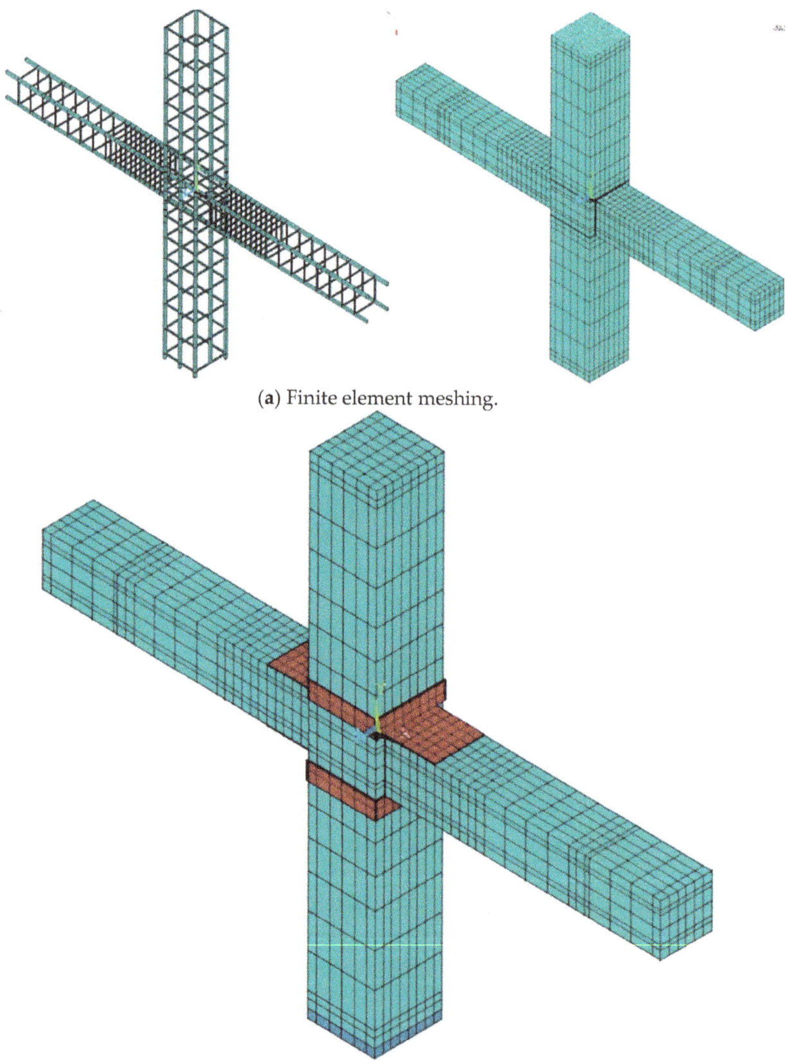

(**a**) Finite element meshing.

(**b**) Retrofitting configuration for plastic hinge relocation off the column.

Figure 6. Finite element model of considered RC joints.

These elements were applied to undo the possibilities of cracks resulting from stress concentration. Eight nodes were assumed for these elements, each having the three DOFs. CFRP sheets were modeled using Solsh190, a solid-shell-type element with eight nodes.

This type of element possesses the geometry of a continuum solid element at each node. Finally, the columns' end was modeled considering suitable boundary conditions to apply an axial compression force at the joint throughout the analysis.

To perform the non-linear pushover analyses, stepwise equal and opposite loads were applied to the beams' ends. The modified Newton–Raphson method was implemented to solve the pushover problem. However, the convergence criteria that were considered were based on displacement, whose tolerance limit was taken equal to the FE software. In order to conduct non-linear analysis, automatic time stepping in FE, which predicted and controls load steps, was used. The FE mesh of the joint retrofitted with the flange-bonded scheme is shown in Figure 6b.

Nearby, 3144 Solid65 elements were used to model concrete elements, while for the steel bars, 1363 Link180 elements were used, and CFRP sheets were modeled with 264 Solsh190 elements. Nonetheless, using the beam element instead of the link element for modeling steel bars may lead to more accurate outcomes by considering the bending behavior of the steel bars.

The mesh sensitivity analysis has been conducted to assess the effect of mesh size and the number of elements on the finite element analysis results. Therefore, the final analysis results were obtained from the most stable model, where changing mesh size did not affect the results by more than +/−1%.

The William–Warnke model was employed to account for the failure criterion of concrete and cracking, as well as possible models of crushing failure [43]. Therefore, for failure surface failure, compressive strength and ultimate uniaxial tensile strength are required. Figure 7 shows a three-dimensional concrete failure surface [32]. It is seen that the stiffness of the cracked concrete vanishes in the FE model; hence, tension cannot be resisted.

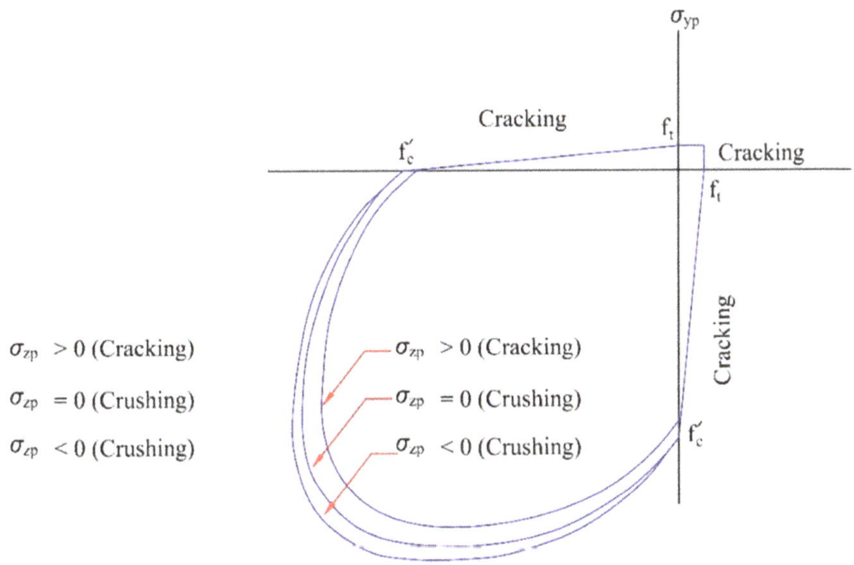

Figure 7. Three-dimensional failure surfaces based on William–Warnke criterion [3].

Since principal stresses are all compressive and lie outside of the failure surface, the concrete starts to crush; thus, the elastic modulus in all directions decreases rapidly. Additionally, β_(t) and β_c are coefficients of shear transfer with values between 0 and 1 introduced for the definition of concrete for open and closed cracks. The value of "0" represents a smooth crack (i.e., complete loss of shear transfer), and "1" stands for a rough crack (i.e., no loss of shear transfers). β_(t) is the shear transfer coefficient, taken as 0.25 for open cracks, and β_c was taken as 0.99 for closed cracks [44,45].

The efficiency of CFRP depends on several parameters, including the number of fibers, mechanical properties, surface preparation, and debonding [46,47]. The latter is a major issue with which numerous CFRP applications are associated. The maximum CFRP strain allowed for the elimination of this mode of failure, which is the result of intermediate cracks, and it is limited to the value suggested by ACI 440.2-08.

4.1. Model Validation

The numerical model was verified by extracting the load–displacement results of the beam through the FE analysis. As shown in Figure 8a, the comparison was made with the experimental testing output, and the results showed very close agreement, confirming the verification for both CS1 and RS1 models. As expected, the FE numerical solution resulted in a stiffer response for both models, showing higher initial stiffness. Following the expected initial discrepancy, trends of the numerical and experimental curves are similar up to 20 KN, followed by a stiffer response from the finite element model. However, the numerical analysis predicts the ultimate capacity relatively accurately, with a slightly higher resistant force, proving that the FE model and the performed analysis were reliable for both considered models (before and after retrofitting).

Accordingly, the results presented in the graph of Figure 8a depict that the maximum strength achieved for both CS1 and RS1 joints through finite element analysis was 45.4 kN and 59.13 kN, respectively. In comparison, the maximum strength capacity for these joints obtained from the experimental tests was 44.8 kN and 56.5 kN, respectively. Therefore, these results validate the FEM simulation with a 1.2% error for CS1 and a 4.3% error for RS1. The higher error for the frame retrofitted with CFRP is mostly due to the assumption of full bonding between CFRP and concrete during FEM modeling, which is unlikely to occur in practical construction operations. To increase the accuracy of the results, it is recommended to insert the action of an adhesive in the model by using proper contact elements or interface elements to simulate the realistic behavior of CFRP and concrete bonding.

4.2. Pushover Analysis of the Retrofitted Connection

Once the numerical model was validated, the CFRP-retrofitted RC joint was analyzed. As stated earlier, the target of this research was to evaluate the ability of the selected CFRP retrofit scheme by shifting the place of the plastic hinges off the column face. However, the selected flange-bonded retrofitting scheme is shown in Figure 6b. In a parametric study, the thickness and number of CFRP sheets required for a successful plastic hinge relocation into the beam were to be determined.

To this end, the non-linear pushover analysis of the joint was performed using different CFRP overlay thicknesses, and for each thickness, the strain state in the beam's tensile reinforcements was considered for locating the true position of the plastic hinge. The joint was then retrofitted with 4, 5, and 7 layers of CFRP laminates, corresponding, respectively, to thicknesses of 0.66 mm, 0.825 mm, and 1.155 mm. As shown in Figure 8b, the maximum strain variation in the longitudinal tensile reinforcements of the beam retrofitted with 4, 5, and 7 layers of CFRP is compared with reinforcement strain variation in the original (non-retrofitted) beam.

Moreover, Figure 8b demonstrates that for the original joint, the maximum strain in steel reinforcement corresponding to plastic hinge location occurs in the vicinity of the column. While the connection is retrofitted using four layers of laminates, there will be two peaks, one close to the column and the other at the end section of CFRP sheets. These two peaks are close in value; it cannot be stated that a safe relocation of the plastic hinge has followed. However, once five layers of CFRP sheets are used to retrofit the joint, the reinforcement strain at the end of the CFRP overlay evidently dominates the strain formed at the column face. This indicates a successful relocation of the hinge over the cover. Higher numbers of CFRP layers would also lead to such relocation, and the retrofitting would then be more expensive. Therefore, for this joint, a five-layer coat is considered as the optimal thickness for successful plastic hinge relocation.

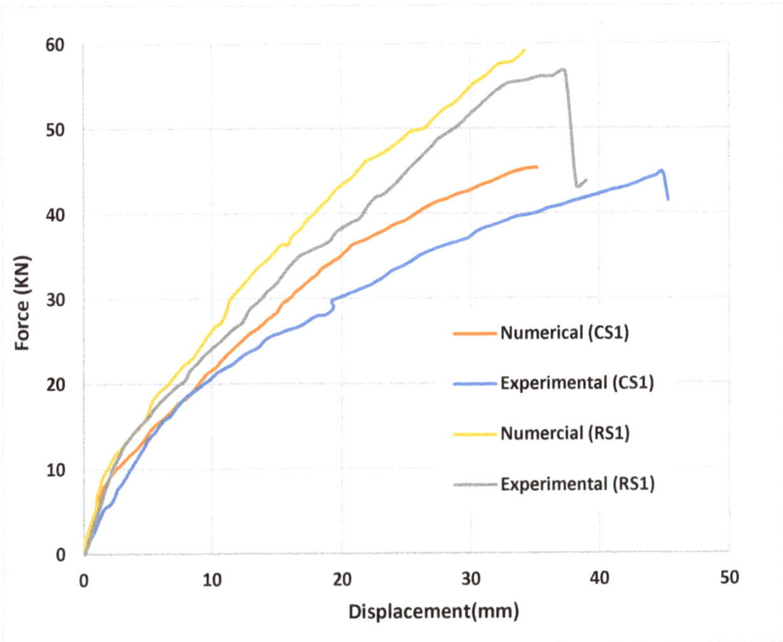

(**a**) Load–displacement curves of the beam tip.

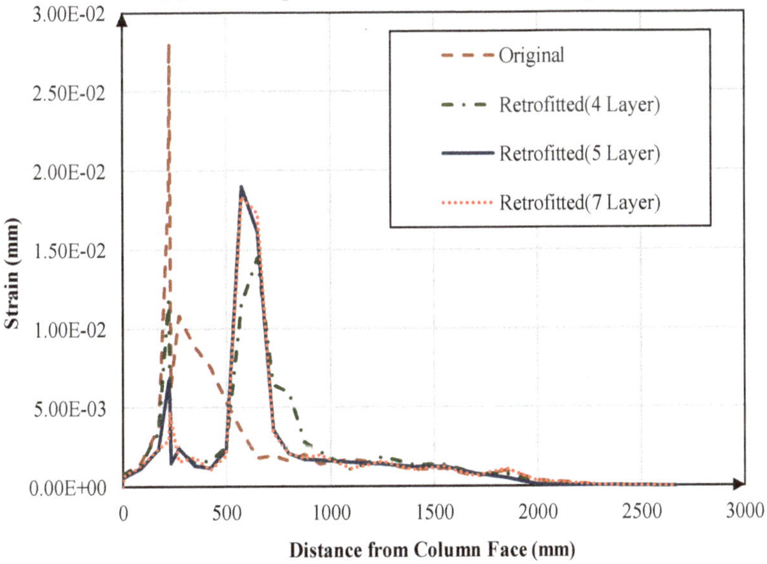

(**b**) Strain variations in the longitudinal tensile reinforcement of the beam.

Figure 8. Load–displacement and strain results.

The joint stress variation and concrete strain contours when the applied load reaches its ultimate level before and after retrofitting are displayed in Figures 9 and 10, respectively. These figures also show the relocation of the plastic hinge due to retrofitting.

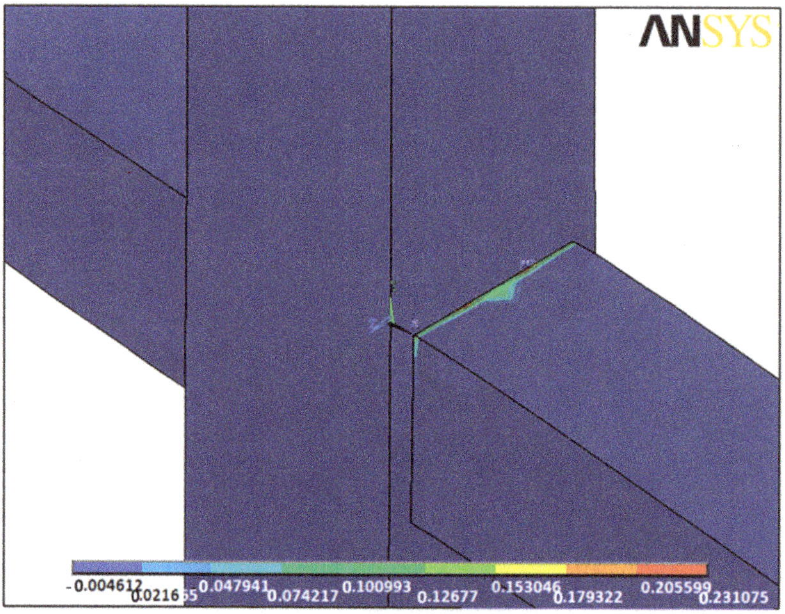

(**a**) Joint without CFRP (CS1).

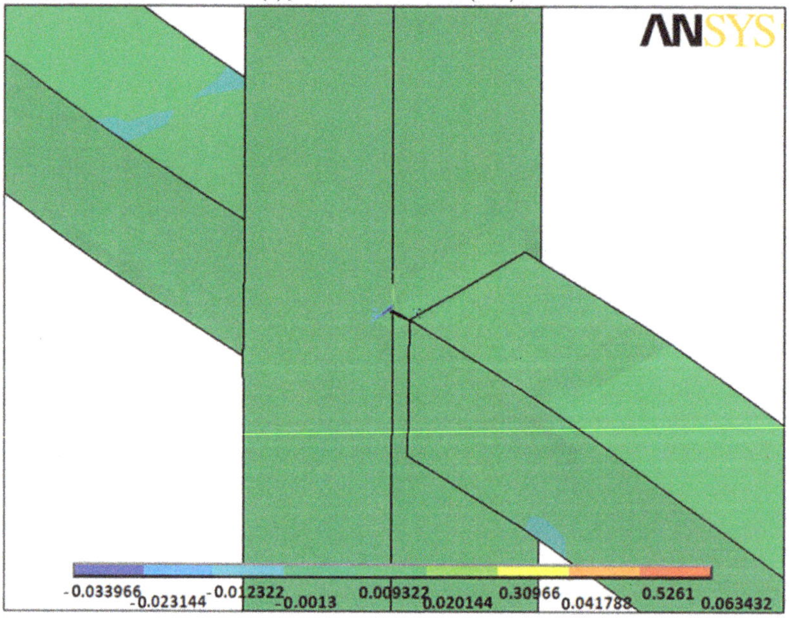

(**b**) Joint with flange-bonded CFRP (RS1).

Figure 9. Stress contours of the joint and successful relocation of maximum stress from joint area to the beam length.

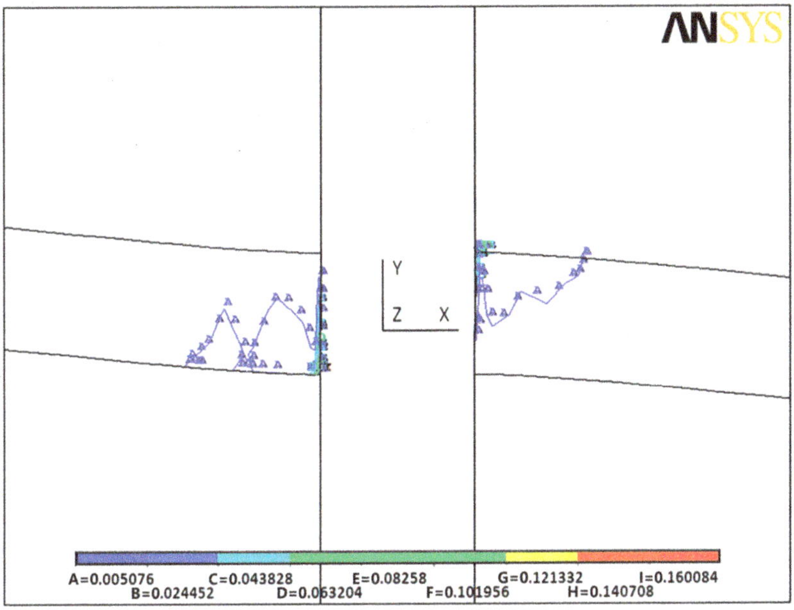

(**a**) Joint without CFRP.

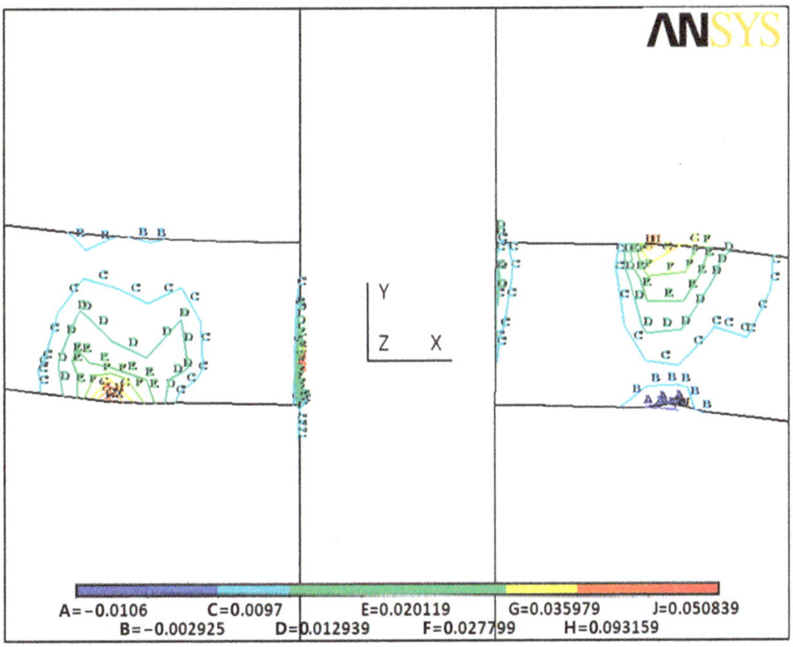

(**b**) Joint with flange-bonded CFRP.

Figure 10. Strain contours of the joint.

In Figure 9, the stress counter indicates that in the control sample (CS1), the maximum stress under applied loads appears at the contact point of the beam–column, specifically at the top part of the beam. However, in the retrofitted joint (RS1), the peak stress occurs in the beam, away from the beam–column connection surface. These results reveal that bonding the flange with CFRP leads to a shift in the position of the peak stress from the beam–column contact point to a location along the beam length, away from the joint.

Similarly, in Figure 10, the strain counter shows that in the CS1 joint, the maximum strain (deformation) appears at the contact surface of the beam–column in the tensile part (top of the right beam and bottom of the left beam). Hence, in the RS1 joint, the peak strain and consequent plastic hinges are observed to occur in the beams, away from the beam–column connection. These results demonstrate that by employing the proposed retrofitting technique, the position of plastic hinge occurrence has successfully shifted from the near beam–column connection face to the beam length at a further distance from the joint area. Therefore, the beam–column joint is able to function during applied loads and provide structural integrity and stability.

However, practically, in order to appropriately design for retrofitting scheme by using flange-bonded CFRP layers, it is recommended to add a large enough number of CFRP layers in the critical length of the beam beside the joint (preferably equal to the depth of the beam) in order to relocate the plastic hinge from the joint area to the length of the beam which is not critical for the stability of the structure. The number of CFRP layers can be estimated based on the amount of imposed shear force and bending moment at the beam and column in the joint.

5. Conclusions

In this study, a new retrofitting scheme involving flange-bonded CFRP was investigated via numerical and experimental investigations of RC beam–column joints. First, the FE method was employed for our numerical study. To ensure the experimental testing setup accorded with real structural conditions under actual loading, a constant load was applied to the top of the column, thus reflecting the dead load on the column of the RC frame. Meanwhile, a couple of forces were incrementally applied at the end of the beams via two jacks to generate a double curvature action in the beam–column joint, which would occur during the application of vertical and lateral loads to the frame. Since the primary goal of this research was to relocate the plastic hinges from the column face to a greater length along the beam, the results of the experimental tests and numerical analyses demonstrate that the proposed retrofitting scheme successfully achieves this objective. By employing this retrofitting approach, plastic hinges are now observed at a considerable distance from the column faces. This significantly enhances the safety and security of the beam–column joint when the structure is subjected to seismic excitations, ultimately ensuring the stability and integrity of the structure. However, the most important experimental and numerical model findings can be summarized as follows.

Experiments and numerical results show that under static loads under a moderate-ductility RC, beam plastic hinges may be present near the column face. Thus, the flange-bonded scheme in this study is practical and can be used in actual 3D frames as it takes the presence of the cross beams into consideration and accounts for the integrated floor slabs.

1. The experimental configuration suggested a significant rise in the connection's ultimate strength (26%).
2. Inelastic deformations took place 300 mm to 400 mm off the column, meaning the required relocation has been achieved.
3. Five CFRP layers were considered as the ideal thickness for successful plastic hinge relocation.
4. Higher numbers of CFRP layers would also lead to such a relocation, and the retrofitting would then be more expensive.
5. The experimental and numerical results confirmed the effectiveness of the suggested grooving method. It is believed that the flange-bonded CFRP retrofitting improves the

capacity of the joints. This proved its ability to send the possible plastic hinge off the joint and further into the beam, leading to a ductility increase and an improvement of the overall RC frame performance.

6. Finally, the results of the successfully conducted experiments and numerical testing indicate that under the condition of seismic loading in a moderate-ductility-reinforced concrete-moment-resisting frame, beam plastic hinges might materialize near the column face.

Author Contributions: Conceptualization, F.H., R.A. and A.A.F.; Methodology, F.H., R.A. and A.A.F.; Software, R.A. and A.A.F.; Formal analysis, R.A. and A.A.F.; Investigation, R.A. and A.A.F.; Resources, F.H.; Data curation, R.A. and A.A.F.; Writing—original draft, R.A. and A.A.F.; Writing—review and editing, F.H.; Visualization, R.A. and A.A.F.; Supervision, F.H.; Project administration, F.H.; Funding acquisition, F.H. All authors have read and agreed to the published version of the manuscript.

Funding: This research received no external funding.

Institutional Review Board Statement: Not applicable.

Informed Consent Statement: Not applicable.

Data Availability Statement: All new data created through this study are already included in this paper. Therefore, data sharing is not applicable to this article.

Conflicts of Interest: The authors declare no conflict of interest.

References

1. Ali, H.T.; Akrami, R.; Fotouhi, S.; Bodaghi, M.; Saeedifar, M.; Yusuf, M.; Fotouhi, M. Fiber reinforced polymer composites in bridge industry. *Structures* **2021**, *30*, 774–785. [CrossRef]
2. Harries, K.A. ASCE 41 Seismic Assessment of FRP-Repaired Concrete Columns. *J. Compos. Constr.* **2021**, *25*, 04021001. [CrossRef]
3. Yazdani, S.; Asadollahi, S.; Shoaei, P.; Dehestani, M. Failure stages in post-tensioned reinforced self-consolidating concrete slab strengthened with CFRP layers. *Eng. Fail. Anal.* **2021**, *122*, 105219. [CrossRef]
4. Liang, S.-L.; Fu, F.; Huang, Z.-Q.; Qian, K. Enhancement of collapse-resistant capacity of non-seismically designed RC frames using various CFRP strengthening schemes. *Eng. Struct.* **2023**, *291*, 116450. [CrossRef]
5. Alkhawaldeh, A.A.; Alrousan, R.Z. Improving cyclic response of heat-damaged non-ductile RC joints using CFRP hybrid systems. *Constr. Build. Mater.* **2023**, *377*, 131150. [CrossRef]
6. Farahbod, F.; Aalam, H.; Jahanmohammadi, A. Seismic behavior of three-dimensional interior reinforced concrete beam–column joints strengthened with CFRP sheets. *Asian J. Civ. Eng.* **2023**, 1–18. [CrossRef]
7. Farhang, K.; Farahbod, F.; Nezamabadi, M.F.; Mansouri, B. Shear strengthening of RC 3D exterior beam-column joints with CFRP sheets an experimental and numerical study. *Asian J. Civ. Eng.* **2023**, *24*, 1599–1619. [CrossRef]
8. Fayaz, Q.; Kaur, G.; Bansal, P.P. Numerical Modelling of Seismic Behaviour of an Exterior RC Beam Column Joint Strengthened with UHPFRC and CFRP. *Arab. J. Sci. Eng.* **2021**, *47*, 4971–4986. [CrossRef]
9. Hajirasouliha, I.; Doostan, A. A simplified model for seismic response prediction of concentrically braced frames. *Adv. Eng. Softw.* **2010**, *41*, 497–505. [CrossRef]
10. Shen, X.; Li, B.; Chen, Y.-T.; Tizani, W. Seismic performance of reinforced concrete interior beam-column joints with novel reinforcement detail. *Eng. Struct.* **2021**, *227*, 111408. [CrossRef]
11. Haridas, A.P.; Mathai, A. Plastic Hinge Relocation in RCC Double-Slotted Beam Connection. In *Advances in Civil Engineering*; Springer: Singapore, 2020; pp. 715–724.
12. Ercan, E.; Arisoy, B.; Ertem, O.B. Experimental Assessment of RC Beam-Column Connections with Internal and External Strengthening Techniques. *Adv. Civ. Eng.* **2019**, *2019*, 2828353. [CrossRef]
13. Ghobarah, A.; Said, A. Shear strengthening of beam-column joints. *Eng. Struct.* **2002**, *24*, 881–888. [CrossRef]
14. Parvin, A.; Granata, P. Investigation on the effects of fiber composites at concrete joints. *Compos. Part B Eng.* **2000**, *31*, 499–509. [CrossRef]
15. Granata, P.J.; Parvin, A. An experimental study on Kevlar strengthening of beam–column connections. *Compos. Struct.* **2001**, *53*, 163–171. [CrossRef]
16. Li, J.; Samali, B.; Ye, L.; Bakoss, S. Behaviour of concrete beam–column connections reinforced with hybrid FRP sheet. *Compos. Struct.* **2002**, *57*, 357–365. [CrossRef]
17. Obaidat, Y.T. Cyclic behavior of interior RC beam-column joints strengthened with NSM-CFRP ropes. *Structures* **2022**, *37*, 735–744. [CrossRef]

18. Farhang, K.; Farahbod, F.; Nezamabadi, M.F. Hysteresis behavior of 3D RC exterior beam-column joints strengthened with CFRP sheets and spike anchors. *Structures* **2022**, *40*, 1065–1077. [CrossRef]
19. Attari, N.; Amziane, S.; Chemrouk, M. Efficiency of Beam–Column Joint Strengthened by FRP Laminates. *Adv. Compos. Mater.* **2010**, *19*, 171–183. [CrossRef]
20. Wang, B.; Wu, X.; Liu, Q.; Wu, Y.; Huang, F.; Xu, L.; Wu, X.; Deng, Y. Effectiveness and Efficiency of Externally Bonded CFRP Sheets for Shear Strengthening of RC Beam-Column Joints. *Polymers* **2022**, *14*, 1347. [CrossRef]
21. De Vita, A.; Napoli, A.; Realfonzo, R. Full Scale Reinforced Concrete Beam-Column Joints Strengthened with Steel Reinforced Polymer Systems. *Front. Mater.* **2017**, *4*, 18. [CrossRef]
22. Antonopoulos, C.P.; Triantafillou, T.C. Experimental Investigation of FRP-Strengthened RC Beam-Column Joints. *J. Compos. Constr.* **2003**, *7*, 39–49. [CrossRef]
23. Le-Trung, K.; Lee, K.; Lee, J.; Lee, D.H.; Woo, S. Experimental study of RC beam–column joints strengthened using CFRP composites. *Compos. Part B Eng.* **2010**, *41*, 76–85. [CrossRef]
24. Niroomandi, A.; Maheri, A.; Maheri, M.R.; Mahini, S. Seismic performance of ordinary RC frames retrofitted at joints by FRP sheets. *Eng. Struct.* **2010**, *32*, 2326–2336. [CrossRef]
25. Maheri, M.R.; Akbari, R. Seismic behaviour factor, R, for steel X-braced and knee-braced RC buildings. *Eng. Struct.* **2003**, *25*, 1505–1513. [CrossRef]
26. Naser, M.Z.; Hawileh, R.A.; Abdalla, J. Modeling Strategies of Finite Element Simulation of Reinforced Concrete Beams Strengthened with FRP: A Review. *J. Compos. Sci.* **2021**, *5*, 19. [CrossRef]
27. Mahdavipour, M.A.; Eslami, A.; Jehel, P. Seismic evaluation of ordinary RC buildings retrofitted with externally bonded FRPs using a reliability-based approach. *Compos. Struct.* **2019**, *232*, 111567. [CrossRef]
28. Huang, Q.; Liu, Y.; Shi, M. Experimental study on seismic performance of RC exterior beam-column joints strengthened by annular CFRP sheets wrapping. *J. Build. Struct.* **2022**, *43*, 76.
29. Seifi, A.; Hosseini, A.; Marefat, M.S.; Zareian, M.S. Improving seismic performance of old-type RC frames using NSM technique and FRP jackets. *Eng. Struct.* **2017**, *147*, 705–723. [CrossRef]
30. Allam, K.; Mosallam, A.S.; Salama, M.A. Experimental evaluation of seismic performance of interior RC beam-column joints strengthened with FRP composites. *Eng. Struct.* **2019**, *196*, 109308. [CrossRef]
31. Jawdhari, A.; Fam, A.; Harik, I. Bond between CFRP rod panels and concrete using cementitious mortar. *Constr. Build. Mater.* **2020**, *235*, 117503. [CrossRef]
32. Basim, S.; Hejazi, F.; Rashid, R.S.B.M. Embedded carbon fiber-reinforced polymer rod in reinforced concrete frame and ultra-high-performance concrete frame joints. *Int. J. Adv. Struct. Eng.* **2019**, *11*, 35–51. [CrossRef]
33. Maheri, M.R.; Torabi, A. Retrofitting external RC beam-column joints of an ordinary MRF through plastic hinge relocation using FRP laminates. *Structures* **2019**, *22*, 65–75. [CrossRef]
34. Akhlaghi, A.; Mostofinejad, D. Experimental and analytical assessment of different anchorage systems used for CFRP flexurally retrofitted exterior RC beam-column connections. *Structures* **2020**, *28*, 881–893. [CrossRef]
35. Zhu, Y.; Chen, Y.; He, K.; Feng, R.; Zhang, X.; Zhu, Q.; Tang, C. Flexural behavior of concrete-filled SHS and RHS aluminum alloy tubes strengthened with CFRP. *Compos. Struct.* **2020**, *238*, 111975. [CrossRef]
36. Arowojolu, O.; Ibrahim, A. Plastic hinge relocation in exterior reinforced beam–column joint and compliance issues to seismic design guidelines—A review. *Struct. Concr.* **2020**, *21*, 1938–1958. [CrossRef]
37. Mosallam, A.; Allam, K.; Salama, M. Analytical and numerical modeling of RC beam-column joints retrofitted with FRP laminates and hybrid composite connectors. *Compos. Struct.* **2019**, *214*, 486–503. [CrossRef]
38. Khan, M.S.; Basit, A.; Ahmad, N. A simplified model for inelastic seismic analysis of RC frame have shear hinge in beam-column joints. *Structures* **2021**, *29*, 771–784. [CrossRef]
39. Al-Rousan, R.Z.; Alhassan, M.A.; Al-Omary, R.J. Response of interior beam-column connections integrated with various schemes of CFRP composites. *Case Stud. Constr. Mater.* **2021**, *14*, e00488. [CrossRef]
40. El-Zeadani, M.; Saifulnaz, M.R.; Hejazi, F.; Amran, Y.M.; Jaafar, M.; Alyousef, R.; Alrshoudi, F. Mechanics-based approach for predicting the short-term deflection of CFRP plated RC beams. *Compos. Struct.* **2019**, *225*, 111169. [CrossRef]
41. Saqan, E.I.; Rasheed, H.A.; Hawileh, R.A. An efficient design procedure for flexural strengthening of RC beams based on ACI 440.2R-08. *Compos. Part B Eng.* **2013**, *49*, 71–79. [CrossRef]
42. Azarm, R.; Maheri, M.R.; Torabi, A. Retrofitting RC Joints Using Flange-Bonded FRP Sheets. *Iran. J. Sci. Technol. Trans. Civ. Eng.* **2017**, *41*, 27–35. [CrossRef]
43. Zarandi, S.; Maheri, M.R. Seismic performance of RC frames retrofitted by FRP at joints using a flange-bonded scheme. *Iran. J. Sci. Technol. Trans. Civ. Eng.* **2015**, *39*, 103.
44. Soudki, K.; Alkhrdaji, T. Guide for the design and construction of externally bonded FRP systems for strengthening concrete structures (ACI 440.2R-02). In Proceedings of the Structures Congress 2005: Metropolis and Beyond, New York, NY, USA, 20–24 April 2005; pp. 1–8.
45. Paulay, T.; Priestley, M.N. *Seismic Design of Reinforced Concrete and Masonry Buildings*; John Wiley & Sons: Hoboken, NJ, USA, 1992.

46. Nguyen-Minh, L.; Phan-Vu, P.; Tran-Thanh, D.; Truong, Q.P.T.; Pham, T.M.; Ngo-Huu, C.; Rovňák, M. Flexural-strengthening efficiency of cfrp sheets for unbonded post-tensioned concrete T-beams. *Eng. Struct.* **2018**, *166*, 1–15. [CrossRef]
47. Sojobi, A.O.; Liew, K.M. Flexural behaviour and efficiency of CFRP-laminate reinforced recycled concrete beams: Optimization using linear weighted sum method. *Compos. Struct.* **2020**, *260*, 113259. [CrossRef]

applied sciences

MDPI

Article

The Effect of Glass Flour on the Microstructure and Properties of Fiber-Reinforced Concrete: Experimental Studies

Gabriela Rutkowska [1,*], Mariusz Żółtowski [1], Filip Chyliński [2], Yuliia Trach [1] and Elżbieta Gortych [1]

[1] Institute of Civil Engineering, Warsaw University of Life Sciences, 166 Nowoursynowska Street, 02-787 Warsaw, Poland; mariusz_zoltowski@sggw.edu.pl (M.Ż.); yuliia_trach@sggw.edu.pl (Y.T.); s200205@sggw.edu.pl (E.G.)

[2] Instytut Techniki Budowlanej, Filtrowa 1, 00-611 Warsaw, Poland; f.chylinski@itb.pl

* Correspondence: gabriela_rutkowska@sggw.edu.pl

Abstract: The introduced limits on carbon dioxide emissions by the European Union encourage experimental work on new-generation materials containing smaller amounts of clinker. Currently, silica fly ash from hard coal combustion is widely used in cement and concrete technology in Europe and Poland. Their wide application is determined mainly by the chemical and phase composition, and in particular by the activity of pozzolanic and its high fineness, like cement. The aim of this study was to assess the effect of glass flour and polypropylene fiber modifiers on the properties of concrete and its microstructure. To analyze the results, samples of reference ordinary concrete and samples with different amounts of glass flour (0–30%) and a constant number of polypropylene fibers (0.025 kg) were used. The obtained test results showed the possibility of producing ordinary concrete with the addition of glass flour. The average compressive strength for concrete containing 10% additive was set at 49.3 MPa, 51.2 MPa, and 53.1 MPa after 28, 56, and 90 days of maturation for a content of 20% of 44.6 MPa, 46.4 MPa, and 48.4 MPa, respectively, and for 30% of 41.5 MPa, 43.8 MPa, and 45.6 MPa, respectively. By modifying concrete with glass flour and polypropylene fibers, a composite resistant to negative temperatures can be obtained. Glass flour shows reactivity with the cement matrix, and in small amounts, it might cause the microstructure to seal and a slight increase in compressive strength.

Keywords: compressive strength; frost resistance; concrete microstructure; glass flour; fibers

1. Introduction

Concrete is undoubtedly the most used composite material among man-made materials and second only to water in the entire complex of materials used, without which modern construction could not function. It owes this position to well-known advantages such as high strength and durability, ease of manufacture and stacking, and low cost of production. However, this material is not perfect. According to the literature, it is susceptible to harmful biological and physical effects. According to estimates, about 3.5 billion m^3 of this material is consumed annually worldwide. It is exploited in increasingly difficult natural conditions, from the hot desert to the ice of the Arctic and the sea floor, from sewers to "skyscrapers". Concrete is a material with a high potential to adapt to specific operating environmental conditions. It is an ecological composite, often made of local raw materials—aggregate, cement, water, admixtures, and possibly mineral additives. It is a safe product that guarantees the stability and load-bearing capacity of a given structure but is also a sustainable, technologically advanced product [1–4].

Looking at concrete from the side of its microstructure, its durability will be shaped by the nature of the cements used, their quantity, admixture properties, proportions of binder hydration products, and water–cement (binder) coefficient [5,6].

For the development of the construction sector, one of the most important issues is striving to make concrete an ecological material that is even more environmentally friendly so that it can meet both the current and future requirements of construction. It should fully meet current social needs and, at the same time, meet new challenges, ensuring the innovativeness of the material. It is also important that it "works" flawlessly, without the need to carry out costly repairs, so that the production of concrete is correlated with the prediction of its behavior during specific operating conditions. For a building material to guarantee economy, environmental friendliness, innovation, or durability, the structure of concrete should be improved or modified. Modification at the structure level changes the physical and mechanical properties of the material. The search for such solutions is indispensable in the case of designing the composition of a concrete mix, whose two components, cement and aggregate, contribute to anthropopressive interactions at the stage of their acquisition and production. In Poland and around the world, there are already areas where obtaining good-quality materials for cement production is a problem. Every year, the world economy needs more and more cement to produce concrete, for which, to this day, no comparable replacement has been found. The big problem is that during the production of 1 ton of cement, 0.5 to 1 ton of greenhouse gases are produced, which is 6–8% of total anthropogenic emissions, according to various data. The concrete production process has a carbon footprint estimated at about 850 kg of CO_2 emitted per ton of clinker needed to produce cement. The European Union's carbon dioxide emission limits (target: 55% reduction in emissions by 2030) encourage research into new generation materials containing smaller amounts of clinker [7–10].

The use of industrial waste in the composition of concrete reduces the amount of cement clinker used and natural aggregates for its production [5]. Nowadays, the materials most widely used to produce concrete are limestone and silica fly ashes and silica dust, referred to as pozzolana [11,12]. Research is being conducted on the possibility of using other components, e.g., fly ash from the thermal treatment of sewage sludge [13–16], bio-ash resulting from the combustion of wood and other plant biomass [17], and shredded glass waste [18–21]. The use of waste provides numerous environmental benefits, such as reduced landfill costs, energy savings, and reduced carbon emissions. In addition, their use can improve the microstructure and mechanical properties of mortar and concrete [22]. Existing environmental requirements are becoming more stringent [23]. In many European countries (Netherlands, Germany, Belgium), the amount of glass collected from consumers exceeds 80% of the glass produced; most of it is recycled.

Crushed glass research is focused on the use of glass as fine concrete aggregates. However, durability concerns over the alkali–silica reaction (ASR) have limited its use as a fine replacement in concrete. Studies have shown that glass behaves pozzolanically if ground finely enough, with a surface area of more than 300 m^2/kg [24,25]. The pozzolanic reaction produces amorphous silica in the SCM, calcium hydroxide (CH) as a byproduct of the cement reaction, and water to form additional calcium silicate hydrate. Glass effect measures such as SCM focus on mechanical and durability properties, and they show an increase in long-term compressive strength, flexural strength, resistance to ASR, and reduction in water sorptivity of concrete containing finely ground glass powder. Moreover, some studies showed that finely ground glass powder had comparable or slightly better mechanical properties at later ages than fly ash and slag, but much less than silica fume [26–28]. Despite all the aforementioned results, few studies have focused on connecting the microstructural properties of cementitious mixtures containing glass powder to the performance characteristics of glass mixtures. Federico [29] performed an extensive study on the glass powder's kinetic and performance properties. However, the effect of curing temperatures on different types of glass cullet reaction kinetics and performance has not been studied.

The aim of the study conducted by Aliabdo et al., was to determine the feasibility of using glass powder as a concrete additive. The pozzolanic activity of glass powder and the effect of this additive as a cement substitute in the range of 0% to 25% on the physical and mechanical properties of the studied composites were evaluated. The test results showed that glass flour is pozzolanic in nature and meets the limits for classes F and C according to ASTM C 618, and the use of glass flour has a negligible effect on the setting time. The use of 10% glass powder as a substitute for cement increased the compressive strength of the mortar by about 9.0%. The use of glass powder in an amount greater than 15.0% as a substitute for cement reduces the 28-day compressive strength of concrete. In order to compensate for the decrease in the compressive strength of concrete, it is necessary to reduce the w/c ratio [30]. Experimental studies conducted by the Kishan Lal Jain and team aimed to evaluate the durability of concrete mixes, containing glass waste flour and granite powder at different levels of substitution. Glass powder (GP) in quantities of 5%, 10%, 15%, 20%, and 25% and granite powder (GrP) in quantities of 10%, 20%, 30%, 40%, and 50% were added to the mixtures as a partial addition to cement and sand, respectively. A significant improvement in the strength properties of concrete containing 15% GP and 30% GrP instead of cement and sand, respectively, was observed. The results show an improvement in water permeability and water absorption by concrete mixed with glass granite [31].

The aim of the research was to learn the impact of adding shredded glass waste and polypropylene fibers (PPFs) during the preparation of the concrete mix of ordinary concrete on its selected technical properties. The obtained results allowed us to determine the strength and frost resistance of the tested material with different contents of glass waste and the same number of polypropylene fibers. In addition, the effect of the addition of glass flour and polypropylene fibers on the microstructure of concrete was evaluated.

2. Materials and Methods

Procedures based on the guidelines contained in current EU regulations and standards were used to carry out the research. Figure 1 shows a conceptual diagram of the research carried out.

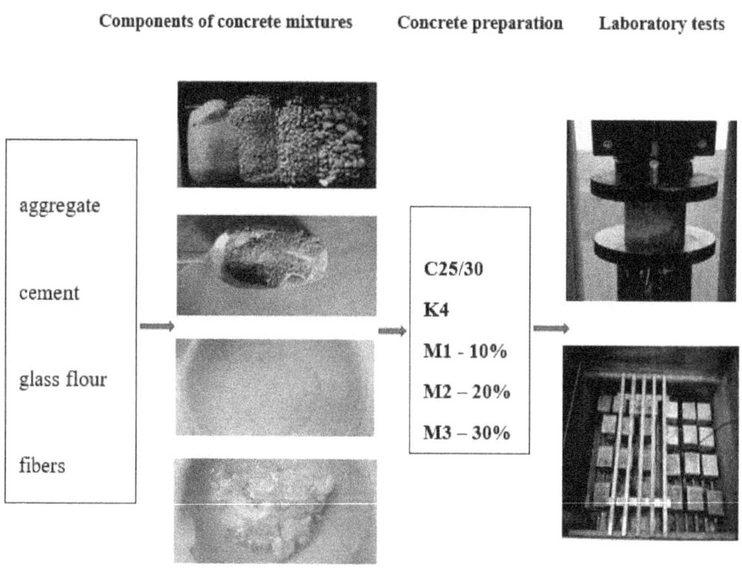

Figure 1. Conceptual diagram of laboratory tests performed.

To determine the effect of glass flour and polypropylene fibers on compressive strength, samples of ordinary concrete of class C25/30 with consistency K4 (semi-liquid consistency) were prepared based on EN 206 + A2:2021-08. To prepare the concrete mix, Portland cement CEM I 42.5 R from the Ożarów cement plant, in accordance with PN-EN 197-1: 2012, was used [32,33]. This cement is characterized by its high early strength. The declared performance of Portland cement is summarized in Tables 1 and 2. The given values are average values, guaranteed by the manufacturer for 2023.

Table 1. Physical properties and phase composition of cement CEM I 42.5 R (data from the manufacturer—own support).

Blaine Specific Surface Area (cm^2/g)	Start of Setting Time (min)	Compressive Strength after 2 Days (MPa)	Compressive Strength after 28 Days (MPa)
3330	218	21.0	49.8
	Share of mineral phases CEM I (% w.)		
C3S—55.54	C2S—14.59	C3A—8.15	C4AF—6.85

Table 2. Chemical properties of cement CEM I 42.5R (data from the manufacturer—own support).

Loss of Ignition (%)	Sulphate Content SO_3 (%)	Cl Chloride Content (%)	Alkali Content as Na_2O_{eq} (%)	SiO_2 (%)
3.19	2.96	0.05	0.76	20.20
Al_2O_3	Fe_2O_3	CaO	CaOw	MgO
4.41	2.42	64.36	1.98	1.98

In all samples, the same granulometric composition of fine aggregate, river sand of the 0–2 mm fraction selected by the sieving method according to EN 933-1:2012 and the same composition of coarse pebble aggregate fraction 4–16 mm selected by the method of successive approximations in three stages—Table 3 was assumed [34]. The bulk density of coarse and fine aggregate was determined by the pycnometric method in accordance with EN 1097-6:2002 [35]. The density for sand was 2.60 g/cm^3 and for gravel, 2.65 g/cm^3. For all laboratory tests, drinking water was used in accordance with EN 1008:2004 [36].

Table 3. Grain composition of aggregate (own version).

Fractions	Fraction Mixing Percentage (for Sand and Gravel)			Particle Size (%)	
	Stage I	Stage II	Stage III	Sand	Gravel
0.0–0.125				1.93	0.60
0.125–0.25				17.82	5.52
0.25–0.50			31	28.62	8.87
0.50–1.0				24.32	7.54
1.0–2.0				27.31	8.47
2.0–4.0		35			24.15
4.0–8.0	45	65	69		20.18
8.0–16.0	55				24.67

As a partial replacement for cement, glass flour was used in amounts of 10, 20, and 30%. Flour is a product that is made by grinding and shredding glass into very fine particles, which gives it a light, powdery consistency but still has a relatively high density compared to some other building materials such as cement or sand. The density of the glass flour is 2.4 g/cm^3. Figure 2 shows the grain size of the glass meal.

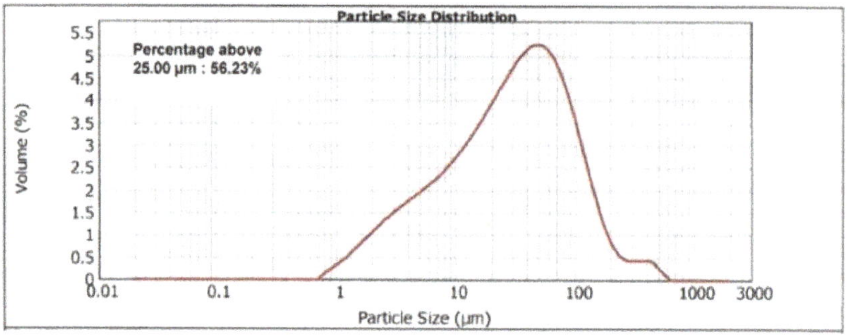

Figure 2. Glass flour grain size curve (manufacturer's data).

Polypropylene fibers were also used to strengthen the concrete structure. Polypropylene fibers extruded from polypropylene granulate/polymeric material ($CH_2CH(CH)_3$), bundled and cut in accordance with EN 14889-2:2006 (Bautech sp.zo.o Piaseczno near Warsaw, Poland) were also used for the tests. In the concrete mix, they form a three-dimensional supporting network resistant to gravity, thanks to which they maintain a constant level of concrete, prevent cracks, reduce plastic shrinkage, and limit the formation of shrinkage scratches [37]. Figure 3 shows the fibers and their characteristics.

Polypropylene fibers

- Tensile strength - 560 MPa.
- Consistency of concrete with 4 kg/m³ of FiberMix 12 mm - VeBe time 7 sec.
- Fiber length - 6mm.
- Content of adhesion. component - 0.8%.
- Chemical and alkali resistance – excellent.
- Distribution in water - 10 sec.
- Moisture content - 1%.
- Density - 0.91 g/cm³.
- Equivalent diameter of single fiber - 0.02mm.

Figure 3. Polypropylene fibers and their characteristics.

The following types of concrete mixes have been prepared for laboratory tests:

- BZ—reference concrete without additives,
- 1M—concrete with 10% glass flour and 4% polypropylene fibers,
- 2M—concrete with 20% glass flour and 4% polypropylene fibers,
- 3M—concrete with 30% glass flour and 4% polypropylene fibers.

Table 4 shows the composition of individual concrete mixes.

Table 4. Concrete mix proportions by weight.

Specification	Components of the Concrete Mix (kg/m^3)				
	Water	Aggregate		Cement	Glass Flour
		Sand	Gravel		
BZ concrete	287.08	488.34	831.49	666.03	-
Concrete with 10% M1 flour	287.08	488.34	831.49	599.43	66.60
Concrete with 20% flour M2	287.08	488.34	831.49	532.82	133.21
Concrete with 30% M3 flour	287.08	488.34	831.49	466.22	199.81

The following tests of fresh concrete mixes were carried out:

- consistency, in accordance with EN 12350-2:2019-08, reflow table method,
- density, in accordance with EN 12350-6:2019-08,
- air content, in accordance with EN 12350-7:2019-08, pressure method [38–40].

The following concrete tests were carried out:

- compressive strength, according to EN 12390-3:2019-07,
- density, in accordance with EN 12390-7:2019-08,
- frost resistance, according to 88/B-06250:2004 [41–43],
- evaluation of the effect of the addition of glass flour and polypropylene fibers on the microstructure of concrete.

The research was carried out in the Building Laboratory at the Faculty of Civil and Civil Engineering, in the Laboratory of Physical Processes at the Water Centre of the Warsaw University of Life Sciences, and in the Concrete Laboratory at the Building Research Institute in Warsaw.

The compressive strength test fc was performed after 28, 56, and 90 days of maturation in the hydraulic testing machine H011 Matest (Italy). Based on the obtained average strengths, the result was converted into cubic samples with a side of 15 cm, and the concrete class was determined.

$$f_{c,cub\ 15} = 0.95 \cdot f_{c,cub\ 10}$$

The frost resistance test was performed on samples measuring 10 × 10 × 10 cm after 28 days of concrete maturation for 150 freeze and thaw cycles. Three criteria were adopted to assess the degree of frost resistance, the fulfilment of which determines the degree of frost resistance achieved:

- no cracks on the samples after all freezing and thawing cycles,
- not exceeding the value of a 5% difference in the weight of the samples soaked in water before and after the frost resistance test,
- a decrease in compressive strength between witness and frozen samples of not more than 20%.

Three types of concrete containing polymer fibers and various amounts of glass flour (samples M1, M2, and M3) were prepared for Scanning Electron Microscope (SEM) examinations. At the first stage, from 10 × 10 × 10 cm concrete cubes, a thick slice from the middle section was cut perpendicular to the trowelling surface for each type of material. In this way, surfaces were grinded and polished. Cut and pre-polished sections of polymer composites are presented in Figure 4.

In the next step for SEM examinations, a chosen region from the pre-polished sections of each sample was cut with surface dimensions of about 20 × 20 mm. After cutting, they were dried in the oven and put into epoxy resin in a vacuum chamber. Microscopic sections were prepared in the same way as in previously published papers [44]. Figure 5 presents images of prepared samples made using an optical microscope. Before SEM examinations, samples were gold-evaporated.

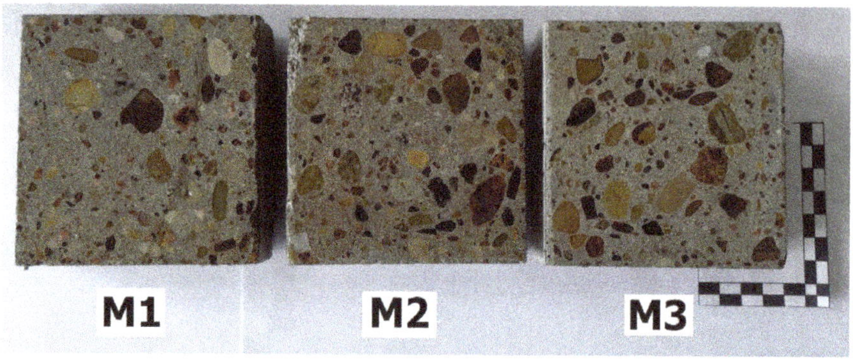

Figure 4. Pre-polished surfaces of concrete samples.

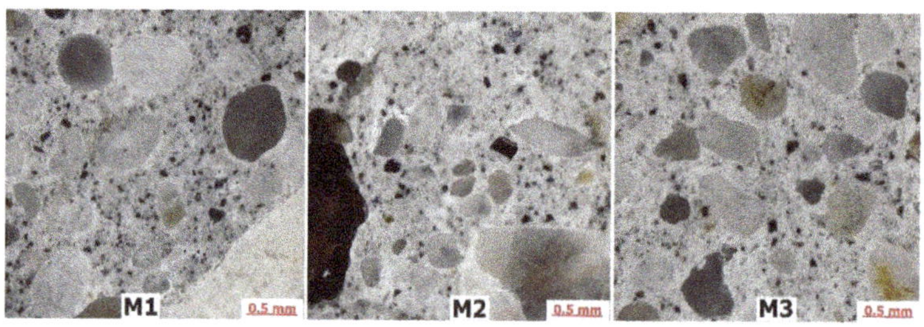

Figure 5. Surface of samples prepared for SEM examinations.

Microstructural analyses were carried out using SEM produced by Zeiss, model Sigma 500 VP (Carl Zeiss Microscopy GmbH, Köln, Germany). Backscattered electron (BSE) images were collected. Phase compositions and mapping were analyzed using the Energy Dispersive X-ray spectroscopy (EDX) detector produced by the Oxford model Ultim Max 40 (Oxford Instruments, High Wycombe, UK).

3. Results

3.1. Results of Concrete Mixes

Figure 6 shows the course of individual tests of concrete mixes, and Table 5 shows the results obtained.

Figure 6. Testing of concrete mix—(**a**) consistency, (**b**) air content, (**c**) density.

Table 5. Test results of concrete mixes (own application).

Specification	Consistency (mm)	Density (kg/m^3)	Air Content (%)
BZ	K4	2370	1.7
M1	K5	2249	2.2
M2	K5	2246	2.7
M3	K5	2247	2.9

Based on the results of tests carried out on concrete mixes, it was found that glass flour and fibers slightly affect its individual parameters—consistency, density, and air content. The results of the consistency tests are consistent for mixtures of M1, M2, and M3, which have obtained a consistency of K5. Glass flour is a product that does not have the ability to absorb water in a manner like traditional building materials. Glass flour is a glassy material with a smooth surface that does not absorb water in a significant way. The density of fresh concrete mix BZ obtained values of 2370 kg/m^3 and for mixtures with flour from 2246 to 2249 kg/m^3. The lowest air content was obtained for BZ samples, equal to 1.7%, while the highest air content was recorded in concrete mixes in which cement was replaced in the amount of 30%, equal to 2.9%.

3.2. Results of Mature Concrete

Compressive Strength

In Figure 7, the course of the compressive strength test is shown, while in Figure 8, the average test results are obtained for individual concretes made based on glass flour with polypropylene fiber.

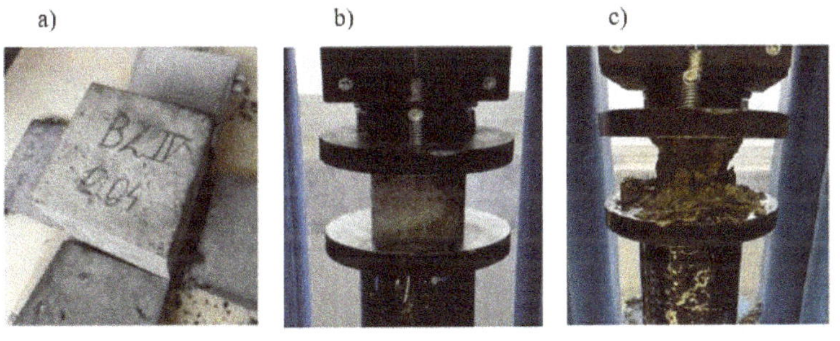

Figure 7. Compressive strength test (**a**) concrete sample, (**b**) sample set in the machine, (**c**) test sample.

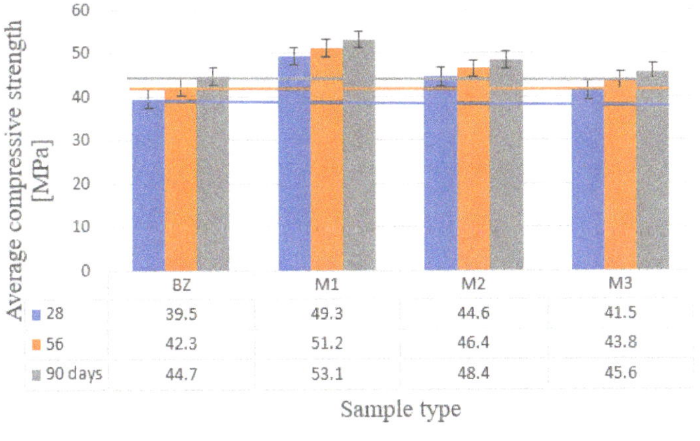

Figure 8. Average compressive strength of concrete.

Analyzing Figure 8, after 28 days of maturation, the highest compressive strength of 49.3 MPa was obtained by concrete in which cement was converted in the amount of 10% to flour, while the lowest strength of 39.5 MPa was achieved by reference concrete. Compared to BZ concrete, the increase was 24.8%. The highest compressive strength after 56 days, equal to 51.2 MPa and after 90 days, equal to 53.1 MPa was also obtained by concrete, in which the flour content accounted for 10%. However, the lowest compressive strength after 56 days of maturation, equal to 42.3 MPa and after 90 days, equal to 44.7 MPa, was obtained by BZ reference concrete. In relation to BZ, the increase in strength amounted to 21.0 and 18.8%, respectively. Replacing cement with glass flour with fibers increased strength over time compared to comparative concrete without an additive.

Figure 9 shows the relationship between the average compressive strength of concrete and the ratio of glass flour to cement in concrete, while Table 6 shows the obtained equations of dependencies.

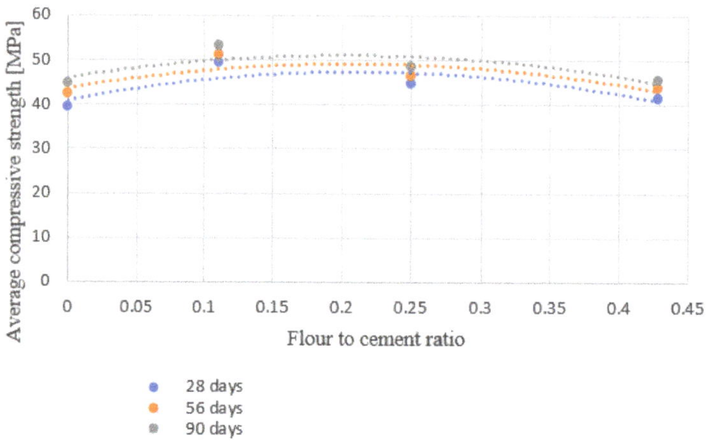

Figure 9. Dependence of average compressive strength on the ratio of glass flour to cement (self-support).

Table 6. Dependency equation for concrete with the addition of glass flour.

Property/Days of Ripening	Dependency Equation x—Ratio of Glass Flour Content to Cement Mass	Coefficient of Determination R^2
Compressive strength (MPa) after 28 days (M1)	$fcm = -144.1x^2 + 61.5x + 40.9$	0.62
Compressive strength (MPa) after 56 days (M2)	$fcm = -126.3x^2 + 52.9x + 43.6$	0.58
Compressive strength (MPa) after 90 days (M3)	$fcm = -122.4x^2 + 50.1x + 46.0$	0.59

Analyzing the graph of functions showing the dependence of the average compressive strength on the ratio of the flour content to cement, the most beneficial from the point of view of strength increase is the proportion of ash to cement in the amount of 0.20. The coefficient of determination for concrete samples ranged from 0.58 to 0.62. The obtained results indicate a low correlation between experimental and computational data.

3.3. Density

Figure 10 shows the results of concrete density tests.

Glass flour is a waste that has a lower density compared to cement, so using it in concrete mixes can reduce the density of concrete. The analysis of the results presented in

Figure 7 indicates that the addition of glass flour and polypropylene fibers causes a decrease in concrete density with a higher flour content. The density ranged from 2275 kg/m^3 for M3 concrete to 2312 kg/m^3 for M1 concrete. The reference concrete obtained a density of 2358 kg/m^3. All concrete can be classified as ordinary concrete, whose density ranges from 2000 to 2600 kg/m^3.

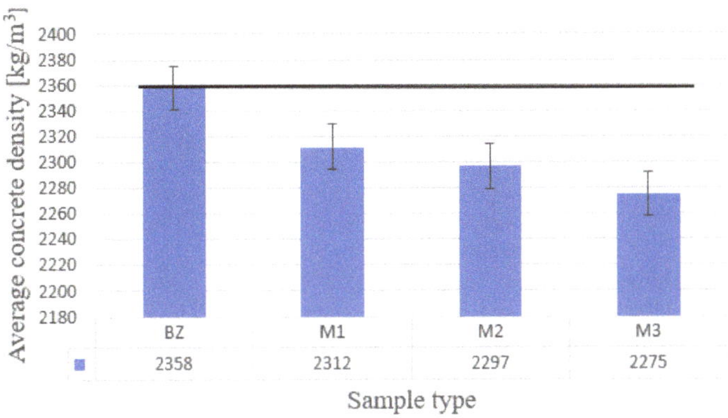

Figure 10. Average density of concrete.

3.4. Frost Resistance

Figure 11 shows a frost resistance test chamber and concrete samples after freezing/thawing cycles. For each type of concrete, 12 samples were prepared (6 reference samples and 6 samples for freezing/thawing cycles), of which averages were calculated (Table 7).

Figure 11. (**a**) A chamber for testing the frost resistance of concrete and (**b**) samples after the test.

The analysis of Table 7 shows that among the reference samples, the highest compressive strength of 52.6 MPa was obtained by M1 concrete, while the smallest was 43.4 MPa in BZ concrete without additives. Considering the compressive strength of samples after 150 freezing cycles, it was observed that the highest average compressive strength of 46.4 MPa was obtained by M1 concrete, in which cement was replaced with glass flour in the amount of 10%. The lowest compressive strength after 150 freezing cycles was recorded for M3 concrete. The average strength in this case was 34.2 MPa. The average decrease in the strength of the samples subjected to freezing in two cases for concrete M2 and M3 exceeded the permissible 20%. The lowest decrease in strength was for BZ concrete samples without additives. The average weight loss after freezing ranged from 0.381% for BZ samples to 0.880% for M3 samples. Samples were subjected to 150 freeze and thawing cycles, in which cement was replaced with glass flour in the amount of 10% and concrete without additives is frost-resistant—F150. The other two types of concrete, in which cement was replaced in the amounts of 20 and 30%, turned out to be non-frost-resistant concrete. The research shows that there is a certain limitation at which the conversion of cement to glass flour causes a decrease in compressive strength after freezing/thawing cycles. In the context of these results, it can be indicated that the optimal content of the additive in the form of glass flour is 15%.

Table 7. Comparison of the average decrease in compressive strength of concrete samples subjected to freezing and the average loss in weight of samples made on the basis of glass flour and polypropylene fibers.

Sample	Average Compressive Strength (MPa)		Average Strength Drop Samples Subjected to Freezing (%)	Average Weight (g)		Average Weight Loss (%)
	Reference Sample	After 150 Freezing Cycles		Before Freezing	After 150 Freezing Cycles	
BZ	43.4	41.2	−5.1	2362	2353	0.381
M1	52.6	46.4	−11.8	2303	2293	0.434
M2	47.3	37.7	−20.3	2280	2265	0.658
M3	44.6	34.2	−23.3	2274	2254	0.880

3.5. Microstructure of Concrete

Figures 12 and 13 show exemplary images of the microstructure of the analyzed reference concrete without the addition of glass flour and polypropylene fibers.

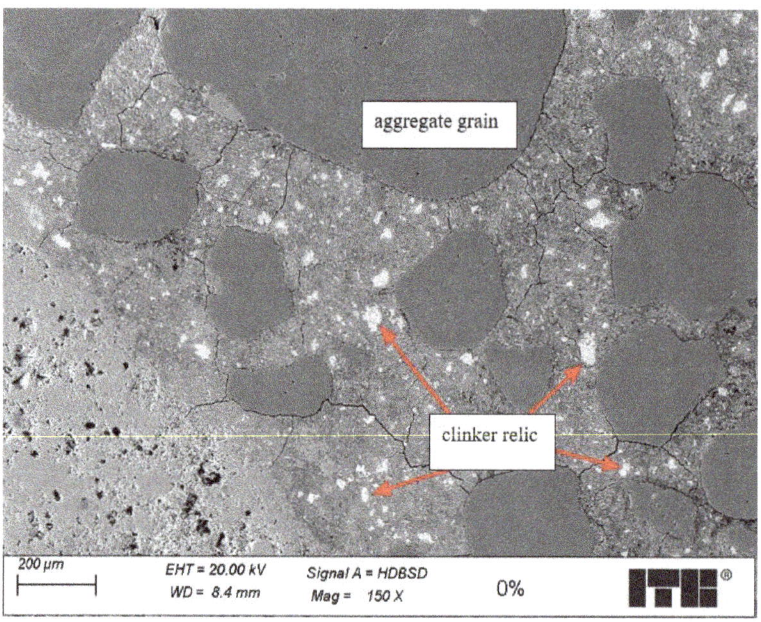

Figure 12. Microstructure of reference concrete (grains of clinker relics and fine quartz aggregate are marked).

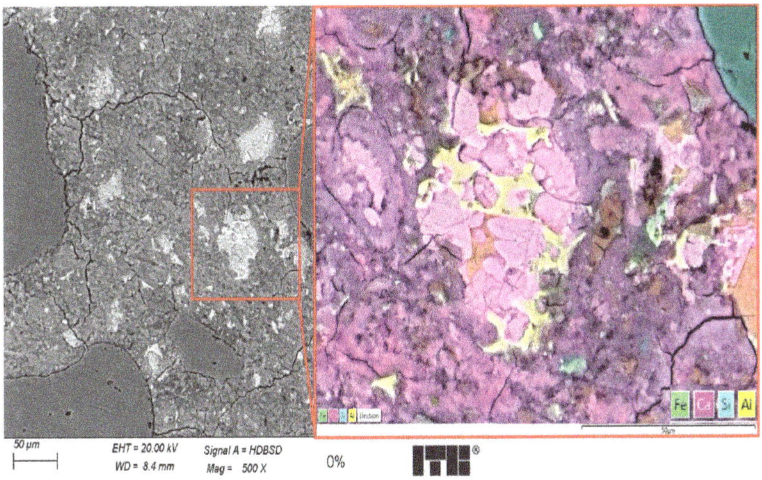

Figure 13. Reference concrete grout microstructure and C-S-H phase mapping around clinker relic grain.

The microstructure of the analyzed reference concrete was compact and tight, without excessive air voids or regular, spherical air pores. Numerous relics of clinker were present. No significant content of other secondary cement components was found, which confirms the use of CEM I cement for the preparation of concrete samples. The C-S-H phase was properly formed with a compact and tight structure. In most cases, the contact zone between the aggregate grains and the cement grout was tight and properly formed. Areas of discontinuous contact zone between aggregate grains and grout were few, which indicates proper mixing and compaction of the concrete mix and proper care of the internal areas of hardened concrete (Figure 13). Cracks that occurred in the cement matrix were caused by the sample preparation due to the drying of C-S-H gel.

The aim of SEM microstructural analysis was to discover the potential causes of the observed increase in compressive strength between the reference sample and the M1 sample and the causes of the observed decrease in compressive strength in samples M1, M2, and M3. Additionally, the microstructure observations might explain the slight increase in freeze–thaw resistance of concretes containing polymer fibers compared to the reference one.

Figures 14–16 present examples of the observed microstructure of concretes M1, M2 and M3.

In the cement matrix of M1–M3 samples, grains of clinker relicts were observed, along with glass flour and polymer fibers with a diameter of about 30 µm. Also, some grains of GGBS (ground granulated blast furnace slag) were observed. Some of the GBBS might be present in the CEM I cement as a minor constituent, according to the EN 197-1 standard. Comparing the microstructure of M1-M2-M3, it was observed that by increasing the content of glass flour in the cement matrix instead of a part of cement, the cement matrix became more porous and cracked. Especially when comparing the M2 and M3 samples. In the M3 sample, there were observed areas with high porosity, probably caused by the difficulties with the compaction of the samples during molding. In the M3 sample, the C-S-H gel was more porous than in other samples, which might be caused by the lower content of clinker in the cement matrix. The C-S-H phase was sealed and well developed in M1 and M2 samples, especially in areas located near the clinker relicts. Some regular spherical air voids were observed in the M1 sample.

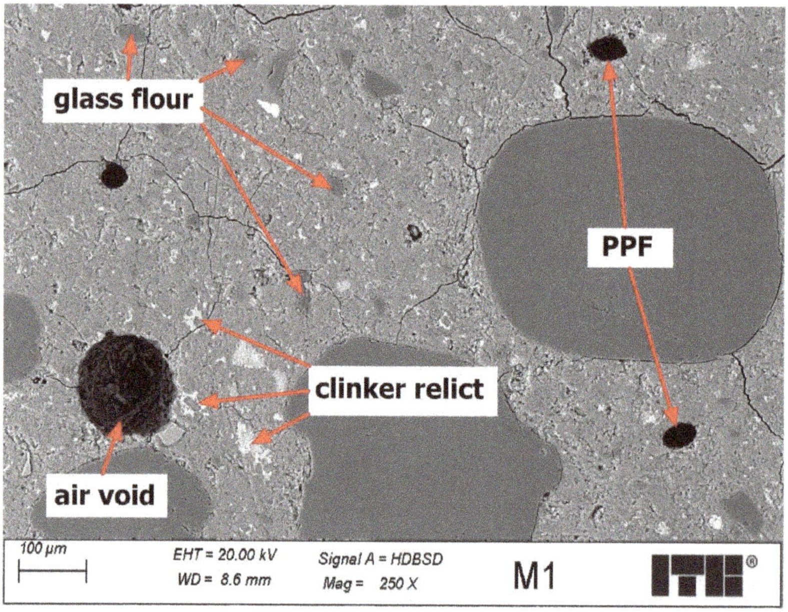

Figure 14. Microstructure of M1 concrete sample.

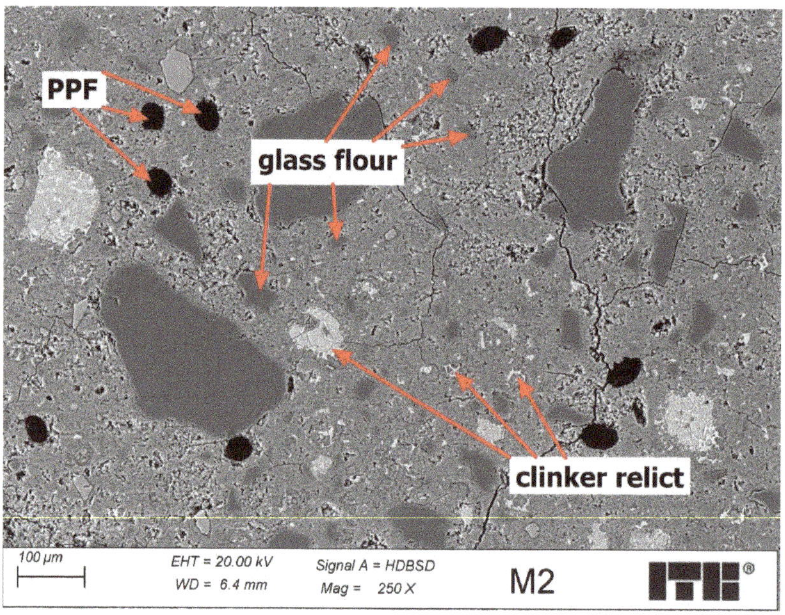

Figure 15. Microstructure of M2 concrete sample.

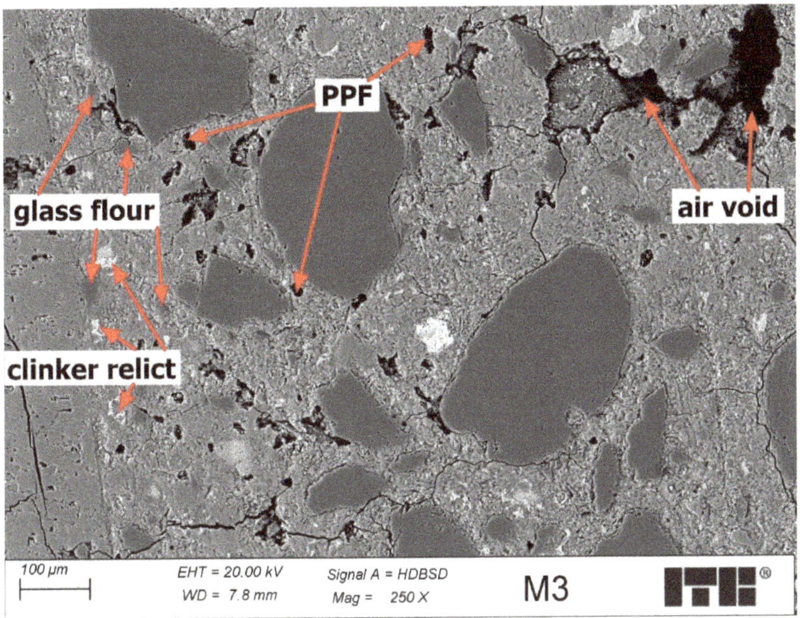

Figure 16. Microstructure of M3 concrete sample.

Figure 17 presents an example of the microstructure of the M2 sample and the EDX mapping of transition zone between grain of glass flour and cement matrix.

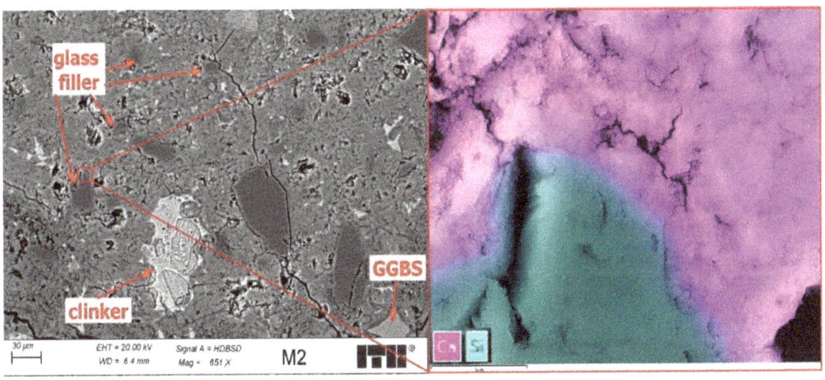

Figure 17. Microstructure of M2 sample and transition zone between glass flour grain and cement matrix.

The observed microstructure of the transition zone between grains of glass flour and cement matrix was mostly sealed with several cracks. The diffusion of silicon ions into the cement matrix and calcium ions in the opposite direction was observed, as shown on Figure 17 as a blue/pink halo.

4. Discussion

Analysis of the results of the compressive strength of concrete in different ripening periods indicates that the strength of concrete increases over time, reaching its highest values for longer periods. This phenomenon is typical for the cement hydration process, in which chemical reactions take place that cause the concrete to harden and develop its strength. The obtained results also indicate that the content of glass flour and a constant number of polypropylene fibers have an impact on the development of strength over time. As the content of the additives used increases, differences in strength can be observed at different time stages. However, there is a limit to the optimal amount of flour to which the compressive strength increases. This suggests that the presence of additives may affect the rate and nature of the cement hydration process because of the interaction between flour, fiber, and cement, which consequently affects the development of concrete strength over time. The research confirms the results of other authors. Khatib and Negim [45] showed that cement partially replaced by glass flour has a beneficial effect on the mechanical properties of the hardened composite. Studies have shown that the maximum increase in compressive strength was achieved for samples with 10% flour, and above this value, there is already a noticeable decrease in this property [45]. Polypropylene fibers have an antispasmodic effect, protect the composite against cracks and scratches [46], and improve the frost resistance of concrete. Research with the addition of polypropylene fibers is described in Michalik and Kupisz [47], which focuses mainly on showing their impact on frost resistance. Reference concrete was compared with concrete containing class Ia (single) and class Ib (fibrillated) microfibers. The authors observed that with an increase in the percentage content of polypropylene fibers, the consistency of the concrete mix decreases, the density of the concrete mix decreases with the increase in the percentage content of single fibers, the addition of single microfibers causes a decrease in the compressive strength of the composite after 28 days, and frost resistance is significantly improved compared to reference concrete [48].

The analyzed microstructure of M1–M3 concrete samples containing additions of polymer fibers and glass flour has shown that the addition of larger amounts of glass flour as a substitute for cement might cause problems with the compaction of the concrete mix, as observed in the M3 sample. This is probably due to the decrease in the workability of concrete mix caused by the increase in water demand from glass flour.

Observed in the cement matrix of the M1 sample, several spherical air voids might be the cause of the observed slight increase in freeze/thaw resistance of this concrete. Although the amount of air voids was too low to make concrete fully resistant to this type of aggression, in the microstructures of other samples (M2 and M3), this type of air void was not observed. In the microstructure of the M3 sample, there were observed areas with high porosity and cracks, which might decrease freeze/thaw resistance even more.

Examining the microstructure of the transition zone between grains of glass flour and the cement matrix, it was discovered that silicon ions from glass migrate into the cement matrix and calcium ions in the opposite direction. It proves that grains of glass flour react with the cement matrix. It might be the cause of the observed slight increase in compressive strength of the M1 sample compared to the reference sample (without glass flour). Although the reactivity of glass flour grains might be the cause of the corrosion of cement composites by the potential alkali aggregate reaction, which needs further investigation.

5. Conclusions

After the tests, it can be concluded that the partial replacement of cement with glass flour and the addition of polypropylene fibers caused:

- increase in compressive strength in the tested ripening periods for samples with 10% glass flour,
- decrease in the density of the concrete mix by approximately 5%,
- deterioration of the quality of the consistency class,
- decrease in the density of hardened concrete from 2 to 4%,
- improving resistance to cyclic freezing/thawing, while obtaining frost-resistant concrete for concrete containing 10% glass flour.

Analyzing the results of microstructure observations of cement composites containing glass filler and polymer fibers, the following conclusions might be drawn.

The addition of polypropylene fibers and a small amount of glass flour causes the presence of several spherical air voids, which might slightly increase the resistance to freeze/thaw corrosion.

Adding a higher amount of glass flour might cause a problem with the workability of the concrete mix due to the increase in water demand.

Glass filler shows reactivity with the cement matrix, and in small amounts, it might cause the microstructure to seal and a slight increase in compressive strength.

Due to the reactivity of glass flour, its potential alkali aggregate reactivity needs to be investigated to prevent the potential risk of corrosion.

Author Contributions: Author Contributions: Conceptualization, G.R.; methodology, G.R.; software, G.R. and F.C.; validation, G.R.; formal analysis, G.R.; investigation, G.R.; resources, G.R.; data curation, G.R. and F.C.; writing—original draft preparation, G.R. and M.Ż.; writing—review and editing, G.R. and M.Ż.; visualization, E.G. and Y.T.; supervision, Y.T. All authors have read and agreed to the published version of the manuscript.

Funding: This research received no external funding.

Institutional Review Board Statement: Not applicable.

Informed Consent Statement: Not applicable.

Data Availability Statement: The data presented in this study are available on request from the corresponding author.

Conflicts of Interest: The authors declare no conflict of interest.

References

1. International Statistics Yearbook. Available online: https://stat.gov.pl (accessed on 12 February 2021).
2. IEA—Shaping a Secure and Sustainable Energy Future for All. Available online: www.iea.org (accessed on 12 February 2023).
3. Deja, J.; Antosiak, B. Degree of progress of the fly ash reaction in alkali-activated fly-ash binders. *Cem. Lime Concr.* **2012**, *17*, 67–76.
4. Giergiczny, Z. *Fly Ash in the Composition of Cement and Concrete*; Silesian University of Technology: Gliwice, Poland, 2013.
5. Masters, L.W.; Brandt, E. Systematic methodology for service life prediction of building materials and components. *Mater. Struct.* **1998**, *22*, 285–392. [CrossRef]
6. Deja, J. *Concrete Technologies and Research Methods*; Association of Cement Producers: Cracow, Poland, 2020.
7. Neville, A.M. *Concrete Properties*; Association of Cement Producers: Cracow, Poland, 2012.
8. Gupta, S.M. Support vector machines-based modelling of concrete strength. *Int. J. Intel. Technol.* **2007**, *3*, 12–18.
9. Kim, J.I.; Kim, D.K. Application of neural networks for estimation of concrete strength. *KSCE J. Civ. Eng.* **2002**, *6*, 429–438. [CrossRef]
10. Lai, S.; Serra, M. Concrete strength prediction by means of neural network. *Constr. Build. Mater.* **1997**, *11*, 93–98. [CrossRef]
11. Pietrzak, A. Pro-ecological technologies in construction on the example of "green concrete". *Constr. Optim. Energy Potential* **2014**, *1*, 8693.
12. Giergiczny, Z. Cements with mineral additives are components of durable concrete. *Eng Constr.* **2010**, *5–6*, 275–279.
13. Maca, P.; Jandekova, P.; Konvalinka, P. The influence of metakaolin addition on the scaling of concrete due to frost action. *Cem. Lime Concr.* **2014**, *1*, 1–7.
14. Yusuf, R.O.; Noor, Z.Z.; Din, M.D.F.M.D.; Abba, A.H. Use of sewage sludge ash (SSA) in the production of cement and concrete—A review. *Int. J. Glob. Environ. Issues* **2012**, *12*, 214. [CrossRef]
15. Rutkowska, G.; Wichowski, P.; Świgoń, K.; Sobieski, P. Testing the properties of concrete with the addition of fly ash from thermal treatment of sludge. *Cem. Lime Concr.* **2017**, *2*, 113–119.
16. Rutkowska, G.; Wichowski, P.; Franus, M.; Mendryk, M.; Fronczyk, J. Modification of Ordinary Concrete Using Fly Ash from Combustion of Municipal Sewage Sludge. *Materials* **2020**, *13*, 487. [CrossRef] [PubMed]
17. Rutkowska, G.; Klepak, O.; Podawca, K. Problems of heat loss in existing-family buildings in the context of construction errors. *Annu. Set Environ. Prot.* **2013**, *15*, 2625–2639.
18. Jura, J.; Ulewicz, M. The impact of bio-ashes on selected properties of cement mortars. *Sci. Rev. Eng. Environ. Manag.* **2017**, *26*, 234–240. [CrossRef]
19. Najduchowska, M.; Różycka, K.; Rolka, G. Assessment of the possibility of using cullet in the construction industry in terms of its impact on the natural environment. *Works Inst. Ceram. Build. Mater.* **2014**, *17*, 46–56.
20. Chen, Z.; Poon, C.S. Comparative studies on the effects of sewage sludge ash and fly ash on cement hydration and properties of cement mortars. *Constr. Build. Mater.* **2017**, *154*, 791–803. [CrossRef]

21. Sikora, P.; Horszczaruk, E.; Skoczylas, K.; Rucinska, T. Thermal Properties of Cement Mortars Containing Waste Glass Aggregate and Nanosilica. *Procedia Eng.* **2017**, *196*, 159–166. [CrossRef]
22. Tamanna, N.; Sutan, N.M.; Yakub, I.; Lee, D.T.C. Strength Characteristics of Mortar Containing Different Sizes Glass Powder. *J. Civ. Eng. Sci. Technol.* **2014**, *5*, 11–16. [CrossRef]
23. Siddique, R. *Waste Materials and By-Products in Concrete*; Springer: Berlin, Germany, 2008.
24. Lotfi, S.; Deja, J.; Rem, P.; Mróz, R.; van Roekel, E.; van der Stelt, H. Mechanical recycling of EOL concrete into high-grade aggregates. *Resour. Conserv. Recycl.* **2014**, *87*, 117–125. [CrossRef]
25. Mohammadreza, M.; Riding, K.A. Effect of curing temperature and glass type on the pozzolanic reactivity of glass powder. *Cem. Concr. Res.* **2014**, *58*, 103–111. [CrossRef]
26. Nassar, R.-U.; Soroushian, P. Field investigation of concrete incorporating milled waste glass. *J. Solid Waste Technol. Manag.* **2011**, *37*, 307–319. [CrossRef]
27. Bajad, M.N.; Modhera, C.D.; Desai, A.K. Effect of glass on strength of concrete subjected to sulphate attack. *Int. J. Civil Eng. Res. Dev.* **2011**, *1*, 2.
28. Meena, A.; Singh, R. Comparative Study of Waste Glass Powder as Pozzolanic Material in Concrete. Bachelor's Thesis, Department of Civil engineering, National Institute of Technology, Rourkela, India, 2012.
29. Federico, L. Waste Glass—A Supplementary Cementitious Material. Ph.D. Thesis, Department of Civil Engineering, McMaster University, Hamilton, ON, Canada, 2013.
30. Aliabdo, A.A.; Abd Elmoaty, A.E.M.; Aboshama, A.Y. Utilization of waste glass powder in the production of cement and concrete. *Constr. Build. Mater.* **2016**, *124*, 866–877. [CrossRef]
31. Jain, K.L.; Sancheti, G.; Gupta, L.K. Durability performance of waste granite and glass powder added concrete. *Constr. Build. Mater.* **2020**, *252*, 119075. [CrossRef]
32. *EN 206+A2:2021-08*; Concrete—Requirements, Properties, Production and Compliance. Polish Committee for Standardization: Warsaw, Poland, 2021.
33. *EN 197-1:2012*; Cement—Part 1: Composition, Requirements and Criteria for Cements for General Use. Polish Committee for Standardization: Warsaw, Poland, 2013.
34. *EN 933-1:2012*; Tests of Geometric Properties of Aggregates—Part 1: Determination of Grain Composition—Sieving Method. Polish Committee for Standardization: Warsaw, Poland, 2012.
35. *EN 1097-6:2022-07*; Tests of Mechanical and Physical Properties of Aggregates—Part 6: Determination of Grain Density and Water Absorption. Polish Committee for Standardization: Warsaw, Poland, 2022.
36. *EN 1008:2004*; Mixing Water for Concrete. Polish Committee for Standardization: Warsaw, Poland, 2013.
37. *EN 14889-2:2006*; Concrete Fibers—Part 2: Polymer Fibers—Definitions, Requirements and Compliance. Polish Committee for Standardization: Warsaw, Poland, 2013.
38. *EN 12350-2:2019-08*; Concrete Mix Tests—Part 2: Consistency Testing Using the Cone Slump Method. Polish Committee for Standardization: Warsaw, Poland, 2019.
39. *EN 12350-6:2019-08*; Concrete Mix Tests—Part 6: Density. Polish Committee for Standardization: Warsaw, Poland, 2019.
40. *EN 12350-7:2019-08*; Concrete Mix Tests—Part 7: Air Content—Pressure Methods. Polish Committee for Standardization: Warsaw, Poland, 2019.
41. *PN-EN 12390-3:2019-07*; Concrete Testing—Part 3: Compressive Strength of Test Specimens. Polish Committee for Standardization: Warsaw, Poland, 2019.
42. *EN 12390-7:2019-08*; Concrete Testing—Part 7: Concrete Density. Polish Committee for Standardization: Warsaw, Poland, 2019.
43. *PN-88/B-06250:2004*; Plain Concreto. Polish Committee for Standardization: Warsaw, Poland, 1988.
44. Chyliński, F.; Michalik, A.; Kozicki, M. Effectiveness of Curing Compounds for Concrete. *Materials* **2022**, *15*, 2699. [CrossRef] [PubMed]
45. Khatib, J.M.; Negim, E.M.; Sohl, H.S.; Chileshe, N. Glass Powder Utilisation in Concrete Production. *Eur. J. Appl. Sci.* **2012**, *4*, 173–176.
46. Zych, T. Contemporary fiber concrete—The possibility of shaping structural elements and architectural forms. *Tech. J. Archit.* **2010**, *107*, 371–386.
47. Michalik, A.; Kupisz, P. Assessment of frost resistance of concrete with the addition of polymer microfibers. *Constr. Insp.* **2020**, *91*, 32–35.
48. Rutkowska, G.; Wichowski, P.; Lipiński, R. The influence of crushed glass waste on selected properties of concrete prepared with their participation. *Sci. Rev. Eng. Environ. Manag.* **2018**, *27*, 463–475.

MDPI AG
Grosspeteranlage 5
4052 Basel
Switzerland
Tel.: +41 61 683 77 34

Applied Sciences Editorial Office
E-mail: applsci@mdpi.com
www.mdpi.com/journal/applsci

www.ingramcontent.com/pod-product-compliance
Lightning Source LLC
Chambersburg PA
CBHW040747250926
51850CB00007BB/1432
* 9 7 8 3 7 2 5 8 5 4 0 7 3 *